Experiment and Natural Philosophy in Seventeenth-Century Tuscany

AUSTRALASIAN STUDIES IN HISTORY AND PHILOSOPHY OF SCIENCE

VOLUME 21

EXPERIMENT AND NATURAL PHILOSOPHY IN SEVENTEENTH-CENTURY TUSCANY

The History of the Accademia del Cimento

LUCIANO BOSCHIERO

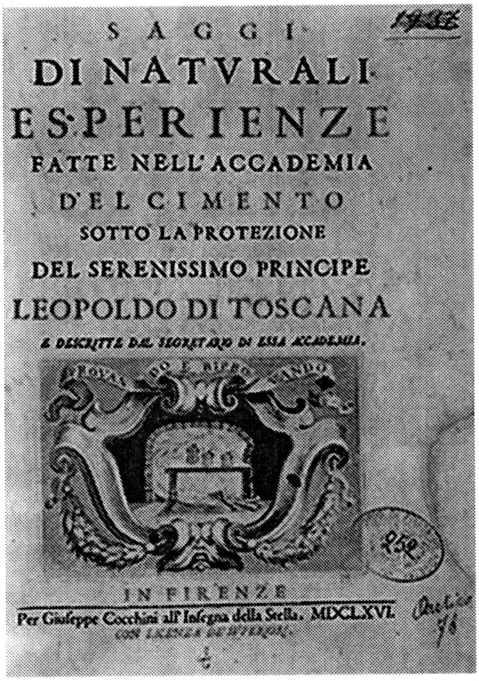
SAGGI
DI NATVRALI
ESPERIENZE
FATTE NELL'ACCADEMIA
DEL CIMENTO
SOTTO LA PROTEZIONE
DEL SERENISSIMO PRINCIPE
LEOPOLDO DI TOSCANA
E DESCRITTE DAL SEGRETARIO DI ESSA ACCADEMIA.

IN FIRENZE
Per Giuseppe Cocchini all'Insegna della Stella. MDCLXVI.

Saggi di naturali esperienze fatte nell'Accademia del Cimento, Frontispiece.
Used with permission of the Istituto e Museo di Storia della Scienza,
Biblioteca Digitale

Springer

A C.I.P. Catalogue record for this book is available from the Library of Congress.

ISBN 978-1-4020-6245-2 (HB)
ISBN 978-1-4020-6246-9 (e-book)

Published by Springer,
P.O. Box 17, 3300 AA Dordrecht, The Netherlands.

www.springer.com

Printed on acid-free paper

TABLE OF CONTENTS

LIST OF FIGURES

ACKNOWLEDGEMENTS

Most of the research required for the completion of this book was carried out during my time as a doctorate student. For this reason I am indebted to the support shown to me by my friends and colleagues at the University of New South Wales, all of whom showed an interest in my work. In particular, I am indebted to John Schuster and David Miller for their invaluable advice and guidance over many years.

For showing support and providing comments at various stages of this project, thanks are due to my wife, Michelle; my parents, Ana and Marino; Katherine Neal; Stephen Gaukroger; Ivan Crozier; John Henry, Simon Schaffer, and several of my postgraduate colleagues during my time at the University of New South Wales and the University of Sydney. Thanks especially to Paolo Galluzzi who offered me guidance when researching this topic in the Italian archives. Galluzzi's support, as well as the assistance of the staff at the Istituto e Museo di Storia della Scienza in Florence (IMSS), and at the Biblioteca Nazionale Centrale di Firenze (BNCF), was invaluable for the preparation of this work. I would also like to acknowledge IMSS Biblioteca Digitale and BNCF for their permission to reproduce the images in this book.

The research presented in this book was also generously supported by research grants from the Italian Foreign Ministry, and the Research Management Committee for the Faculty of Arts at the University of New South Wales.

INTRODUCTION

The aim of this book is to explore and understand the activities undertaken by the Florentine Accademia del Cimento, one of Europe's first scientific societies. The Cimento operated for ten years, between 1657 and 1667, and during that time performed many experiments and observations in physics and astronomy, rivalling the achievements of the Royal Society of London and the Parisian Acadèmie Royale des Sciences. This book will attempt to sift through the available primary evidence, as well as secondary accounts of the Cimento's activities, in order to examine the intellectual concerns that the individual academicians acquired throughout their careers and that they pursued while carrying out and interpreting their experiments for the Cimento and the Tuscan Court. Those interests will also shed some light on the ways in which the academicians performed and used experiments.

Inspired by Galileo's success with experiments and instruments during the first half of the seventeenth century, the Cimento academicians developed an experimentalist approach to their natural inquiry that attempted to eliminate any dependence on theoretical presuppositions and preconceptions. The group's purported aim was to rely solely on the senses to accumulate knowledge of nature. This experimental philosophy framed the way in which historians have since viewed the Cimento's practices.

This book will not, however, be an attempt to trace the early modern origins of scientific institutions, or of the experimental philosophy that is believed by many historians to have been the newest form of scientific inquiry prevalent in academies during the mid to late seventeenth century. Many historians have already examined these topics in great detail and have devised varying theories about the foundations and workings of the early Royal Society of London, the Parisian Acadèmie Royale des Sciences and the Cimento.

The Accademia del Cimento formally began on 19 June 1657, when Prince Leopoldo de' Medici (1617–1675) invited nine of his courtiers and experts in natural philosophy to the Pitti Palace in Florence. This group included: Giovanni Borelli (1608–1679), Vincenzio Viviani (1622–1703), Carlo Rinaldini (1615–1698), Alessandro Marsili (1601–1670), Francesco Redi (1626–1697), Carlo Dati (1619–1676), Alessandro Segni (1633–1697), Candido del Buono (1618–1676), and Antonio Uliva (d. 1668). Under the patronage of the Medici

Court, these men reportedly committed themselves to making experiments and observations. The academicians' dedication to experimentalism, it would seem, is typified in their motto, 'Provando e Riprovando', referring to the rigorous 'testing and retesting' of their own experiments as well as those performed previously by other natural philosophers of the period.[1] Yet the best testimony to the Accademia del Cimento's supposedly strict experimentalist approach to researching nature was the publication of their work in 1667, titled *Saggi di naturali esperienze*. This text was devoted to the narration of the experiments performed in the Accademia del Cimento during its first few years in operation and stated the academicians' intentions never to stray into speculative arguments, but simply to report the experiments they performed.

Indeed, the author, and the Accademia's secretary after 1660, Lorenzo Magalotti (1637–1712), expresses this aim clearly in the Preface to the *Saggi*:

> ... if sometimes in passing from one experiment to another, or for any other reason whatever, some slight hint of speculation is given, this is always to be taken as the opinion or private sentiment of the academicians, never that of the Academy, whose only task is to make experiments and to tell about them.[2]

The reason Magalotti gave for this experimentalist and non-speculative approach to producing natural knowledge, was that experiments were believed to provide the only true descriptions of nature. For too long, claims Magalotti, natural philosophers had been relying on the authority of past writers and had been reaching false conclusions about the causes of nature's structure and movement.[3] Therefore, Magalotti asserts in the Preface, although geometry provided some possibility for arriving at the truth, the only way of completely avoiding theoretical speculation about causes of natural phenomena was through the use of experiments: '[T]here is nothing better to turn to than our faith in experiment.'[4] It must be made clear that Magalotti did not suggest that the academicians completely abandoned any intentions to search for causes. On the contrary, as philosophers of nature, they were still determined to find causal descriptions of natural phenomena. The point is simply this: that experiments were purported to be the only way of properly 'fitting effects to causes and causes to effects'.[5]

Magalotti created the impression for his readers that the members of the Accademia del Cimento never engaged in theoretical and speculative discussions, and that instead they were accumulating factual knowledge regarding the causes

[1] This phrase is mentioned in the Preface to the Accademia's publication written by their secretary, Lorenzo Magalotti. *Saggi di naturali esperienze fatte nell'Accademia del Cimento sotto la protezione del serenissimo principe Leopoldo di Toscana*, Florence, 1667, 84. All references to the *Saggi*, and pages given, are from its publication in G. Abetti and P. Pagnini (eds.), *Le opere dei discepoli di Galileo Galilei. Edizione Nazionale. I. L'Accademia del Cimento. Parte Prima*, Florence, 1942.

[2] '... se talora per far passaggio da una ad un'altra esperienza, o per qualunque altro rispetto, si sarà dato qualche minimo cenno di cosa specolativa, ciò si pigli pur sempre come concetto o senso particolare di accademici, ma non mai dell'Accademia; della quale unico istituto si è di sperimentare e narrare'. Magalotti, *idem.*, 86–87. All translations of passages from the *Saggi* are from W.E.K. Middleton, *The Experimenters: a study of the Accademia del Cimento*, Baltimore, 1971.

[3] Magalotti, *idem.*, 84.

[4] *Ibid.*

[5] *Ibid.*

of natural phenomena using only experiments. In fact, as we shall see later, Magalotti intentionally excluded the academicians' debates about theory in order to create this appearance of a non-speculative and uncontroversial academy, adding greater credibility and authority to the Cimento's work, and therefore helping to boost the status and reputation of the academicians, as well as their princely patrons.

Soon after the Cimento was founded, other European institutions began to produce the same type of reports of experimental knowledge-making. The best-known early modern institutions to have used a similar experimental rhetoric were of course the Royal Society of London and the Acadèmie Royale des Sciences in Paris. The statutes drawn up for these institutions upon their foundations, declared their intentions only to report experiments without offering any theoretical interpretations.[6] So the statement from the *Saggi*, quoted above, is an example of the experimentalist rhetoric that appears to have been sweeping across Europe during the latter half of the seventeenth century.

With regard to the Accademia del Cimento, the story is particularly powerful, since the Cimento appears to have been the first academy in Europe to be founded on this philosophy. More specifically, it is supposed that the Florentine academicians were the first group of thinkers in the seventeenth century to adopt an organised form of knowledge-making based on an inductivist method of experimentation.[7] Such a method may be termed 'atheoretical' since it was claimed that no theoretical suppositions entered the procedure and that only this procedure could provide sound theory, or causal explanation.[8] For this reason, the activities inside the Accademia del Cimento have been a focal point for these traditional historiographies of Italian science that attempt to trace the origins of modern science.[9] In fact, as we shall see in Chapter One, early accounts of seventeenth-century Italian science, beginning with those written immediately after Galileo's

[6] In 1663, Robert Hooke drew up the statutes for the Royal Society, and laid down the following rules for the reporting of experiments. 'In all reports of experiments to be brought to the society, the matter of fact shall be barely stated without any prefaces, apologies, and rhetorical flourishes; and entered so in the register book by order of the society'. C.R. Weld, *A History of the Royal Society*, 2 vols., New York, 1975, ii, 527. In the case of the Parisian Academy, one of its leading members, Christian Huygens, wrote the following words in a memorandum to his fellow academicians in 1666. 'The principal aim and most useful occupation of this Assembly should be, in my view, to work on a natural history more or less according to the plan of Bacon .... One must distinguish chapters in this history and amass to it all observations and experiences which pertain to each particular'. C. Huygens, *Oeuvres Complètes*, 22 vols., The Hague, 1888–1950, vi, 95–96. As cited by R. Hahn, *The Anatomy of a Scientific Institution: The Paris Academy of Sciences, 1666–1803*, Los Angeles, 1971, 25.

[7] Such a universally applicable experimental method has often been seen as the essence of modern science. See J.A. Schuster and R.R. Yeo, 'Introduction', in *The Politics and Rhetoric of Scientific Method* (eds. *idem.*), Dordrecht, 1986, x.

[8] This supposed detachment of theory from fact, was also discussed by Paul Feyerabend, *Against Method*, London, 1975.

[9] As we shall see in Chapter One, with regard to the Accademia del Cimento, these historians include Giovanni Targioni Tozzetti, Giovanni Batista Clemente Nelli, Raffaello Caverni, and Antonio Favaro. In more recent times, authors such as Martha Ornstein, Eugenio Garin, Rupert Hall, and Roger Emerson, have also discussed the rise of an experimental method amongst the members of the Tuscan, English, and French Courts.

death in 1642, have been written almost purely with this theme in mind. These historiographies have considered the early seventeenth-century reports about Galileo's experimental exploits, especially Viviani's account of his teacher's work, and have reshaped those reports into stories about the rise of a modern experimental science. They have claimed that Galileo came up with a loosely articulated experimental method that was exploited and perfected by his students and followers to the point of providing a standard of research recognisable as 'modern science'. These historiographies will be referred to here as 'traditional', since it is a story that has been adopted time and again and has remained virtually unchanged even until the end of the twentieth century.

More recently, 'cultural' historians have focused on the social and political circumstances which contributed to the foundation of the Accademia in mid-seventeenth-century Florence, and the reasons for the academicians' purported devotion to the new experimental philosophy. Thanks to the work of such erudite scholars as Jay Tribby, Mario Biagioli, Paula Findlen, and Marco Beretta, we now have a thorough understanding of the proclaimed experimental programme adopted by Tuscany's early modern thinkers and sponsored by the Medici Court.[10] In fact, these authors have argued that the Cimento's experimental philosophy, much like the experimental science that Shapin and Schaffer describe in their writings regarding the early Royal Society of London, was aimed at producing atheoretical matters of fact: this is, experiments with no natural philosophical arguments attached, thus keeping clear of intellectual conflicts.[11] Therefore, as we are told by Tribby, Biagioli, Findlen, and Beretta, the members of the Accademia del Cimento and their Medici patrons maintained courtly etiquette and gentlemanly decorum, as well as a social standard for gaining legitimacy, both for the individual thinkers amongst their scientific colleagues and for the Medici Court amongst the wider European community of royal courts. In short, some authors identify this type of rhetoric as the beginnings of a loosely articulated, theory-neutral method for accumulating matters of fact. Such an experimentalist-courtly culture supposedly replaced natural philosophical concerns and conflicts, establishing the factual and gentlemanly origins of experimental science.

These types of historiographies will be referred to here as 'cultural' studies since they have certainly helped us to understand some of the court culture and political circumstances surrounding the foundation of the Accademia del Cimento. The focus of this literature on issues of courtly patronage, etiquette, and

[10] J. Tribby, 'Dante's Restaurant: The cultural work of experiment in early modern Tuscany', in *The Consumption of Culture. 1600–1800* (ed. A Bermingham and J. Brewer), London, 1991, 321; M. Biagioli, 'Scientific revolution, social bricolage, and etiquette', in *The Scientific Revolution in National Context* (ed. R. Porter and M. Teich), New York, 1992, 11–54; P. Findlen, 'Controlling the experiment: rhetoric, court patronage and the experimental method of Francesco Redi', *History of Science* (1993), **xxxi**, 39–40; M. Beretta, 'At the source of western science: the organisation of experimentalism at the Accademia del Cimento (1657–1667)', *Notes and Records of the Royal Society of London* (2000), **54** (2), 131–151.

[11] Shapin's and Schaffer's best-known works on this topic include: S. Shapin and S. Schaffer, *Leviathan and the Air-Pump*, New Jersey, 1985; S.Shapin, 'The House of Experiment in Seventeenth-Century England', *Isis* (1988), **79**, 373–404. Shapin, *A Social History of Truth: Civility and Science in Seventeenth Century England*, Chicago, 1994.

social legitimation is particularly valuable for our understanding of the academicians' aims and interests, and the negotiation of the plausibility of some of their claims. However, although the focus has shifted in the recent literature towards the wider social and political circumstances that contributed to the Cimento's foundation and workings, there still seems to be an implicit acceptance of the traditional historiography discussing the birth of modern experimental science.

In both 'traditional' and 'cultural' historiographies, the implications behind the use of the term 'experimental science', or 'experimental method', in association with these institutions, are that Aristotelian natural philosophy had been replaced in the seventeenth century by the birth of an organised, atheoretical, inductivist method of the type purportedly used by the Accademia del Cimento. Chapter One will explore in detail what terms such as 'experimental science' or 'experimental method' imply in the 'traditional' and 'cultural' historiographies. I will attempt to discard the notion that the foundation of institutions such as the Cimento established the origins of a modern science by first rejecting the pursuit of natural philosophy and by second substituting an 'experimental method'. In place of these historiographies I will formulate a new model for understanding the activities of these so-called experimentalists. I will begin by challenging the notion that it was indeed an 'experimental method' that Galileo and his successors in Tuscany had developed and refined. Galileo and his students did use different types of experiments to validate their work against Aristotelianism, but they did not adopt an experimental method in their knowledge-making of the type that various historians believe originated in the Cimento and in their allegiance to Galileo. In fact, considering the theory with which experiments are laden, according to philosophical and sociological analysts of scientific knowledge, it is difficult to imagine that any such method even existed.[12] So it will be argued here that if we focus solely on such method rhetoric in the presentational techniques of the academicians, it will distract us from the broader intellectual aims and interests that were being pursued inside the Tuscan Court.

Accordingly, we shall find through a careful analysis of the works by Galileo, Evangelista Torricelli, Viviani, Borelli, as well as the other members of the Cimento, that experiments played a subsidiary role in their work. As historians Naylor, Clavelin, Segre, Drake, and Settle have established, experiments were used as a tool of persuasion for the wider-reaching natural philosophical skills, commitments, and agendas of Galileo and his students.[13] In other words, rather than

[12] Philosophers of science, including W.V. Quine and Pierre Duhem, have been making this argument since the early twentieth century. Sociological analysts of scientific knowledge who have borrowed from the philosophical works on this subject include Trevor Pinch, Harry Collins, Barry Barnes, and the earlier works of Steven Shapin.

[13] R.H. Naylor, 'Galileo's experimental discourse', in *The Uses of Experiment: Studies in Natural Science* (ed. D. Gooding, T. Pinch and S. Schaffer), London, 1990, 117–134; M. Clavelin, *The Natural Philosophy of Galileo: Essay on the Origins and Formation of Classical Mechanics* (tr. A.J. Pomerans), Cambridge, 1974; M. Segre, 'The Role of Experiment in Galileo's Physics', *Archive for History of Exact Sciences* (1980), **23**, 227–252; S. Drake, *Galileo at Work: His Scientific Biography*, Chicago, 1978; T.B. Settle, 'Galileo's Use of Experiment as a Tool of Investigation', in *Galileo: Man of Science* (ed. E. McMullin), New York, 1967, 315–337.

finding merely traces of an 'experimental method', this book will reveal that the Cimento academicians were still far more committed to verifying and propagating their respective natural philosophical beliefs. The lives of most of the academicians included educations grounded in natural philosophical practices of an anti-scholastic tenor, with strong commitments to the linking of natural philosophy to findings and techniques of mathematics and mechanics. This came about as a result of the lessons passed on to Galileo's school of students and followers, including the members of the Accademia del Cimento. Most of the academicians were devoted to the mathematical arts, or mixed mathematics as it was also known to Aristotelians, but with the additional aim of addressing wider natural philosophical concerns. This indicates that these so-called experimental scientists were actually interested in the much broader field of natural philosophy and within it preferred an approach which, following some contemporaries and some modern historians, may be termed 'physico-mathematics'.[14] This is part of the culture of natural philosophising that dominated seventeenth-century Italian thought and that will be examined in Part One by grasping the type of natural philosophising that Galileo pursued, and the aims and interests that each of the academicians attempted to fulfil throughout their careers.

The reason why so much time and effort will be afforded to the analysis of these individuals' natural philosophical commitments before they entered the Accademia, is to show exactly what intellectual skills and agendas they took to the tasks of constructing and interpreting the group's experiments. More specifically, Part One will demonstrate that the debates inside the Cimento were not based on clashes of egos or attempts to grab the Prince's attention, or even mere opinions about how an experiment should be carried out. Instead, we will be seeing that each academician was educated and trained according to the natural philosophical debates that pervaded the colleges, universities, and courts of seventeenth-century Europe. At this time, scholastics, that is, university scholars who were teaching and practicing recently refurbished versions of Aristotelian natural philosophy, were defending the efficacy of their beliefs against the new and varying ontological and cosmological views of Neoplatonists and mechanists. Therefore, rather than study only the courtly setting of the Accademia del Cimento, or simply their rhetorical use of experiments, my aim is to show that the group's activities were situated within the wider culture of natural philosophising.[15]

[14] Peter Dear defines the term 'physico-mathematics' as an expression coined in the seventeenth century to denote the use of mathematics in the study of physics, including the natural philioopshical search for physical causes. Recently, Gaukroger, Schuster, and Sutton have also identified the use of the term by René Descartes in his attempts to find mathematical expressions of physical causes. It is with this definition in mind that I use the term at various times throughout this book, especially with regard to the rise of the mechanical philosophy in Chapter One. P. Dear, *Discipline and Experience: The Mathematical Way in the Scientific Revolution*, Chicago, 1995; P. Dear, *Revolutionizing the Sciences: European Knowledge and its Ambitions, 1500–1700*, Basingstoke, 2001, 199; S. Gaukroger, J. Schuster, and J. Sutton, 'Introduction', in *Descartes' Natural Philosophy* (eds. *idem.*), London, 2000.

[15] The natural philosophical culture in early modern Europe has been the subject of John Schuster's recent treatment of the Scientific Revolution, with which my own thesis concurs. See J.A. Schuster, 'L'Aristotelismo e le sue Alternative' in D. Garber (ed.), *La Rivoluzione Scientifica*, Rome, 2002, 337–357.

In Part Two we will be turning to the case studies. Most of the Cimento's irregularly scheduled meetings during its first five years of existence were centred on the resolution of questions regarding the pressure of air, the creation of a vacuum, the freezing process of liquids, and the properties and effects of heat and cold. In each of these fields, almost all of the academicians made contributions. But these were not simply suggestions for new experiments that could provide 'matters of fact'. Instead they were experiments that had been specifically suggested and contrived either to support or negate important natural philosophical claims. The *Saggi*'s author never made any references to any of the academicians, but letters and manuscripts reveal that there existed a culture of debate within the Cimento based on theoretical disputes that were framed according to the competing natural philosophies of Aristotelians and corpuscularian mechanists within the group. Once the academicians decided to embark on studies of these various fields of experimental inquiry, they wished to incorporate the beliefs and intellectual concerns that had dominated each of their careers up until that point, including their work in the disciplines that they were pursuing inside the Cimento. In fact, despite Magalotti's efforts in the style and rhetoric of the *Saggi* to provide the greatest possible reputation for the Accademia's members and patrons as reliable producers of natural knowledge, the text still contains traces of the natural philosophical contestation entangled in each of their experiments.

To begin with, as we shall see in Chapter Five, the academicians investigated the pressure of air and the creation of vacuous spaces through the barometer that Torricelli constructed in 1643. But rather than this being a demonstration of the Italians' dedication to innocent play with instruments and experimentalism, leading to atheoretical 'matters of fact', the construction of the barometer and its various uses throughout Europe during the 1640s and 1650s, indicate the presence of wider-reaching issues. Torricelli constructed an instrument for measuring the weight of air, so that he could apply his knowledge of mathematics to the physical world, and just as importantly, so that he could also refute the theories offered in previous decades regarding the question of whether air has any weight, and whether it is possible to create a vacuous space. The question was an important one for scholastics who vigorously argued that nature abhorred the production of a void. This was a cornerstone of their natural philosophical beliefs, since it upheld the cosmology of five elements that moved according to their natural tendencies. Atomists challenged this view in the sixteenth century, but an anti-scholastic position did not become a significant part of the natural philosophical landscape until the wider incorporation of mathematical and physical demonstrations. That is, various advocates of a range of Neoplatonic and mechanical views weighed into the discussion of air pressure and the void, leading to Torricelli's barometric contribution, and the physico-mathematical concerns that ran through the issue as it was discussed first in Paris and then Florence. So, by the time the Cimento decided to study pneumatics, physico-mathematical and mechanistic natural philosophical concerns had already been well established in that discipline.

A similar story underlies the Accademia's experiments on the freezing process of water, and the properties and effects of heat and cold. Once again, there are very few indications in the *Saggi* that there was any theorising occurring during these experiments, or indeed that any of the academicians pushed for certain natural philosophical interpretations. Yet a closer look at their work in this field will reveal, first, that corpuscularian beliefs were incorporated into the construction of the experiments, and that members, such as Borelli and Viviani, were intent on finding shortcomings in the scholastic opinions on the topics. Second, we shall see that the interpretations of the experiments made by some academicians involved the use of mixed mathematical skills derived from statics and the accumulation of quantified data that they believed represented the dynamical force of the expansion of freezing water, a typically physico-mathematical concern with deriving natural philosophical results. Finally, it will be revealed that even Leopoldo was participating in the construction and interpretation of experiments that supported the mechanistic world view. Freezing appeared to be Leopoldo's favourite topic and his heavy involvement in the creation of natural philosophical theories during the construction and interpretation of these experiments indicates that he was not enforcing a theory-free experimental method on his academicians during their first five years in operation, before they embarked on the publication of their work.

After establishing the natural philosophical issues that the academicians contested inside the Cimento during their first five years in existence, our attention in Part Three will turn to the subsequent presentation of their works. This is where we come to appreciate the political circumstances behind the rhetorical framing of the *Saggi*. When Leopoldo decided to publish a collection of the Cimento's experiments, he had to decide what such a publication should set out to achieve. Leopoldo and the Grand Duke of Tuscany Ferdinando II, clearly desired to revive the glory of patronising outstanding natural philosophical work that they had experienced with Galileo, but it would seem that on this occasion, they preferred to keep clear of any controversial claims. Such a strategy would protect the academicians from any religious confrontations and would also provide them with the image of non-speculative and non-theoretical experimentalists. This would not only explain the rigorous editing and censoring process behind the *Saggi*, but it also gives us an indication of why they preferred not to publish their vast amount of work regarding the controversial field of astronomy. So the Medici were looking to capitalise on their association with the Galilean school, and improve their status amongst the other European Courts.

Furthermore, as we shall also see in Chapter Seven, as well as in the case study regarding astronomy in Chapter Eight, this 'geo-political' pressure deflected the personal political ambitions of the academicians. Undoubtedly, they would have preferred to publicise their individual contributions to the knowledge produced by the Cimento. But since they were not able to do this, they each were still trying to position themselves for favouritism inside the Court. They did this by maintaining their natural philosophical contributions to the knowledge produced inside the Tuscan Court, and by hoping that they would be justly credited for it.

Therefore, the theoretical significance of the experiments discussed in Parts One and Two reflects the disciplinary and natural philosophical concerns of the academicians. That is to say, the Cimento's members constructed knowledge claims in disciplines that were part of a natural philosophical domain that was, furthermore, recognised and pursued all over seventeenth-century Europe. Additionally, how these concerns were used and presented inside the Tuscan Court to the royal family, and by the Court to the rest of Europe, reflects the issues of courtly status and reputation discussed by the 'cultural' historians mentioned earlier. Behind the convenient rhetoric of experimental 'matters of fact' was a deep concern with natural philosophical inquiry.

In summary, this book will be aimed at gaining an understanding of the natural philosophical skills and commitments that were a part of the careers of each of the academicians and the disciplines they studied. Meanwhile, we shall find that experimental science, in the way it has been presented by many historians as the pure, factual, and inductivist practice of an experimental method in the controlled environments of royal courts such as in Tuscany, did not, in fact, play a role in the Cimento's knowledge-making process, their construction and interpretation of claims. This is not to say, however, that experiments or courtly culture were not an important part of the landscape of natural philosophising in the mid- to late-seventeenth-century Tuscan Court. In fact, throughout this analysis, we shall see evidence of the persuasive and authoritative role of experiments for practical knowledge-making, and for maintaining the relationship between natural philosophers and their patrons. The rigorous use of experiments, or the published devotion to an experimental programme of some sort, strengthened public perception that one was appealing to an approach to making natural knowledge that was detached from theoretical convictions or presuppositions. This meant that an individual's or an institution's credibility depended on the perception from fellow natural philosophers and thinkers across Europe, that some type of experimental method was being used. This is precisely why experimental rhetoric was so valuable to the presentation of claims.

This is to suggest that there was a distinct difference between what the Cimento academicians presented in their publication, and what they were actually discussing in their meetings between 1657 and 1662, before they decided to publicise their work. But this does not mean that a non-rhetorical realm existed before they embarked on the publication process, or that I am attempting to read the minds of the academicians to find out what they were thinking during the first five years of the Cimento's existence. Rather, the aim here is to show that for political and presentational reasons, the natural philosophical concerns the academicians actually pursued when constructing their experiments had to be suppressed from public consumption. Fortunately, those concerns were preserved in the academicians' manuscripts and letters. Therefore, an understanding of the political concerns of the Medici Grand Duke and Prince Leopoldo, along with the intellectual concerns and conflicts amongst the Cimento's members, will assist us greatly in coming to terms with the actions and the pursuits of this small group of thinkers and how the *Saggi* were compiled.

PART ONE

GALILEO AND BEYOND

CHAPTER ONE

350 YEARS OF COMING TO GRIPS WITH THE EXPERIMENTAL ACTIVITIES OF GALILEO AND HIS FOLLOWERS

Most early accounts of seventeenth-century Italian science, beginning with those produced immediately after Galileo's death in 1642, were written almost purely with an experimentalist image in mind. They project the theme that Galileo and his students and followers in seventeenth-century Tuscany pursued a form of inquiry that included only the performance of experiments and the inductive collection of matters of fact. They claim that Galileo came up with a loosely articulated experimental method that was exploited and perfected by his students and followers to the point of providing a standard of research recognisable as 'modern science'.[1] More recently, several 'cultural' historians have provided more valuable contributions, associating experimental philosophy with the social and political context of the Tuscan Court. The aim of this chapter, then, is to evaluate the validity of the different traditional and cultural accounts of Italian early modern science. This will be done by also examining the issues that these historiographies have failed to mention, in particular, the culture of natural philosophical theorising and contention that existed throughout all of Europe during the seventeenth century, and that included a clash between Aristotelians and the new and changing versions of mechanism. We begin with the traditional accounts regarding the supposed experimental method of Galileo and the academicians who followed him.

1. EARLY UNDERSTANDINGS OF GALILEO'S AND HIS STUDENTS' EXPERIMENTALISM

The fascination with the life and works of Galileo began immediately after his death in 1642. During the second half of the seventeenth century, Galileo's followers conducted long and thorough searches for his letters and manuscripts

[1] The obvious exception to this type of historiography is, of course, the work by Alexandre Koyré, who argued that Galileo did not even perform experiments. See *Galileo Studies* (tr. John Mepham), Hassocks, 1978.

scattered all over Europe. The aim of these searches, led by Galileo's last student, Vincenzio Viviani, and supported by Tuscany's ruling Medici family, was to preserve the memory of the life and works of the Pisan natural philosopher and mathematician. Indeed, this was the aim behind Carlo Manolessi's 1655 Bolognese publication of the collection of Galileo's works.[2] In support of this publication, Prince Leopoldo asked both Viviani and another Galilean follower, Niccolò Gherardini, to write biographies of Galileo. These were originally intended for Manolessi's publication, but for some reason were never included. In any case, Viviani's efforts to recover Galileo's papers, were tireless and eventually resulted in the vast collection of Galilean manuscripts now held at the Biblioteca Nazionale Centrale in Florence.[3]

Since this early movement to preserve the memory of Galileo, he has been the subject of numerous biographies and at the centre of countless stories about the Scientific Revolution. His experiments and claims about celestial and terrestrial motion, directed against the traditional values and beliefs of the scholastics, have always attracted the interests of Galilean historians and admirers.[4] Most of these historiographies have been steadily creating an image of Galileo and his 'school' as representing the origins of experimental science.

Aside from his controversy with the Catholic Church, Galileo is of course well remembered for his recordings of countless astronomical telescopic observations relating to the problem of planetary motion and his experiments regarding terrestrial mechanics. The enduring images of Galileo include his enticing of others to peer at the stars through his telescope, his dropping of heavy objects from the Leaning Tower of Pisa, and his observations of the pendulum-like movements of the lamp inside Pisa's cathedral. With the exception of his telescopic observation sessions with his supporters and critics, these events were never recorded by Galileo. In fact, Viviani was the only one to have annotated these experiments after Galileo's death.

In 1654, Viviani wrote *Racconto istorico della vita del Sig. Galileo Galilei*, in the form of a letter to Prince Leopoldo de' Medici.[5] This was the biography intended for publication in Manolessi's *Opere di Galileo Galilei*. Later Viviani also composed *Vita di Galileo*, published posthumously in 1717.[6] In both works he begins by documenting Galileo's early observations in Pisa, leading into the first accomplishments of his long career. Viviani first tells his readers about Galileo's steps towards the invention of the pendulum clock. After casually observing a swinging

[2] This edition of Galileo's works did not include, of course, the banned *Dialogue*. C. Manolessi, *Opere di Galileo Galilei*, 2 vols., Bologna, 1655–1656.

[3] Viviani's collection of Galileo's papers is documented by A. Favaro, *Documenti inediti per la storia dei manoscritti galileiani nella biblioteca Nazionale di Firenze*, Rome, 1886.

[4] For some indication of the bibliographical richness of Galileo's life and works, see M. Segre, *In the Wake of Galileo*, New Jersey, 1991, 36.

[5] Antonio Favaro has since published this in a collection of Galilean works.V. Viviani, 'Racconto Istorico di Vincenzio Viviani', in *Le Opere di Galileo Galilei, Edizione Nazionale* (ed. A. Favaro), 20 vols., Florence, 1890, xix, 597–632.

[6] V. Viviani, *Vita di Galileo* (ed. L. Borsetto), Bergamo, 1992. See also V.Viviani, *Vita di Galileo* (ed. F. Flora) Milano, 1954.

lamp inside the Duomo in Pisa, Galileo tested 'the equality of its oscillations, when it occurred to him to apply its use to medicine and measuring the beat of one's pulse'.[7]

There is considerable doubt regarding the validity of Viviani's narration of this event. The moment of revelation inside the Duomo supposedly occurred in 1583, yet the lamp to which Viviani refers was not installed in the cathedral until 1587.[8] Nevertheless, this story quite clearly, and possibly deliberately, creates the impression that Galileo gained access to a part of nature through his brilliant intellect and an enlightening moment of experience and observation. With this alone, any reader of this biography could be impressed by Galileo's experimentalist tendencies at a young age. But the impressive stories do not end there.

In possibly the best example of how Galileo supposedly obtained knowledge through direct experience of nature, Viviani reports how Galileo had devised his theory of the uniform acceleration of falling bodies using 'repeated experiments carried out from the height of Pisa's bell tower in the presence of other lecturers, philosophers and all the students'.[9] This event, although not mentioned in any other source, has become a prime example of Galileo's experimental philosophy. It is questionable whether these events actually ever took place, since Viviani was the only person to have recorded them. In any case, regardless of whether Viviani's stories are true or not, they would have certainly been sending out a powerful message to all seventeenth- and eighteenth-century thinkers who read through either Viviani's manuscript or his posthumous publication of the biography of Galileo. Indeed, Segre notes that Viviani's work was typical of 'the Renaissance practice of turning biography into hagiography' by exaggerating and even inventing stories about moments of inspired brilliance.[10]

In addition to these observational and experimental exploits, as reported by Viviani, Galileo was also a key member of the Accademia dei Lincei, the so-called lynx-eyed natural philosophers, under the protection of a young Roman noble by the name of Federico Cesi (1585–1630). Until its demise in 1630 this group, working according to Cesi's interests and directions, consisted of very keen practitioners of the telescope and microscope.[11] During its existence, it also supported the publication of Galileo's *Letters on Sunspots* (1613) and the *Assayer* (1623), as well as Cesi's *Apiarium*, a recording of observations made with the microscope published in 1625.

[7] 'Con la sagacità del suo ingegno inventò quella semplicissima e regolata misura del tempo per mezzo del pendulo, non prima da alcun altro avvertita, pigliando occasione d'osservarla dal moto d'una lampada, mentre era un giorno nel Duomo di Pisa; e facendone esperienze esattissime, si accertò dell'egualità delle sue vibrazioni, e per allora sovvennegli di adattarla all'uso della medicina per la misura della frequenza de' polsi'. Viviani, 'Racconto Istorico', 603; Viviani, *Vita* (ed. F. Flora), 30.

[8] Segre, *In the Wake*, 36; Viviani, *Vita* (ed. F. Flora), 23, 165.

[9] 'con replicate esperienze, fatte dall'altezza del Campanile di Pisa con l'intervento delli altri lettori e filosofi e di tutta la scolaresca', Viviani, 'Racconto Istorico', 606; Viviani, *Vita* (ed. F. Flora), 34.

[10] M. Segre, 'Viviani's Life of Galileo', *Isis* (1989), **80**, 208–209, 228.

[11] J.E. McClellan III, *Science Reorganized. Scientific Societies in the Eighteenth Century*, New York, 1985, 3, 44. See also D. Freedberg, *The Eye of the Lynx: Galileo, His Friends, and the Beginnings of Modern Natural History*, Chicago, 2003.

Galileo's association with the empirical exploits of the Lincei, in addition to his own experimental achievements as presented in Viviani's story, have been enough to convince many historians that Galileo's career was based on a strong adherence to a pure experimental method with little or no role for theorising. Indeed, one need only look through some early to mid-twentieth century texts to see the results of such simple evaluations of Galileo's life.[12] Viviani's influence in these is particularly evident. For instance, Viviani's allegedly fictional account of the experiments on falling bodies at the Tower of Pisa is used by Martha Ornstein to conclude that Galileo was 'the originator of the scientific method'.[13] Abetti and Pagnini, early twentieth-century editors of Galileo's and the Cimento's works, also conclude that being part of Italian natural philosophy during the second half of the seventeenth century meant 'putting into practice the experimental method of research' first used by Galileo.[14] So these historiographies assert that the Pisan natural philosopher initiated the use of the experimental method, a practice that was perfected by the following generation of thinkers. Clearly then, almost 300 years after Viviani wrote Galileo's biography, he has still managed to convince some historians that Galileo was on the verge of creating modern experimental science. This is why we may refer to these historiographies as traditional – they accept Viviani's stories about Galileo's experimental work and continued to regard Galileo and his students as the first modern experimental scientists to produce atheoretical, factual knowledge of nature by use of a unique, efficacious method. Meanwhile, they make no effort to analyse the mathematical and mechanical principles Galileo used to construct his claims. Instead, just like Viviani, they rely on inaccurate evidence to create an image of the great Pisan natural philosopher that ties in with today's notions of reliable and trustworthy scientific enterprise.

Of course, not all historians have taken up this position. For example, far from arguing that Galileo used an experimental method, Alexandre Koyré claims that the Pisan natural philosopher performed no actual experiments at all. While perhaps this view opposing the traditional historiographies is a slight exaggeration of how Galileo used, or did not use, experiments, other historians, such as Stillman Drake and Thomas Settle, also question the 'traditional' image of Galileo as experimental scientist.[15] We shall return to these shortly. In the meantime, the

[12] For a sophisticated discussion regarding the need to replace 'traditional' historiography of Galileo with more contextual studies, see J. Renn (ed.), *Galileo in Context*, Cambridge, 2001.

[13] M. Ornstein, *The Role of Scientific Societies in the Seventeenth Century*, Chicago, 1938, 24–26.

[14] 'mettendo in valore il metodo di ricerca sperimentale'. Abetti and Pagnini, 'Premessa', in *Le opera dei discepoli di Galileo Galilei. Edizione Nazionale. I. L'Accademia del Cimento. Parte Prima.* Florence, 1942, 8.

[15] According to Koyrè, by the time Galileo started to work on natural motion, the Aristotelian method of observation was not always utilized. Instead, Koyrè refers to the "Archimedisation" of physics under the sixteenth-century practical mathematicians, as the main impetus behind Galileo's own work: 'An Archimedean physics means a deductive and 'abstract' mathematical physics, and it was just such a physics that Galileo was to develop at Padua. A physics of mathematical hypotheses; a physics in which the laws of motion ... are deduced 'abstractly', without ... recourse to experiments with real bodies. The 'experiments which Galileo, and others after him, appealed to, ... were not and could never be any more than thought experiments'. Koyrè, *Galileo Studies*, 37.

traditional experimentalist stories do not end with Galileo. His achievements in so-called experimental science supposedly served as an example for his students and followers who took up the task of preserving and enhancing Galileo's experimentalist image. We have seen that Viviani's biography of Galileo may be interpreted as an example of this – how one of Galileo's students managed to show us the origins of experimentalism in seventeenth-century Italy. Yet Viviani's work is not the only model example for such experimentalist historiographies.

In 1644, Evangelista Torricelli (1608–1647), Galileo's most famous student and his successor in the prestigious position of First Mathematician in the Medici Court, performed the famous experiment with mercury that resulted in the construction of the first barometer. Despite Torricelli's commitment to mathematical and geometrical problems throughout his career, and his natural philosophical agenda discussed in the following chapters, this achievement has stood out as his most important, and also significantly, as a great indication of a purely experimentalist mentality that Galileo had supposedly instilled in his students. Raffaello Caverni even goes so far as to suggest that Galileo was not a full-blown experimentalist, but his student Torricelli certainly was. In fact, according to Caverni, Torricelli was the starting point for the art of experimental philosophy in the seventeenth century.[16]

Furthermore, some writers have regarded Galileo's followers, including the members of the Accademia del Cimento, virtually as the ambassadors of Galileo's experimental philosophy during the mid to late seventeenth century. One may get this impression from reading Giovanni Targioni Tozzetti's preface to his famous eighteenth-century publication regarding the academicians' activities, *Notizie*. Here Targioni Tozzetti is interested in exploring the progress of Tuscan celestial and terrestrial science. This includes the productive discoveries and diligent experiments carried out firstly by Galileo, then his disciples, and finally the Accademia del Cimento.[17] Meanwhile, Giovanni Batista Clemente Nelli, also writing during the eighteenth century, believes that Viviani clearly guided the natural philosophical interests of the Medici Grand Duke into the field of Galilean experimental science.[18]

This image of the birth of experimental science in Tuscany established even firmer roots in twentieth-century writings. In 1903, Stefano Fermi wrote about how Galileo's students freed themselves from traditional natural philosophical theorising and adopted an inductive experimental method.

> The common characteristic of the followers of the Galilean school ... is the spirit that pushes them to observation, to reckoning, to experience, to the inductive method, and sways them from metaphysical deductions, from subtle discriminations and from the a priori demonstrations of the stale philosophical and scientific school.[19]

[16] R. Caverni, *Storia del Metodo Sperimentale in Italia*, 6 vols., Florence, 1891–1900 (reprinted Bologna, 1970), i, 177.

[17] G. Targioni Tozzetti, *Notizie degli aggrandamenti delle scienze fisiche accaduti in Toscana nel corso di anni LX del secolo XVII*, 3 vols., Florence, 1780, i, 5.

[18] G.B.C. Nelli, *Saggio di storia letteraria fiorentina del secolo XVII scritta in varie lettere*. Lucca, 1759, 111.

[19] S. Fermi, *Lorenzo Magalotti. Scienziato e Letterato (1637–1712)*, Piacenza, 1903, 87.

It is presumably this 'spirit' that Gustavo Barbensi also claims to identify in the practices of the Cimento. According to Barbensi, the academicians perfected the 'experimental method of the Galilean School'.[20] Martha Ornstein also asserts that early modern thinkers in Italy after Galileo gave prime importance to their experiments, and since universities refused to sponsor these experimentalist activities, they turned to scientific societies such as the Cimento: 'It is superfluous to say that they made every effort to foster the cause of experimental science. This was the keynote, the charter of their existence, the motive underlying their every activity.'[21] Finally, Rupert Hall, Roger Emerson, and Gaetano Pieraccini also describe the use of Galilean or Baconian experimental method as the sole impetus behind the Cimento's work.[22]

Once again, we see the traditional image of the birth of experimental science in seventeenth-century Tuscany being exploited by some historians. Admittedly, there are greater grounds for projecting an experimentalist image for the Cimento than there are for Galileo. At least the academicians left us published declarations of their intentions to perform only experiments during their meetings, this being the rhetoric of the *Saggi*. Yet with little more than the *Saggi* as evidence, is it fair to say that 'experimental science' underlay the Cimento's 'every activity'; that the academicians virtually lived and breathed purely inductivist experimental knowledge from 1657 to 1667? Later I shall be suggesting how manuscript evidence provides for quite a different story; how there was a significant difference between what the academicians said in public, and what they actually did – the natural philosophical issues about which they contended. For now, however, it is important to note how other recent writers have also suggested that the situation was more complex than the pure adaptation of an experimental method that the *Saggi*'s author and editors were prepared to reveal, or what the early twentieth-century historiographies suggest. More specifically, 'cultural' historians have linked early modern Tuscan experimentalism to certain rules of behaviour and etiquette for investigating nature inside the princely court. In particular, for Jay Tribby, Paula Findlen, and Mario Biagioli, the focus has shifted to the social and political aims and interests of the Cimento's patron, the Medici family. This has provided us with some valuable material regarding the cultural context of the Cimento's existence and will aid us in our understanding of the academicians' activities and the construction of their knowledge claims. Nevertheless, the continued references in some of these works to the birth of experimental science carry some serious implications. Those implications will be analysed here after revising some of the issues of courtly culture that these authors have brought to light.

[20] G. Barbensi, *Borelli*, Trieste, 1947, 19.

[21] Ornstein, 259.

[22] A.R. Hall, *The Scientific Revolution 1500–1800: The Formation of the Modern Scientific Attitude*, 2nd edn., Boston, 1966, 38; R. Emerson, 'The organisation of science and its pursuit in early modern Europe', in *Companion to the History of Modern Science* (eds. R.C. Olby, et al.), London, 1990, 964; and G. Pieraccini, *La stirpe de' Medici di Cafaggiolo*, 3 vols., Florence, 1925, ii, 603.

2. MEDICI PATRONAGE OF SEVENTEENTH-CENTURY NATURAL PHILOSOPHY

During the fourteenth and fifteenth centuries, the economic and political rise of the Medici House in Florence coincided with the emergence of great Florentine writers, sculptors, and artists of the Italian Renaissance, including among others, Dante, Boccaccio, Petrarch, Brunelleschi, Ghiberti, and Lippi.[23] The Medici were not always involved in the patronage of these artists, yet from their cultural and political dominance of the period, they had always closely associated themselves with them and their works and the wealth they brought to the region during that golden period of Tuscany's history. Yet according to Jay Tribby, the strong relationship that is thought to have existed between the Medici and the great Renaissance artists, was actually largely a product of the Medici's propaganda during the sixteenth and seventeenth centuries.[24]

During the early 1500s, the Medici family was exiled from Florence following invasions from the French and Spanish. But nearing the middle of the sixteenth century, they returned to the Florentine political scene and eventually acquired a dominance all over Tuscany. The new Medici Grand Dukes then sought to regain prestige and power by appealing to the social ambitions of Tuscany's nobles and bureaucrats. According to Tribby, they did this by reliving the memory of their Renaissance heroes and challenging Tuscany's artists and thinkers to emulate their predecessors: '[T]his Renaissance was dangled before the members of these groups as the key to their social success within the new culture of the court.'[25] In other words, the Medici offered members of the Tuscan Court the promise that they could become part of Europe's intellectual, political and cultural elite.

Apart from the literary and artistic experts who continued to expand upon the work of their early Renaissance predecessors, during the late sixteenth and early seventeenth centuries mathematicians also began to join the ranks of Tuscany's cultural elite. During the Renaissance, in Tuscany and across most of Europe, the retrieval and reform of Ptolemy's and Archimedes' mathematical works, against the background of the recovery of Plato's works, eventually led to the application of such classical mathematics to practical fields such as navigation and engineering, considered valuable to regions like Tuscany with military and trade interests.[26] Also, by the beginning of the seventeenth century, classical mathematics even began to be applied by some Neoplatonic philosophers to the causal inquiry of physical phenomena. As Jim Bennett argues, these mathematical sciences were positioning the status and reputation of

[23] This was under the guidance of Cosimo Il Vecchio (1389–1464) and Lorenzo Il Magnifico (1449–1492).
[24] Tribby, 'Dante's Restaurant', 319–320.
[25] *Ibid.*, 320.
[26] P. Galluzzi, 'Il mecenatismo mediceo e le scienze', in *Idee, Istituzioni, Scienza ed arti nella Firenze dei Medici* (ed. C. Vasoli), Florence, 1980, 191–192.

mathematicians as valuable members of Europe's intellectual elite during the late Renaissance.[27]

So the Medici gained an interest in natural philosophy and mathematics as a means of boosting the social status of the Tuscan Court in Europe. This is not to suggest that the Medici House in Florence had been a protector of the mathematical sciences since the fifteenth century. Rather it is only to give us an indication of how the Medici Court had long involved itself with the elite artists and thinkers of this community, and this was to lead eventually to the Medici patronage of seventeenth-century natural philosophers and mathematicians.

Thanks to recent historiographies focused on courtly culture, we can identify how early seventeenth-century Tuscan mathematicians such as Galileo, legitimated their work within the region's princely court. Inversely we can examine how the Court also used these new representatives of elite culture to raise their own status as patrons of the latest intellectual activity.[28] Therefore, in order to analyse how the Medici rulers intertwined their political aims and interests with the natural philosophical and mathematical practices of the early modern period in Italy, we must begin with Galileo's employment in the Medici Court and his relationship with his patron, the Grand Duke Cosimo II.

For many years, a prominent position inside the Medici Court meant an attractive lifestyle and high wages. For Galileo and the many late Renaissance artists and thinkers working in Medicean Florence, the way of acquiring such a position was to present a gift to the Grand Duke that could bring the Medici further honour and prestige, helping to establish the Tuscan royal family as amongst the most powerful in Europe.[29] This gift could have been an impressive artistic or literary work, but in Galileo's case, it was his use of the telescope that provided him with entry into the Tuscan Court. When he observed four of Jupiter's satellites he immediately named them the 'Medicean Stars', a dedication to the Grand Duke and his family. Galileo would have been particularly pleased to find exactly four stars; it was not only symbolic of the four Medici brothers governing at that time, but it also associated their power in Tuscany to the eternity of the heavens.[30] So Galileo's observations became a very symbolic natural monument to the Medici rulers – a gift from client to patron that guaranteed Galileo a highly prestigious position, that of 'Court Mathematician and Philosopher'. As Biagioli argues, the observations of Jupiter's moons with the use of a new instrument was

[27] J.A. Bennett, 'The Challenge of Practical Mathematics', in *Science, Culture, and Popular Belief in Renaissance Europe* (ed. S. Pumfrey, P.L. Rossi, and M. Slawinski), New York, 1991, 176–179. See also E. Garin, 'La cultura filosofica fiorentina nell'età medicea', in *Idee, istituzioni, scienza ed arti nella Firenze dei Medici* (ed. C. Vasoli), Florence, 1980, 83–86.

[28] Tribby, 'Dante's Restaurant', 324.

[29] P. Findlen, 'The Economy of Scientific Exchange in Early Modern Italy', in *Patronage and Institutions: Science, Technology, and Medicine at the European Court, 1500–1750* (ed. B.T. Moran), Suffolk, 1991, 5–7; W. Eamon, 'Court, Academy, and Printing House: Patronage and Scientific Careers in Late-Renaissance Italy', in *Patronage and Institutions: Science, Technology, and Medicine at the European Court, 1500–1750* (ed. B.T. Moran), Suffolk, 1991, 39.

[30] R.S. Westfall, *The Construction of Modern Science*, Cambridge, 1977, 19–20.

a 'status-carrying' gift for the Medici, and they repaid Galileo with their patronage for the rest of his life, assuring him a tremendous rise in social status.[31]

However, during the 1640s, the opportunities for the Medici to continue promoting natural philosophical thought had almost come to an end. After Galileo's death in 1642, Evangelista Torricelli continued to produce significant work in the areas of mathematics and pneumatics, but he, along with two other well-respected Tuscan mathematicians, died in 1647, leaving a lack of established talented thinkers in Florence.[32] Nonetheless, a new generation of Galilean followers soon emerged under the new Grand Duke of the mid to late seventeenth century, Ferdinando II (1610–1670) and his brother Prince Leopoldo, both also eager 'disciples' of Galileo.[33] This is where we may begin to appreciate the issues of princely patronage involved in seventeenth-century Tuscan natural philosophy. Galileo's impressive observations, as well as his relationship with his Medici patrons, were to set the pattern for the Accademia's foundation 15 years after his death. In other words, Galileo's instruments and experiments, provided the prestige to the Tuscan Court that the Medici were seeking. Therefore, after Galileo's death, they promoted his practices in order to extract maximum advantage from his reputation. This is particularly evident in their strong support for Viviani's search for Galileo's letters and manuscripts, as well as the 'heroisation of Galileo', as Segre describes it, in Viviani's biography of his teacher.[34]

Furthermore, in the decade leading up to the Accademia's first recorded meeting in 1657, Ferdinando is believed to have dedicated quite a bit of his time to the promotion of natural philosophical activity in Tuscany. He even contributed personally to several fields of study. During the 1640s, the Tuscan Grand Duke took part in the development of accurate and useful thermometers, hygrometers, and hydrometers, all instruments used later by the academicians. In 1649, he was the first to suggest the use of mercury in thermometers, rendering it a much more practical instrument.[35] During this time Ferdinando possibly even supervised an informal experimentalist academy, the predecessor to the official Accademia del Cimento. Although there is very little evidence supporting the existence of this academy, there is no doubt that during this period Ferdinando called upon the expertise of Torricelli, Viviani, Renieri, and the del Buono brothers, to carry out some investigations on the freezing of various liquids, the melting of ice, the growth and nourishment of plants, and the speed and movement of sound and light.[36] Furthermore, there is reason to believe from the existence of two diaries of experimental activity in 1657, that Ferdinando continued to supervise

[31] M. Biagioli, 'Galileo's System of Patronage', *History of Science* (1990), **28**, 18–19; M. Biagioli, *Galileo's Instruments of Credit: Telescopes, Images and Secrecy*, Chicago, 2006, 130–131.

[32] The other two were Buonaventura Cavalieri (1598–1647) and Vincenzio Renieri (1606–1647).

[33] Ferdinando took over the regency from his mother (Cosimo II's wife) Maria Maddalena of Austria, in 1628.

[34] Segre, *In the Wake*, 122–126.

[35] Targioni Tozzetti, i, 150.

[36] *Ibid.*, ii, 163–180.

experimental sessions parallel to those of the Accademia del Cimento.[37] The indication that we get from the diary thought to belong to Ferdinando's group is that the same academicians participated in both sets of meetings, and much of the time the same issues were investigated through rigorous experimentation.[38]

In the meantime, Leopoldo, possessing the same intellectual interests as his eldest brother, also sought to practice natural philosophy, establishing the Accademia del Cimento in 1657. Leopoldo's genuine interests in natural philosophy are not only evident, as we shall see later, in his contributions to the Cimento's concerns with the freezing process of liquids, but can also be seen in a letter written by Magalotti to his friend in Rome, Ottavio Falconieri (1646–1676), in July 1664. Here Leopoldo was said to have been 'satisfied to act as an academician, and not as a Prince'.[39] So Leopoldo was not only interested in natural philosophical thought, but also even contributed to the discussions and practices carried out in the Court.[40]

In addition, Magalotti also noted that the Prince still took his role as protector and promoter of the Accademia quite seriously. In fact, Leopoldo gave the commands and decided on the actions of the organisation regarding the fields they worked on; when they would meet; or how their experiments would be presented in publication. After all, Leopoldo, like his ruling ancestors, was well aware of the importance of the work that was produced under his and Ferdinando's patronage. It was crucial that the Cimento's achievements project the political power and cultural identity of Florence and the Medici rulers.[41] This political aim implanted at the Cimento's very foundation is particularly evident in Leopoldo's eagerness to extract a publication from his academy. As Middleton's research on the *Saggi*'s publication process shows, the Prince hinted at the possibility of a publication to a correspondent as early as 1660. This suggests that Leopoldo saw a publication as the ultimate goal of the Accademia, a publication that, as Michelangelo Ricci wrote to Leopoldo in 1660, would so impress the world that they would 'return the applause that is merited by the talent and diligence of these gentlemen [the academicians], and first of all by the magnanimity of Your Highness'.[42] It is thus clear from Ricci's words that a certain prestige

[37] Both these diaries will be referred to regularly in Part Two. They are manuscripts kept among Galilean papers in the Biblioteca Nazionale Centrale in Florence. The diary believed to be a copy of the official Cimento diary is in Ms. Gal. 262, while the diary thought to belong to Ferdinando's academy is in the folder labelled Ms. Gal. 261. There are also three other diaries related to the Cimento, the first (BNCF, Ms. Gal. 260, ff. 2r–32r) appears to be an incomplete copy of the official Cimento diary, while the second (BNCF, Ms. Gal. 260, ff. 34r–39r) is believed to belong to Rinaldini. The last diary is in Viviani's handwriting (BNCF, Ms. Gal. 260, ff. 226r–281r).

[38] Much of the analysis of the separate diaries is due to: Middleton, *The Experimenters*, 46–7; See also Targioni Tozzetti, i, 161–162.

[39] 'si contenta di far da Accademico, e non da Principe'. Fabroni (ed.), *Delle lettere familiari*, i, 86.

[40] In particular, in his experiments for the Cimento, Leopoldo demonstrated his leanings towards a mechanical natural philosophy. It will be shown in Chapter Six how he was quite determined to implement his own skills and commitments as a natural philosopher when performing experiments that he himself requested from his academicians.

[41] Targioni Tozzetti, i, 93; Tribby, 'Dante's Restaurant', 321.

[42] Fabroni, *Delle lettere familiari*, ii, 110. As cited by Middleton, *The Experimenters*, 66.

awaited the Medici should they make the academicians' work public. Indeed, if we are to believe the dedicatory letter to Ferdinando II at the beginning of the *Saggi*, this was precisely the aim of the publication:

> The printing of the first samples of the experiments in natural philosophy that have been made for many years in our Academy ... carry to those regions of the world in which virtue shines most brightly, new evidence of the great munificence of Your Highness and call back towards you with a new sense of gratitude the true lovers of the fine arts and the most noble sciences.[43]

So most certainly, Ferdinando's and Leopoldo's heavy participation in the experiments performed inside the Court during the mid-seventeenth century, begins to support the emphasis that Tribby, Findlen, and Biagioli place on the importance of Medici patronage in seventeenth-century Tuscan experimental philosophy. The benefits of having founded the Cimento were soon evident for the Medici family. This institution, along with the Medici's support for the search for Galileo's lost documents and the publication of his life story, kept alive Galileo's advancements in natural philosophy, making the most of the prestige associated with his work. Furthermore, as Biagioli states, employment by the Medici Court also gave natural philosophers, such as each of the academicians, greater status among their peers and a secure working environment.[44] Therefore, the academicians were obliged to act within this courtly setting, praising the Prince's decisions at every opportunity and yielding to Leopoldo's opinions on any natural philosophical issue.

Thus, there are benefits to be gained from analysing the courtly culture in Tuscany from the Renaissance until the seventeenth century. First, we become aware of the relationship between the Medici and their court mathematicians and philosophers. Second, we recognise the social and political agenda behind the foundation of the Accademia. Undoubtedly, therefore, our understanding of these issues of courtly interests in the Italian context aids us immensely in our studies of the Accademia del Cimento. At this point, however, it must be noted that it is easy to get carried away with such 'cultural' historiographies. They can be used effectively to examine the broad political context of early modern science, but at times, these writings slip into the type of discussions typical of 'traditional' historiographies. That is, they wind-up claiming that the birth of an experimental method occurred in institutions such as the Accademia del Cimento.

[43] 'Il pubblicar con le stampe i primi saggi delle naturali esperienze, che per lo spazio di molti anni si son fatte nella nostra Accademia ... è una cosa stessa che recar nuova testimonianza a quelle regioni del mondo dove la virtù più risplende, dell'alta munificenza dell'A.V. e richiamare verso di lei a nuovi sensi di gratitudine i veri amatori delle bell'arti e delle scienze più nobili'. Magalotti, 82. The publication process finally got under way in 1662 when Magalotti composed a draft of the *Saggi* and passed it on to Borelli, Viviani, and Rinaldini, for editing. The entire writing and editing process of the text will be discussed in greater detail in Part Three.

[44] Biagioli, 'Scientific Revolution', 18; see also Findlen, 'Controlling the Experiment', 39–40.

3. SURVEY OF RECENT HISTORIOGRAPHIES OF THE EXPERIMENTAL LIFE IN EARLY MODERN COURTS

In recent years, Steven Shapin and Simon Schaffer have produced extensive studies of the experimental life of the early London Royal Society.[45] Their works have focused on the rise of 'the new empirical science of seventeenth-century England' in place of traditional natural philosophical interests.[46] According to Shapin and Schaffer, Robert Boyle (1627–1691) made it clear to his colleagues in England that the only certain way of acquiring knowledge was through a 'programme' of experimental fact-making. Furthermore, Shapin and Schaffer claim that the success of this experimental science depended on the trustworthiness of the experimenters to produce matters of fact: that since they reported their experimental findings to each other according to codes of civil and honest gentlemanly behaviour and discourse. All players in this gentlemanly, courtly game could trust and build on each other's reports.[47] This practice of reporting matters of fact supposedly replaced natural philosophical concerns that were previously pursued in the early seventeenth century.

In all fairness to Tribby, Findlen, and Biagioli, their intentions are not to present this type of origin story for the Accademia del Cimento. As we have seen, their work helps to provide a strong link between the rise of experimental practices in seventeenth-century Italy, and the social and political ambitions, and customs of the Tuscan Court. However, what distinguishes their position from Shapin's and Schaffer's is that they claim not to be interested in seeking the origins of experimental science as a product of courtly culture. In fact, they openly and strongly construct their arguments on the basis of their interest in 'cultural' history. As Tribby puts it:

> My reading of experiment under Ferdinando II is less concerned with the place of the Cimento within the history of science than with the place of the *cimento*, the wide range of tests through which individuals displayed their social capacities and purchased their social status, within the culture of the court.[48]

Similarly, Findlen expresses her wish to show how the presence of the Cimento academicians in the Tuscan Court 'contributed significantly to the portrayal of the Medici as wise and munificent rulers who patronised and participated in the development of new forms of scientific knowledge'.[49] Finally, Biagioli claims that while his work does not challenge Shapin's and Schaffer's views on the early Royal Society and its use of experiments, he believes that the practice of experiments in the grand duchy of Tuscany depended almost exclusively on its use as a source of 'social legitimation' of its practitioners and its patrons.[50] Therefore, we may conclude that while maintaining a distance from experimentalist origin stories, these

[45] See Introduction, note 11.
[46] Shapin, *Social History of Truth*, xxi.
[47] Shapin and Schaffer, 22.
[48] Tribby, 'Dante's Restaurant', 326.
[49] Findlen, 'Controlling the Experiment', 39.
[50] Biagioli, 'Scientific Revolution', 37–38.

historians of early modern Italy actually provide valuable material for a contextual account of Tuscany's seventeenth-century experimental practices.

Nevertheless, while we may be willing to praise these writings for their erudite work on 'cultural' history, Tribby and Findlen still make some allusions to the history of science and the supposed birth of atheoretical experimental knowledge inside the Cimento. For example, Findlen often describes Francesco Redi's career inside the Tuscan Court as the beginnings of an 'experimental method', or 'scientific method'.[51] Meanwhile, Tribby describes:

> [T]he emergence of a new vocational category within the court, that of the experimenting courtier who, in contrast to the philosophising courtier, relies on these new, narrowly conceived activities known as *esperienze* to keep his feet – and his thoughts – ... far away from the speculative work that had ruined the career of another Medici courtier, Galileo, just a few decades earlier.[52]

Regardless of his insistence that he does not deal with science or philosophising in his analysis of the Tuscan Court, Tribby still makes the point in this passage that the 'experimental life' inside the Court in the mid to late seventeenth century, signalled a rupture from the theoretical and speculative work produced by natural philosophers such as Galileo. The new 'experimenting courtiers', as Tribby calls them, all of a sudden began producing atheoretical knowledge claims thanks to a perfected inductive method of research. This is where the works by Tribby and Findlen carry some serious implications. While they claim to be doing 'cultural' history, any spin-offs from their arguments into history of science could mean once again slipping into stories about the origins of modern experimental science of the type produced either by traditional historians or by Shapin and Schaffer. An example is Marco Beretta's recent work on the Cimento.[53]

Uninterested in the Accademia's links to Renaissance culture and politics, Beretta clings to the type of history of the birth of experimental science to which Findlen and Tribby allude and that has evidently survived since the eighteenth century. With the assistance of Shapin and Schaffer's analysis of early modern 'experimental life', Beretta arrives at the conclusion that the Cimento academicians were, like the members of the Royal Society, breaking away from natural philosophical theorising in order to produce factual experimental knowledge. Beretta claims that the Cimento's emphasis on experimental science signalled the emergence of a society completely different to the Renaissance academies: 'As a matter of fact', states Beretta, 'the foundation of the Accademia del Cimento sanctions the birth of a new way of confronting science'.[54] Beretta does not go so far as to mention the role of the Tuscan Court's gentlemanly culture in maintaining a 'matter of fact' investigation of nature. Nevertheless, he clearly insists that

[51] Findlen, 'Controlling the Experiment', 43–45. Findlen's analysis of Redi's career inside the Medici Court will be discussed in more detail in Chapter Four.

[52] J. Tribby, 'Of Conversational Dispositions and the *Saggi*'s Proem', in *Documentary Culture Florence and Rome From Grand Duke Ferdinand I to Pope Alexander VII: Papers from a Colloquium held at the Villa Spelman, Florence, 1990* (eds. E. Cropper, G. Perini, and F. Solinas), Bologna, 1992, 386.

[53] Beretta, 131–151.

[54] *Ibid.*, 134.

the Accademia broke away from traditional natural philosophy, including Galileo's emphasis on mechanics and mathematics, to be the first institution to practice experimental science, providing 'the birth of a new form of scientific knowledge'.[55] By taking as literal reporting the experimentalist rhetoric of the *Saggi*, Beretta claims that the 'Accademia del Cimento remained neutral, adhering faithfully to the mere description of facts'.[56] This historiography is thus reminiscent of the traditional twentieth-century authors relying on the *Saggi* in order to reach the same conclusions about the origins of experimental science. Furthermore, Beretta makes a loose reference to how 'unpublished manuscripts and laboratory diaries also confirm the general tendency of the academicians to proceed on a purely experimental ground'.[57] Unfortunately, he does not tell us to which manuscripts he is referring. The problem here, as we shall see in the upcoming chapters, is that the academicians surviving correspondence and manuscripts actually provide crucial evidence showing that they were concerned with much more than producing purely atheoretical matters of fact.

Marco Beretta, like Tribby and Findlen, does not take at all seriously the academicians' natural philosophical concerns, but while the latter two at least locate the supposedly atheoretical pursuits of the Accademia in their presumed cultural context, Beretta does not provide much indication that he appreciates the social and political value of experiments to the Tuscan Court. Therefore, Beretta's work seems to endorse the traditional stories about experimental method that we have seen in the early twentieth-century writings. By doing this, Beretta leaves his account hostage to the following point, to be established during the course of the following chapters: despite the experimentalist rhetoric of the *Saggi*, the manuscript evidence provides valuable clues regarding how the Accademia del Cimento constructed experiments and knowledge claims according to their natural philosophical concerns. We shall see that rather than search for the origins of 'experimental science' in the courtly traditions of civility and gentlemanly behaviour when accumulating factual accounts of nature, we should understand that the use of an experimental programme, and the gathering of 'matters of fact', were not the central concerns of these early modern thinkers. In other words, while the decision-making and action-taking processes of the Cimento were undoubtedly linked with, and partially shaped by, the cultural and political interests and traditions of their Medici patrons, this does not mean that the thinkers employed by Ferdinando and Leopoldo were willing to abandon the natural philosophical concerns they had been pursuing throughout their entire careers. Instead, besides their allegiance to the Court's social and political traditions and ambitions, they were also concerned with much deeper and pertinent issues related to natural philosophical debates which spanned the entirety of Western Europe. With this in mind, it is now time to turn our attention to an alternative story to the traditional accounts regarding the origins of experimental science, and we may begin by coming to grips with the natural philosophical issues of the seventeenth century.

[55] *Ibid.*, 148.
[56] *Ibid.*, 141.
[57] *Ibid.*, 137.

4. SEVENTEENTH-CENTURY MECHANICAL NATURAL PHILOSOPHY, PHYSICO-MATHEMATICS, AND EXPERIMENT

Natural philosophy since the writings of Plato and Aristotle, had been based on determining the answers to the following four questions: What kind of matter does nature consist of? How is that matter organised into a cosmos? How and why do changes and motion in nature occur (the question of causation)? And what are the best ways to verify one's answers to the first three points (the question of method)? These are the types of questions that framed natural philosophical concerns during the Scientific Revolution.[58] Furthermore, seventeenth-century thinkers faced additional tasks while answering these questions. First, their discourses had to engage with the theological, political, and pedagogical issues of the period. Second, actors faced the task of linking the significance of their experimental hardware and results to their natural philosophical beliefs – their answers to the above four questions.[59] In other words, the process of constructing and interpreting knowledge claims included packaging those claims within a certain natural philosophical discourse. Finally, this type of intellectual environment where the meaning of certain theories and disciplines were contested amongst contrasting social and natural philosophical beliefs, created an evolving subculture of competing natural philosophical discourses. Seventeenth-century Aristotelians were forced to argue the soundness of their claims against varieties of Neoplatonism and against the new changing versions of mechanism.[60]

The four questions mentioned above, as well as the cultural settings they encompass have been examined recently by authors such as Schuster and Watchirs and Schuster and Yeo, and are used by these authors to argue that early modern experimental method rhetoric, such as that found in the *Saggi*, should not distract historians from the wider intellectual interests that existed behind the experimentalist facade.[61] So, as will be argued here in the case of Tuscan natural philosophy, from Galileo to the Accademia, talk of experimental method provides historians with few clues about the conceptual interests pursued by the Italian natural philosophers.[62]

Instead, as we shall see throughout our analysis of the Cimento and its members, there existed a field of contrasting and competing natural philosophical beliefs in seventeenth-century Tuscany. Furthermore, it will be evident how natural

[58] J. Schuster, 'The Scientific Revolution', in *Companion to the History of Modern Science* (eds. R.C. Olby, G.N. Cantor, J.R.R. Christie, and M.J.S. Hodge), London, 1990, 225.

[59] J.A. Schuster and G. Watchirs, 'Natural philosophy, experiment, and discourse: beyond the Kuhn/Bachelard problematic', in *Experimental Inquiries: Historical, Philosophical and Social Studies of Experimentation in Science* (ed. H.E. LeGrand), Dordrecht, 1990, 14.

[60] Schuster and Watchirs, 15.

[61] Schuster and Yeo, xii.

[62] J.A. Schuster and A.B.H. Taylor, 'Blind trust: The gentlemanly origins of experimental science', *Social Studies of Science* (1997), **27**, 1–34. See also J.A. Schuster, 'Whatever should we do with Cartesian method: Reclaiming Descartes for the History of Science', in S.Voss (ed.), *Essays in the Philosophy and Science of René Descartes*, New York, 1993, 195–223.

philosophers studied and framed their understandings of several disciplines, including pneumatics, hydrostatics, and astronomy, according to their wider conceptual beliefs regarding nature's structure, organisation, and movements, as well as the explanation of causes. Finally, adding to the complexity of seventeenth-century natural philosophising, we shall see how other subordinate disciplines, such as those of mixed mathematics, or proclivities such as experimentation, were made to take on different roles in these competing natural philosophies. We are going to find that the leading and preponderant group in the Cimento consisted of post-Galilean corpuscular mechanists. So, an historical analysis of the academicians' mechanical, and corpuscularian views regarding the structure, organisation, and movement of all matter, and the question of causation, would begin to explain the emergence of Florentine theories about issues such as the existence of the void and the weight of air, the freezing process of liquids, and much more. The academicians produced their theories on a variety of disciplines, within the dominant mechanical and corpuscularian natural philosophical discourse of that time and group and in opposition to the traditional and still widely entrenched Aristotelian beliefs. Evidence of natural philosophical conflict about these issues in the Accademia will also, therefore, be forthcoming.

The rise of mechanism as such a viable challenger to Aristotelianism, and how this alternative natural philosophy gained such importance for mid to late seventeenth-century thinkers such as our academicians, can be traced back to the 'scientific humanist' movement during the sixteenth and early seventeenth centuries.[63] I have already noted the increased interest during the Italian Renaissance, in the recuperation and commentary of classical texts. In particular, by the middle of the sixteenth century, ancient mathematical treatises were increasingly used in practical fields such as engineering and navigation. Furthermore, the mathematical knowledge gained from classical sources helped early modern scholars to strengthen the efficacy of such mixed mathematical fields as astronomy, optics, and mechanics, that consisted of natural phenomena less obvious to the senses.[64] For example, as Schuster argues, until realist versions of Copernican theory began to be discussed by some natural philosophers as a viable challenger to Aristotelian natural philosophy, astronomy was merely a discipline for the application of geometrical models, such as Ptolemy's. Aristotelian scholars did not actually believe that these models were true reflections of the material and causal principles related to the celestial realm. Similarly, optics when analysed according to geometrical principles was nothing more than a

[63] Dear, *Revolutionizing the Sciences: European Knowledge and its Ambitions, 1500–1700*, Basingstoke, 2001, 30–48

[64] Dear, *Discipline and Experience*, 194–195. Mechanics, in this case in the scholastic context, is not about mechanistic natural philosophy as has so far been discussed in this section, but about the analysis of dynamics and forces as in simple machines – the study of forces on solids, liquids, and air at rest or in motion that for contemporaries made up disciplines such as statics, hydrostatics, aerostatics and kinematics. As we shall see in the following chapters of Parts One and Two, these were disciplines that were also used by the academicians in their mathematical demonstrations that formed the basis of their mechanical natural philosophy.

mathematical exercise, far from a qualitative description of the reality of the actual physical nature of light.[65] So the increased use and practical application of the mathematical sciences, or mixed mathematics, helped elevate the status of mathematicians and the mixed mathematical fields in universities and royal courts, but this does not mean that they were being used in natural philosophy. In fact, according to scholastics, the mixed mathematical fields were still subordinate to natural philosophical inquiry. That is, according to them, these fields could not be used to find the material and causal structures of the cosmos, that was a matter to be properly dealt with only by Aristotelian natural philosophy.

However, by the seventeenth century, the increased use and practical application of mathematics, eventually raised the status of mathematicians to the point where, for many, their knowledge-making became elevated to the level of natural philosophising. First, the geometry of Euclid, Archimedes, and Apollonius, began to appeal to natural philosophers such as Galileo, Descartes, and Kepler who were already using realist interpretations of Copernican astronomy to challenge traditional Aristotelian natural philosophy. In other words, across Europe, the mixed mathematical fields such as astronomy, optics, and mechanics were not only being used to construct practical knowledge, but they were also becoming a tool for providing an alternative set of explanations for the structure, organisation, and movements of nature and the causal explanations for physical phenomena. This was the new form of inquiry known by many contemporaries, and referred to here, as 'physico-mathematics', which helped to establish the emerging versions of mechanical natural philosophy during the 1630s and 1640s. Second, navigational, engineering and military interests provided a foundation for a mathematical philosophy to ally itself with mechanical causality, complimenting the emerging mechanical tradition in natural philosophy.[66]

So, during the mid-seventeenth century, the increased use and status of the mathematical disciplines, previously regarded by scholastics as subordinate to natural philosophy, began to change the field of natural knowledge. As an example of this emerging physico-mathematical culture of natural philosophising, Peter Dear analyses the work of René Descartes (1596–1650). Cartesian metaphysics was based on the concept that all sense experiences are deceptive and provide only questionable and uncertain knowledge of nature. Therefore, the only means by which true and reliable knowledge is attainable is through the use of the only tool available that provides sound and irrefutable claims, mathematics. So, as Dear puts it, 'Descartes recognizes as physical phenomena nothing except the behaviours of mathematically defined matter; there is nothing else'.[67] Indeed, Gaukroger, Schuster, and Sutton point out that after Descartes met Isaac Beeckman in 1618 their discussions regarding mathematics and natural philosophy even led

[65] J.A. Schuster, 'L'Aristotelismo e le sue Alternative', in D. Garber (ed.), *La Rivoluzione Scientifica*, Istituto della Enciclopedia Italiana, Rome, 2002, 338.
[66] J.A. Bennett, 'The mechanics' philosophy and the mechanical philosophy', *History of Science* (1986), **24**, 5.
[67] P. Dear, *Revolutionizing the Sciences*, 89.

them both to the use of the term 'physico-mathematics' to describe their mechanical philosophy, an alternative to Aristotelianism.[68]

But Dear's analysis of the culture of physico-mathematics did not end with the example of Descartes. He goes on to examine the relationship between this culture and early modern experimental philosophy. According to Dear, the various ways in which experiments were carried out and reported during the first half of the seventeenth century, also reflected the rise of the mixed mathematical arts to the level of natural philosophising. However, for Dear, natural philosophy denotes only Aristotelianism, including the scholastic reliance on sense experience and the belief that the only way of accumulating reliable knowledge of nature is through general accounts, or 'universal statements' about how things happen. As Dear puts it: 'Such statements appeared in already generalized form, rather in the form of singular experiences referring to historically specific events. One did not say "this heavy body fell when I dropped it"; one simply said that all heavy bodies fall.'[69]

According to Dear, the increasing use of the mathematical sciences during the seventeenth century eventually led to the triumph of a more 'modern' scientific practice, Newton's mathematical singular event experiments.[70] In other words, Dear argues that a rupture occurred in the history of early modern science when experiments were not only aided by mathematical skills that allowed for the quantification of nature, but when they were also carried out and reported 'scientifically', as singular events. Dear claims that before arriving at Newton's most superior form of knowledge-making – before even arriving at the 'matters of fact', inductivist experimental philosophy of the Royal Society of London, the Parisian Academy of Sciences and the Accademia del Cimento – the authors of innovative experiments during the first half of the seventeenth century still compiled only general reports about their observations of nature.[71]

Dear's favourite example of how the scholastic tradition of reporting experiments remained until the mid-seventeenth century, is in the experimental reports by Pascal and Roberval during the 1640s and 1650s regarding the measurement of the pressure of air and the creation of a vacuum.[72] These two men insisted on using their skills in the mathematical sciences in order to carry out their experiments and to describe a physical phenomenon, air pressure. We shall see later that

[68] Gaukroger, Schuster, and Sutton, xvii. For an example of such a development in detail, see S. Gaukroger and J. Schuster, 'The Hydrostatic Paradox and the Origins of Cartesian Dynamics', *Studies in History and Philosophy of Science* (2002), **33**, 535–572.

[69] Dear, *Revolutionizing the Sciences*, 132.

[70] Dear, *Discipline and Experience*, 180.

[71] Dear has not examined the work carried out inside these societies in great detail, except for claiming that the members of the Royal Society, led by Robert Boyle, were uninterested in the mathematical sciences. In accordance with Shapin's and Schaffer's claims, Dear believes that the societies were only producing experimental, 'matters of fact'. This was still far from the triumphant Newtonian mathematical-experimental approach to natural knowledge, but was, nonetheless, a step in the direction of modern science: 'Boylean experimental philosophy was not the high road to modern experimentalism; it was a detour'. *Ibid.*, 3.

[72] The experiments carried out by Pascal and Roberval on this topic using Torricelli's barometer, will be discussed in more detail in Chapter Five.

those experiments were, in turn, subject matter for natural philosophical claims and debates. However, Dear ignores this issue and instead claims that their experiments were not singular events, describing what happened on one occasion that an experiment is performed and rendered reliable by witnessing. Rather, Dear suggests that Pascal and Roberval were simply creating 'unchanging universals of experience' of the type scholastics relied upon in order to guarantee the certainty of natural knowledge. In other words, according to Dear, Pascal, and Roberval appealed 'to universalized Aristotelian experiences' for knowledge-making, signifying that while they were capable mathematicians, they had not yet arrived at the scientifically superior practice of carrying out more reliable single experimental events.[73]

Dear's analysis of the rise of the mathematical arts to the point where they were regarded by many as 'physico-mathematics' – no longer a subsidiary to natural philosophy, but actually a part of making natural knowledge – will be extremely useful in the following chapters of this thesis. Yet we cannot continue without offering a clarification and critique of the role he has assigned to experimental knowledge in the seventeenth-century investigations of nature. Despite Dear's enlightening illustrations about how experiments were reported and witnessed, he has not accounted for how experiments of all types are laden with the theoretical interests and agendas of the experimenter. As Schuster and Taylor point out, we should not ignore the lessons from Gaston Bachelard that experimental hardwares are produced according to the conceptual structures of the experimenter. That is, that the construction and interpretation of an experiment embodies certain theoretical aims.[74]

What this means for our examination into the rise of mathematics and experiment in the seventeenth century is that no such thing as an applicable, efficacious experimental method, free from the theoretical constraints of the experimenter, even existed. Indeed, this has been the conclusion of several insights into the notion of inductive scientific method, whether Baconian or Newtonian, that have emerged within the field of Sociology of Scientific Knowledge (SSK) since the 1970s.[75] To suggest that some type of modern experimental method was actually being practiced by seventeenth-century natural philosophers, or that institutions such as the Accademia del Cimento or the Royal Society of London, were part of a rupture in the history of science that contributed to the origins of modern experimental practices, not only overlooks the broader intellectual and cultural traditions of the period as we have argued, but on SSK and related post-Kuhnian principles simply asserts the impossible.

[73] Dear, *Discipline and Experience*, 197.

[74] J. Schuster and A. Taylor, 'Seized by the Spirit of Modern Science', *Metascience* (1996), New Series Issue Nine, 18.

[75] In particular, see B. Barnes, *T.S. Kuhn and Social Science*, London, 1982; P. Feyerabend, *Against Method*, London, 1975; H.M. Collins, *Changing Order: Replication and Induction in Scientific Practice*, London, Sage, 1985; T. Pinch, 'Towards an Analysis of Scientific Observation: the externality and evidential significance of observational reports in physics', *Social Studies of Science* (1985), **15**, 3–36.

Experiments did, however, have an important role in the presentation of knowledge claims. They were used to create the impression for readers of the *Saggi*, and other such contemporary texts narrating the exploits of an individual or an institution, that true knowledge of nature was attained by the use of an unbiased and objective experimental programme, free from contentious speculations and theories. In other words, the role of experiments in seventeenth-century natural philosophy was as an authoritative and persuasive tool; a rhetorical device adding credibility to the claims being presented.[76]

Therefore, the emerging tradition of physico-mathematics was part of a culture of contrasting and competing natural philosophies within which experiments were performed in order to persuade one's readers of the validity of one's theoretical/natural philosophical beliefs. Pascal and Roberval may have been formulating general, 'universal' reports about their experiments, but the literary practices of these thinkers were a subsidiary concern to their natural philosophical skills, commitments, and agendas, which were based on their abilities as mixed mathematicians. So the key issue underlining our definition of physico-mathematics is that the competition for widespread acceptance between the contrasting natural philosophical perspectives, including Aristotelianism, and varying versions of both Neoplatonism and Mechanism, was based largely on how these competing perspectives on matter and causes dealt with and incorporated the developing mathematical disciplines into their processes of making natural knowledge.

We shall see in the case studies in Part Two that this culture of physico-mathematics – the use of mathematical skills to describe matter and causes in nature and to provide for an alternative to traditional Aristotelian natural philosophy – is also recognisable in the work carried out by the Accademia del Cimento. While clearly not Cartesian mechanists, most of the academicians were still using a mechanical natural philosophy based on their skills in mixed mathematics. This crucial point underlines the Cimento's participation in this emerging culture of physico-mathematics and mechanical natural philosophising. The early careers of Viviani and Giovanni Borelli in particular show how they refined their knowledge of classical sources and later used that knowledge in their understanding of physics and even physiology. We will see from our case studies of the Cimento's work that it is misleading to claim that this group was practicing some type of atheoretical experimental method detached from natural philosophical concerns, or that it was merely doing 'experimental science' as part of the gentlemanly culture of court etiquette.

In the meantime, classical atomism went through a resurgence also through the recuperation and reinterpretation of classical writings by philosophers such as Leucippus, Democritus, and Epicurus. Pierre Gassendi (1592–1655), who was highly respected by the Cimento's members, worked particularly closely with the idea of the atomistic structure of the universe.[77] Buonaventura Cavalieri's belief in

[76] Schuster and Watchirs, 20.

[77] Gassendi was mentioned several times in the *Saggi* and his writings also featured in a reading list Rinaldini compiled for Prince Leopoldo in 1656. See Middleton, *The Experimenters*, 4. For a thorough analysis of Gassendian natural philosophy see B. Brundell, *Pierre Gassendi: From Aristotelianism to a New Natural Philosophy*, Dordrecht, 1987.

indivisible particles of nature also captured the interest of the Tuscan Court members. Torricelli used Cavalieri's work to calculate the movement of the cycloid, and the academicians also drew on their own conception of atomism to formulate a mechanistic and corpuscularian argument for the expansion of freezing water.[78] For our Tuscan protagonists, as elsewhere in Europe, corpuscularian beliefs and impetuses towards physico-mathematical approaches in natural philosophy were coalescing in a mechanistic and anti-Aristotelian natural philosophy.

In summary, the central tenets of the mechanical philosophy were that nature consisted of corpuscles, or atoms, and that the organisation and movement of these atoms were as in a machine, requiring an understanding through the application of mathematics and mechanics. Finally, method rhetoric focused on ideas of observation and experiment, adding authority to these natural philosophical claims.[79] These conceptions answered the key natural philosophical questions that were mentioned at the beginning of this section and that the Italian seventeenth-century thinkers would have been asking themselves. So although mechanism took on different forms during the seventeenth century, the issue continually at stake for Galileo, Descartes, Gassendi, and many other seventeenth-century thinkers, apart from wider social and political linkages, was that Aristotelianism could no longer account for nature's structure, organisation, and movements, while mathematics, geometry, and mechanics provided the more efficient and accurate tools for understanding the universe.

5. GALILEO, NATURAL PHILOSOPHY, AND EXPERIMENT

Turning now specifically to the Tuscan setting, one may argue that Galileo was certainly not a systematic mechanical philosopher, since the boundaries of mechanism were more formally and widely established during the 1640s and 1650s.[80]

[78] One of the strongest indications that our early modern natural philosophers were beginning to invest quite a bit of interest in the ancient atomists, is in the career of the seventeenth- century Tuscan poet and mathematician, Alessandro Marchetti (1633–1712). During the Cimento's ten years in existence, Marchetti, also employed by the Medici Court, was working on a translation of *De rerum natura*, by Lucretius, a Roman poet and atomistic philosopher in the first century BC. Lucretius' poem argued in accordance with the theories of Democritus and Epicurus that the universe is an infinite extent of empty space and consists of an infinite number of irreducible particles of matter differing only in shape, size, and weight. Marchetti's translation was published posthumously, but his work still had an impact on the academicians, which is particularly reflected in his collaboration with Borelli. For a detailed account of Marchetti's life and work, see M. Saccenti, *Lucrezio in Toscana*. Firenze, 1966.

[79] Schuster and Watchirs, 20.

[80] What is meant here by systematic mechanical philosophers is those who adopted a system of natural philosophy based purely on mechanical principles. That is, when force and motion are analysed strictly on the basis of the collision and contact of atoms, and all bodies in either the celestial or terrestrial realm function like machines. An ideal mechanist would not entertain the possibility of immaterial causes in nature that are not treatable mathematically. While there were few such strict mechanical natural philosophers – Descartes comes close to fitting the ideal type, as does Giovanni Borelli, about whom we shall hear more in Chapter Three – mechanism was still a generic term for the new philosophers of the mid to late seventeenth century, replacing Aristotelianism.

Nevertheless, his aims had always been based firmly within the natural philosophical field of contention. Writers such as Koyré, Clavelin, and Drake have helped us to understand the mathematical, mechanical, and geometrical issues that ran through Galileo's works on terrestrial motion and his intentions to discredit Aristotelianism. Galileo's use of mathematics and geometry was thus the central factor behind his debate with the scholastics, who had different ideas about the structure, organisation and movements of nature.

However, it was not a question of mathematics alone, for experiments came to hold a crucial position in Galileo's attempts to prove the soundness of his work. He performed experiments of different kinds, including 'thought' experiments contrived to support his theories, and he presented them with more than one aim in mind; to persuade his readers of the reliability of his claims, and to access natural phenomena in a way which would allow his mathematical principles to be incorporated into wider natural philosophical concerns. As both Clavelin and Koyré note, the experimental proof Galileo provided was often a justification of previously established theories, and even if he had not actually performed the experiment, he could provide the mathematical explanations of what would happen in a hypothetical experiment.[81] According to Naylor, this helped to persuade Galileo's readers that his conclusion could only be true. This is the rhetorical role experimentalism served in two of Galileo's publications, *Dialogue* (1632) and *Two New Sciences* (1638). In these texts Galileo used thought experiments as a persuasive device, promoting a new, alternative view of nature in place of traditional, Aristotelian thought.[82] This would seem to subordinate the role of experiments in Galileo's natural philosophy, but we should not believe that he did not regard his experiments as efficacious. On the contrary, either through thought experiments or in those that he actually performed, Galileo was providing universal knowledge claims, that is, irrefutable and unchanging natural phenomena that could easily be explained with the certainty and regularity provided by mathematical principles.[83] Through experiment, therefore, came a vehicle for the mathematical treatment of physical problems, such as terrestrial motion.[84] Galileo's sophisticated combination of mathematics and experiment was thus crucial to his constitution of terrestrial mechanics in the presentation of his work, but his use of experiments was still subservient to his mathematical, geometrical, and anti-Aristotelian natural philosophical agenda. Its most significant role was as an authoritative tool, used to persuade the reader to refute Aristotelianism and support a mechanical, Archimedean physics.

These tensions surrounding Aristotelian and Galilean natural philosophy, continued to be played out amongst the Accademia's members; Galileo's 'followers'.

[81] The best known Galilean thought experiment would have to be the dropping of a cannonball from the mast of a ship. Although this was probably never actually performed, he was able to use his skills in mathematics to demonstrate what would happen. Clavelin, 27; A. Koyré, *The Astronomical Revolution* (tr. R.E.W. Maddison), London, 1980, 470.

[82] Naylor, 124.

[83] Dear, *Discipline and Experience*, 124–126.

[84] S. Gaukroger, *Explanatory Structures: A Study of Concepts of Explanation in Early Physics and Philosophy*, Sussex, 1978, 210–220.

In other words, the central concerns of the period were not so much to accumulate disparate 'facts' or knowledge claims through an experimental programme. Indeed, it is dangerous to assume that any such programme or method is possible, considering as sociologists of scientific knowledge have done, the theoretical skills and commitments that are carried into the construction and interpretation of experiments. Instead, the aim was to prove the validity of theories within a field of natural philosophical contention,[85] with the difference that the Galileans also had as a resource, systematically developed statements of mechanistic natural philosophies in Gassendi and Descartes.

We shall soon see how this culture of natural philosophising, featuring contention between scholastics and mechanists, played through the Cimento's work, and indeed how the academicians' themselves on both sides, used their respective natural philosophical concerns to establish and support their positions in a variety of disciplines. But before we dive into an analysis of this academy, it will be worth our effort to investigate the lives and works of the Cimento's members. It is important to remember that the Cimento was a very small private institution, and the intellectual interests of its two most influential contributors, especially Viviani and Borelli, were crucial in determining the group's activities and directing their discussions. In other words the biographies of these two, as well as an understanding of the intellectual aims and interests of the other lesser-known academicians, will assist us immensely in our understanding of the Cimento's construction of knowledge claims by showing us the natural philosophical skills and commitments that each academician contributed to the Cimento experiments.

[85] This crucial point is well argued in Schuster and Watchirs, 21; see also Schuster and Taylor, 'Blind trust', 515.

CHAPTER TWO

VINCENZIO VIVIANI (1622–1703): GALILEO'S LAST DISCIPLE

Almost all of the Cimento's members maintained long and prestigious careers, either in Tuscany or in other European courts, which stretched well beyond the contributions they made to the short-lived Accademia del Cimento. Nevertheless, their careers have traditionally been remembered in connection with the experimentalist image of the seventeenth-century Italian Galilean school. Marco Beretta and other historians have assumed that the strict experimentalist programme imposed on the Cimento academicians meant that they all abandoned their mathematical and natural philosophical interests in favour of producing atheoretical experimental knowledge claims. I hope to demonstrate here that this was not so. Once they became members of the Cimento, the academicians could not have found it an easy task to abandon the intellectual and natural philosophical interests that they had pursued throughout their careers. Indeed, beginning with an analysis of Viviani's career, I intend to show that they could not, and did not wish to abandon those natural philosophical concerns when constructing knowledge claims for the Medici Court.

This chapter therefore offers a brief biography of Viviani – a sketch of his intellectual pursuits drawn from his publications, manuscripts, and letters. It will become evident that his career involved much more than what many traditional and 'cultural' historians have suggested, the employment of rules of gentlemanly behaviour for the pursuit of atheoretical and factual knowledge. While we will find Viviani's courtly appointment to be of vital importance to the manner in which his career unfolded, we shall also see that he had deep natural philosophical concerns deriving from his education and training in mathematics and mechanics.

1. VIVIANI THE STUDENT

Viviani's early education in geometry was under the guidance of Clemente Settimi, a follower of Galileo and mathematics teacher at the Pious Schools (a congregation founded in 1597).[1] Settimi was also a student of Famiano Michelini, a court mathematician and mathematics teacher to Grand Duke Ferdinando II's two youngest brothers, Leopoldo and Giovanni Carlo.[2] Once Viviani began to show his tutors great potential in his studies, this network of mathematicians connected with the Tuscan Court provided him with a passage to what was to become a prestigious career in mathematics and natural philosophy. In 1638, Michelini found himself in Livorno where the Medici Court was staying temporarily. He mentioned Settimi's promising student to the Grand Duke, who immediately insisted on meeting Viviani.[3] During the long trip from Florence, he studied the first three books of Euclid's *Elements*.[4] Once he met the Grand Duke, gathered with his court philosophers, he explained eloquently the first 16 propositions of Book I of the *Elements* and comfortably responded to a mathematical problem put forward by Michelini under Ferdinando's insistence.[5] Viviani gave a good demonstration of his talents to the Medici Court, and Ferdinando was so impressed by his 'magnanimity' that he offered Viviani a monthly salary to help him continue in his studies. He also gave Viviani the opportunity to meet Galileo, who was still under house arrest in Arcetri.[6]

So Viviani was placed on the Medici payroll and we may assume that he was being groomed as a future court mathematician. Indeed his duties for the Medici came to hold a crucial place in his long life inside the Tuscan Court. For much of his career, he was expected to carry out several laborious responsibilities assigned to him by the Grand Duke, and as Court Mathematician, the title he eventually acquired in 1666, he was required to respond to the Grand Duke's questions and demands. These issues of courtly etiquette, studied thoroughly by Mario Biagioli, were an indispensable part of how Tuscan natural philosophers, starting with Galileo, went about making natural knowledge. We have already seen how courtly life affected Galileo's career; this was certainly not all that was happening in the intellectual movements of mid to late seventeenth-century Tuscany. In Chapter One, I mentioned Galileo's natural philosophical aims in the disciplines of mechanics and astronomy. Despite activity that often is believed to have come close to fulfilling a 'programme' of experimental knowledge-making, his central concerns lay with constructing solid mathematical foundations, at least, for an

[1] Details of Viviani's early life are recorded in some letters and manuscripts now held in the Galilean collection of the Biblioteca Nazionale Centrale in Florence. These include a document written by Viviani's nephew recounting his uncle's life, and transcribed by historian Giovanni Batista Clemente Nelli in 1758 (BNCF, Ms. Gal. 155, ff. 1r–4r), as well as an autobiographical letter written to Abate Marquis Salviati in 1697, also transcribed by Nelli (BNCF, Ms. Gal. 155, ff. 5r–23r); and published by A. Fabroni, *Lettere inedite d'uomini illustri*, 2 vols., 1775, ii, 6. The best secondary sources to deal with Viviani's education under Galileo are: A. Favaro, *Amici e Corrispondenti di Galileo*, 3 vols., Florence, 1983, ii, 1007–1163; and M.L. Bonelli, 'L'ultimo discepolo: Vincenzio Viviani' in *Saggi su Galileo* (ed. C. Maccagni), 2 vols., Florence, 1972, ii, 656–688; and Targioni Tozzetti, i, 321.

[2] R. Galluzzi, *Istoria del Granducato di Toscana sotto il Governo della Casa Medici*, 6 vols., Milano, 1974, iv, 127.

[3] Bonelli, 660.

[4] BNCF, Ms. Gal. 155, f. 1r.

[5] Favaro, *Amici e Corrispondenti*, 1015; Bonelli, 661; BNCF, Ms. Gal. 155, f. 2r.

[6] BNCF, Ms. Gal. 155, f. 2r.

anti-Aristotelian and proto-mechanistic position in natural philosophy. It is with this in mind that we should explore the intellectual interests that Viviani picked up under Galileo's guidance for three years and how Viviani's courtly life was intertwined with his natural philosophical interests.

2. ARCETRI: 1638–1641

In 1638, nearing the end of his life, Galileo was heavily restricted in his movements and in his pursuit of knowledge. This was not only due to his loss of sight, but was also because of the condemnation of his *Dialogue* by the Holy Office, restricting Galileo to his house and thereby limiting his contact with colleagues and his participation in anti-Aristotelian teachings. Nevertheless, he was still interested in working diligently on those issues of motion and mechanics that had interested him his entire career. In particular, he was still greatly motivated by his friends and colleagues to provide mathematical demonstrations supporting the notion of accelerated motion and extend his work in terrestrial mechanics. All this we are told by Viviani, Galileo's self-proclaimed 'last disciple', in a biography of Galileo written 12 years after his death. That text, published posthumously in 1717, not only provides us with some insight into the last years of Galileo's life, but more importantly for our purposes here, it will show us the foundations of Viviani's physico-mathematical interests, and the grounding of the natural philosophical skills and commitments that he was to carry with him throughout his career, including his years of participation inside the Accademia del Cimento. For this reason it will be important to examine here how Viviani assisted Galileo to demonstrate geometrically how falling bodies accelerate uniformly. The scholium that Viviani added to Galileo's *Two New Sciences* on this topic will show Viviani's skills as a mathematician and that his thoughts were far from focused on so-called modern experimental science and atheoretical knowledge-making.

By directing Viviani to Arcetri, Ferdinando was introducing a possible new assistant to Galileo, still the Grand Duke's first mathematician and philosopher. Since 1636, Galileo had been complaining about the near total loss of his sight and the impediment this created for analysing the thoughts and concepts that had remained unwritten up until then.[7] Galileo needed a capable assistant who was prepared to listen to him, comment on his arguments, and help him compose his works. This job required someone who could understand his principles, intelligently describe them with illustrations and figures, and even make observations and experiments. Viviani, although very young, seemed to fit this role perfectly because of his background and enthusiasm in the practice of mathematics. So the two met towards the end of 1638 and Viviani recalled the first few months of his acquaintance with Galileo:

[7] Galileo expressed this in a letter written in 1637 to Father Fulgenzio Micanzio in Venice. In A. Favaro (ed.) *Le Opere di Galileo Galilei, Edizione Nazionale*, *Nazionale*, 20 vols., Florence, 1890, xvii, 125–126.

> Soon after this unexpected publication [*Two New Sciences*], Signor Galileo allowed me into his villa in Arcetri where he was staying. I was able to benefit from our intelligent conversations and his precious teachings and he was content that in the study of mathematics, which I had only recently begun, I could turn to his own voice for the solution to those doubts and difficulties that I often found through the natural weakness of my intellect.[8]

From this passage we find the next clue pointing to the foundations of Viviani's education. We have already seen a suggestion of his early command over Euclidean geometry. In the presence of Galileo, there was still no mention of learning about some type of inductivist experimental method or how to produce experimental knowledge inside the Medici Court. Instead, as indicated in his correspondence, Viviani saw the situation as a further opportunity to strengthen his skills in what scholastics called mixed mathematics, but what in the anti-Aristotelian context of Galileo's agenda easily qualified as physico-mathematics.

It is particularly pertinent to note that just as Viviani was settling into his role of amanuensis to Galileo, copies of *Two New Sciences*, published in Leiden, were only starting to become available in Italy, placing terrestrial motion firmly in the interests of many Italian natural philosophers. So during the last months of Galileo's life, as Viviani embarked on a career inside the Tuscan Court, they worked together in order to strengthen the role of mathematics in natural philosophy. In the process, they would assist in the construction of a physico-mathematical tradition that was both anti-Aristotelian and bound to become quite mechanistic. The topic they worked on together to achieve this was the geometrical demonstration of accelerating falling bodies. Galileo's and Viviani's combined efforts to illustrate accelerated motion on inclined planes, resulting in the scholium Viviani added to the Third Day of *Two New Sciences*, became a major part of Viviani's education, shaping the natural philosophical skills, commitments, and agendas he was to use during his entire career.

Galileo had always been interested in empirically and mathematically exploring dynamical terrestrial motion.[9] In his early work in *De motu* (c.1590), he was concerned with the force with which a body falls along planes of different degrees of inclination. In this text, Galileo not only considered the topic to be a question of why bodies move at different speeds along differently inclined planes, but also what the ratios of speeds are at these various inclinations.[10] This was how Galileo created a dynamical analysis of falling bodies.

Galileo returned to this topic many years later in 1638. On this occasion, as we may see in the Third Day of *Two New Sciences*, Galileo added the notion of the uniform acceleration of falling bodies. Additionally, to complete his dynamical

[8] V. Viviani, *Vita di Galileo* (ed. L. Borsetto), Bergamo, 1992, 105; Favaro, *Amici e Corrispondenti*, 1017.

[9] In particular, as is shown by Wolfgang Lefèvre, Galileo examined scholastic issues related to practical mathematics, such as kinematics and hydrostatics. But since the beginning of his career, as a student in Pisa, he had been attempting to search for dynamical solutions, incorporating theories on force and causes, to problems in motion. W. Lefèvre, 'Galileo Engineer: Art and Modern Science', in *Galileo in Context* (ed. J. Renn), Cambridge, 2001, 11–27.

[10] G. Galilei, *On Motion and On Mechanics* (eds. and trs. S. Drake and I.E. Drabkin), Madison, 1960, 63.

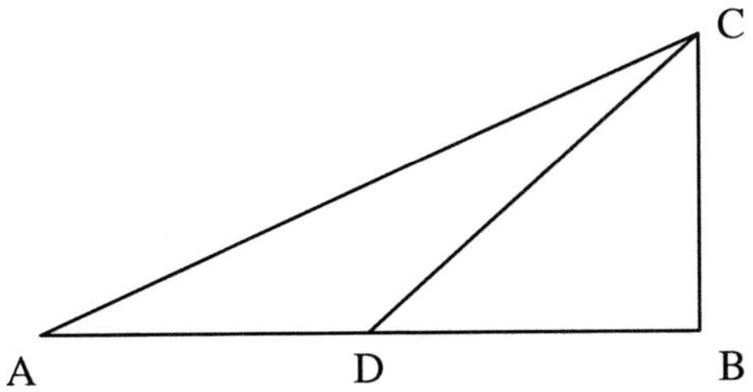

Figure 1. Reproduction of diagram used by Galileo in *Two New Sciences,* to describe the final velocity reached by a body falling along an inclined plane.

analysis of the ratios of speeds of bodies falling down differently inclined planes, he wrote the following postulate; a critical part of his mathematical demonstration of the physical phenomenon of accelerating falling bodies (Figure 1):

> The degrees of speed of the same moveable, descending along the inclined planes CA and CD to points A and D, are equal, because their height is the same CB; and the like is also to be understood of the degree of speed that the same body falling from the point C would have at B.[11]

In other words, according to Galileo, the body always reaches the same speed, or final velocity, at the bottom of each plane. For Galileo, this is the final piece in the puzzle for understanding the speeds of falling bodies. In his attempt to verify this postulate, Galileo relied upon an experiment with a pendulum, described in the text by Salviati, the interlocutor representing Galileo. Supposing that all impediments deriving from the medium are removed, a pendulum will rise to the same height from which it is dropped. Salviati then proposes that when the string of the pendulum is stopped by a nail, the pendulum will still rise to the same height as when it is allowed to fall freely. Regardless of where the nail is placed, the momentum required for the pendulum to reach its original height is always the same, provided of course, that it is always dropped from the same height. Similarly, according to Galileo, the momentum of the body falling down variously inclined planes of the same height is always equal, generating the same final velocity.[12]

Galileo assumed that this experiment was sufficient proof of the accuracy of the earlier postulate. In order to verify the claim mathematically, he formulated two additional propositions and two corollaries supporting his theory on the ratios of time, distance, and speed. He applied a measuring technique used by medieval scholastics to describe several geometrically constructed instants or units of distance on a plane, and to measure how the speed increases uniformly along each instant.

However, when Viviani came across this topic in his readings of *Two New Sciences* during the early months of 1639, he doubted that Galileo's pendulum experiment, and his additional geometrical corollaries, provided a convincing explanation of the postulate. Viviani became so interested in this important section

[11] G. Galilei, *Discourses and Mathematical Demonstrations Concerning Two New Sciences Pertaining to Mechanics and Local Motion* (ed. and tr. S. Drake), Madison, 1974, 162.

[12] Galileo, *Opere*, viii, 205–206.

of Galileo's work that he turned the focus of his discussions with his teacher solely upon the topic of demonstrating the natural motion of heavy falling bodies, and their final speeds when dropped along inclined planes. Viviani described in the biography of his mentor, how in their discussions he made it clear that he did not doubt the truth of the notion that heavy bodies falling along inclined planes of equal heights would reach equal final speeds. He was simply concerned with whether it had been satisfactorily proved.[13] Viviani, therefore, insisted that Galileo provide more convincing demonstrations of the postulate. However, it must be made clear that these demonstrations did not refer to any crucial experiments, such as Galileo's observations of the pendulum. Rather, Viviani was not satisfied that Galileo had employed convincing geometrical and dynamical principles in order to support his views regarding accelerated motion. This is particularly clear when Viviani discusses the first steps that he and his teacher took towards the discussion about Euclidean 'same ratios' that eventually came to be published in further editions of *Two New Sciences*. He recalls:

> One day I asked him for clearer confirmation of that principle [of bodies reaching equal final speeds along variously inclined planes], and during one of the following nights ... he rediscovered the Geometrical Mechanical demonstration, deduced from the doctrine that he demonstrated against a proposition made by Pappus of Alexandria, made in his old treatise on mechanics.[14]

In *Le Meccaniche* (c.1594) Galileo used the claims by Pappus of Alexandria as a springboard for his own discussion regarding ratios of force and weight on inclined planes. Galileo was, in other words, concentrating upon the dynamics of falling bodies, the forces causing the velocities of heavy bodies rolling down inclinations. Pappus claimed that regardless of the degree of inclination of different planes of equal heights, only one set force could prevent the heavy body from moving. That is to say, as the inclination of the plane varies, the opposing force needed to resist the body from falling would always be the same. Through a series of geometrical demonstrations, Galileo denied that Pappus was right and instead concluded that heavy bodies 'have greater resistance to being moved upon variously inclined planes, according as one is more or less tilted than another'.[15] In other words, the force needed to prevent a heavy ball from rolling down a slope becomes greater as the inclination also increases; thus a force is required that is proportional to the inclination of the plane.

Following this recall of Galileo's previous reflections concerning inclined planes, he and Viviani discussed Eudoxian proportion theory as propounded by Euclid in Book V of the *Elements*.[16] They were seeking a geometrical and

[13] 'Appena ebbi scorsi i primi Elementi, che impazziente di vederne l'applicazione, passai alla scienza de'moti naturali nuovamente promossa da Galileo, e che allora appunto era uscito in luce: et arrivato a quel principal supposto, che le velocità de'mobili naturalmente descendenti per piani d'una medesima elevazione sieno uguali tra loro, dubitai, non già della verità dell'assunto, ma dell'evidenza di poterlo supporre come noto.' Viviani, *Vita* (ed. L. Borsetto), 215.

[14] *Ibid.*, 216.

[15] G. Galilei, *On Motion and On Mechanics* (tr. S. Drake and E. Drabkin), Madison, 1960, 171–172.

[16] Viviani, *Vita* (ed. L. Borsetto), 105–106.

mechanical demonstration of the notion that bodies descending along differently inclined planes, but from equal vertical heights, reach the same speed at the bottom of the plane. That is, that two bodies descending a vertical and an inclined plane accelerate uniformly, but at different rates, so that the greater distance needed to cover the inclined plane is proportional to the time needed to reach the same speed as in the vertical. This way, through their search for a geometrical demonstration of the postulate, Galileo and Viviani were establishing a new dynamical theory that relied upon Galileo's early thoughts as expressed in *Le Meccaniche*, as well as a reconsideration of Euclid's version of Eudoxus' proportion theory.

As Drake argues, the use of ancient geometers, whose works had come to light as crucial to natural philosophical studies only during the sixteenth century, 'lies at the basis of most of Galileo's applications of mathematics to physics'.[17] As a result, they also formed the basis of Viviani's education in natural philosophy. Galileo and his young student were particularly interested in exploiting Euclid's concept of 'same ratios', as given in Book V, Definition 5 of the *Elements*.[18] How this definition is used in the Galilean arguments concerned with natural motion along inclined planes, and in particular, Galileo's postulate regarding final speeds, is clear from the scholium composed by Galileo with Viviani's assistance after 1638, and added by Viviani to subsequent editions of *Two New Sciences*.[19] In fact, this scholium addresses Galileo's postulate, supposedly adding the evidence validating Galileo's proposition that heavy bodies reach equal final speeds when dropped from variously inclined planes.[20]

Galileo's work and Viviani's contribution, therefore, were based largely on the ancient readings regarding Eudoxian proportion theory. This was a rigorous mathematical and geometrical exercise which had occupied a great deal of Galileo's career since his beginnings in Pisa. Thanks largely to Viviani's enthusiasm for further exploring this notion of ratios between weight and force on inclined planes, Galileo arrived at a more convincing demonstration for the postulate. As Galileo himself confirmed on 3 December 1639, in a letter to another Galilean disciple in Rome, Benedetto Castelli (1578–1643), Viviani's curiosity and constant questioning led to Galileo's last investigations on the topic.

> Objections made to me many months ago by this young man [Viviani] who is now my guest and disciple, against that principle postulated by me in my treatise on accelerated motion ... made me think about this again in such a way as to persuade him that

[17] S. Drake, 'Galileo Gleanings – XXIII Velocity and Eudoxian Proportion Theory', *Physis* (1973), **15**, 51.

[18] T.L. Heath, *The Thirteen Books of Euclid's Elements*, 3 vols., New York, ii, 113.

[19] Viviani narrated these demonstrations, including the arguments based on the refutation of Pappus, in a dialogue form to suit Galileo's style of presentation and inserted an additional section into the crucial Third Day of the text for subsequent editions.

[20] Galileo and Viviani used Euclid's notion of 'same ratios' to conclude 'that the time along the incline has to the time along the vertical the same ratio that the incline has to the vertical'. Favaro (ed.) *Le Opere*, viii, 218–219. This is the theorem concluding the added scholium. It is intended to demonstrate Galileo's postulate in *Two New Sciences* and in the process, to provide a new dynamical solution to a problem in kinematics.

> that principle might be conceded as true. Finally, to his and my great delight, I succeeded in finding a conclusive demonstration.[21]

Viviani, therefore, made a valuable contribution to Galileo's last years of physico-mathematical natural philosophising by assisting him to formulate much more rigorous support for a key postulate in his mathematical theory of accelerated free fall. That theory was of great natural philosophical relevance, and Galileo and Viviani freely indulged in explicit dynamical and causal analysis in the course of working on the scholium. Clearly, moreover, Viviani's early education and collaboration with Galileo was centred strongly on a respect for ancient mathematicians and geometers, and an intention of firmly establishing Galilean terrestrial mechanics. In Chapter One we came to appreciate the mathematical and geometrical interests that dominated Galileo's natural philosophy. As we have now begun to see through these early years of Viviani's education, those interests set the intellectual foundations for Galileo's 'disciples', including Viviani, one of the most dominant members of the Accademia del Cimento.

3. TORRICELLI'S ARRIVAL IN ARCETRI

This interest in advancing Galileo's previous accomplishments by further exploring the work of the ancient mathematicians and geometers eventually attracted the attention of Evangelista Torricelli (1608–1647) in Rome, and his teacher Castelli.[22] In April 1641, Castelli visited Galileo at Arcetri and brought Torricelli's manuscript on the motion of projectiles. This was part of the dissertation that Torricelli eventually published in 1644 under the title *Opera Geometrica*.[23] Castelli, considering Galileo's ill health, recommended that Torricelli assist the Tuscan Court's mathematician and philosopher during his exile in his villa.[24] Galileo was impressed by Torricelli's sample of work on projectiles and he immediately invited Castelli's student to Arcetri where he arrived in October.[25] By this stage, Viviani had been staying in Galileo's villa for quite some time, and although helpful, he was still very young and Torricelli's more experienced contributions would have been welcomed, especially with regard to Galileo's interests in projectile motion.

Torricelli died only five years after Galileo, so obviously he did not directly contribute to the Cimento's foundation, but through his work and his friendship with Viviani he was to leave a lasting legacy that was undoubtedly part of the post-Galilean generation and that came to form the intellectual foundations of

[21] Favaro (ed.) *Le Opere*, xviii, 126. As translated by S. Drake, *Galileo at Work: His scientific biography*, Chicago, 1978, 405. See also Favaro, *Amici e corrispondenti*, 1017.

[22] Torricelli was a mathematics student at the Jesuit school of Faenza. Showing promise in geometry, he was sent to Rome in 1626 to study under the guidance of Castelli, the mathematics lecturer at the Sapienza and a supporter of Galileo.

[23] E. Torricelli, *Opera Geometrica*, Florence, 1644.

[24] Viviani, *Vita* (ed. L. Borsetto), 217; Bonelli, 662.

[25] Drake, *Galileo at Work*, 416–417, 419.

that institution. For this reason, it is worth briefly examining how Torricelli constructed and presented his natural philosophical claims regarding projectile motion, the topic he pursued just before and after meeting Galileo. This case study demonstrates the natural philosophical skills and commitments of Galileo and his students, as well as their ability to appeal to the social and political interests of their Medici patrons.

Galileo's theory of projectiles and its application to ballistics is explained in the Fourth Day of *Two New Sciences*. Treating the simplest case, a body projected horizontally, parallel to the earth's surface, Galileo argued that 'all impediments being put aside', not considering any possible interference from wind resistance and making purely mathematical and geometrical calculations, one must assume that a constant horizontal speed imparted to the projectile combined with the acceleration of a naturally falling body (according to the squared ratios of the times as was shown in Day Three and summarised earlier), will create a perfect parabolic line.[26] Furthermore, should the projectile not be halted by the earth's surface, then the moveable would continue its parabolic trajectory.

For the Medici Grand Dukes, ballistics had long been an important matter open to research. It was believed in the European courts of the sixteenth and seventeenth centuries that the ability to measure the potential of artillery and mortar cannons, would be valuable knowledge for the construction of military fortifications. This certainly was a topic of interest for Galileo while he was in Padua, writing two treatises about military fortifications and architecture, and later a manual for his geometrical and military compass used to measure the elevation of projectiles.[27] It is clear from these writings, including *De motu*, also written during these early years of Galileo's career, that he placed much importance on the practical application of his geometrical and mathematical principles, especially for firing guns, arrows and all artillery in general.[28] Indeed, this seems to be the reason for the tabulation of his estimations of the degrees of elevation of parabolas in *Two New Sciences*.[29] Furthermore, his knowledge in this discipline would not have done any harm to his employment prospects with the Tuscan

26 'Secondo la proporzion duplicata de i tempi, e che tali moti e loro velocità, nel mescolarsi, non si alterino perturbino ed impedischino.' *Ibid.*, viii, 273.

27 The two early writings remained unpublished during Galileo's life. They are *Breve Istruzione all'Architettura Militare*, written circa 1592–1593, and the second from around the same period, titled *Trattato di Fortificazione*. The third was a publication dedicated to Cosimo de' Medici in 1606; *Le Operazioni del Compasso Geometrico et Militare di Galileo Galilei Nobil Fiorentino Lettor delle Matematiche nello Studio di Padua*. All three have been published in Favaro (ed.), *Le Opere*, ii, 15–75, 77–146, 365–424.

28 During his years in Padua, Galileo believed that the trajectory of projectiles was a straight line, and only began to curve once the heavy body lost its initial impetus. After the curve it was supposed that it took on another straight-line trajectory in its fall. This point is highlighted by Drabkin in his translation of *De motu* in: Galilei, *On Motion and On Mechanics*, (tr. Drake and Drabkin), 9, n.1. For an analysis of the scholastic grounding of Galileo's early work on projectile motion, see J. Renn, P. Damerow and S. Rieger, 'Hunting the White Elephant: When and How did Galileo Discover the Law of Fall?', in *Galileo in Context* (ed. J. Renn), Cambridge, 2001.

29 Galileo, through Sagredo, mentions how the maximum range of shots depends on the angle of elevation of the artillery. The three interlocutors then go on to discuss the angles which provide greatest distance and elevation, and provide the table. Favaro (ed.), *Le Opere*, viii, 304–307.

Court. As Paul Lawrence Rose and A.R. Hall both point out, in 1637 Galileo wrote a letter to Elia Deodati, a literary and legal consultant in Paris, noting the purpose of the tabulation in the text:[30]

> I wish for now to close the treatise with a table which I have proved and calculated for artillery and mortar trajectories, showing their flights and with what proportion they increase and diminish according to the various degrees of elevation. The practice of this table will be useful to gunners, its theory of great delight to philosophers.[31]

It is evident that when compiling information on this topic, Galileo was not only demonstrating his knowledge of projectiles to a natural philosophical audience, but he was also appealing to the practical benefits of his work and therefore proving his value to the Medici Court. This was the type of work that was required of natural philosophers inside the Tuscan Court as they debated intellectual principles with their colleagues while establishing their reputations with their patrons, where they earned their wages. These were therefore the challenges that Galileo and Torricelli faced when carrying out their work on projectile motion, showing us just how important mixed mathematical skills and commitments were to natural philosophising in the mid-seventeenth century, and how vital it was to present one's claims appropriately for the Court.

According to Michael Segre, many of Galileo's own students and colleagues were critical of his theory of parabolic trajectories. Mathematicians Buonaventura Cavalieri, Antonio Nardi, and Giovanni Batista Renieri, maintained that Galileo's claim could be easily challenged through some simple experiments and observations.[32] This criticism also came from Paris, especially from Mersenne, Descartes, and Roberval.[33] In particular, as Rose points out, Descartes seemed doubtful about the applicability of Galileo's theory; whether the tables in *Two New Sciences* could be used for measuring the range of all artillery or just slow projectiles as Galileo seemed to suggest.[34] Finally, according to John Guilmartin's analysis of sixteenth-century warfare at sea, gunners would have seen no practical value at all from quadrants and theoretical measurements. Instead, their best judgement regarding the range of a mortar cannon came from their own experience.[35]

[30] A.R. Hall, *Ballistics in the Seventeenth Century*, Cambridge, 1952, 91; Rose, 156.

[31] Favaro (ed.), *Le Opere*, viii, 156. As translated by Paul Lawrence Rose, 'Galileo's Theory of Ballistics', *British Journal for the History of Science* (1968), **4**(14), 156.

[32] Segre, *In the Wake*, 96. According to Segre, these claims were particularly consistent from Renieri against Torricelli's defence of the theory.

[33] *Ibid.*, 93.

[34] Favaro (ed.), *Le Opere*, 276. Galileo made a distinction between fast and slow projectiles; bodies that are projected 'supernaturally' by gunpowder are subject to more variations than arrows shot by slings and bows that are still within the limits of terminal speed, 'the maximum that such a heavy body can naturally attain through air'. *Idem.*, 279. This issue is discussed by Rose, 'Galileo's Theory of Ballistics', 156–159.

[35] J.F. Guilmartin Jr., *Gunpowder and Galleys: Changing Technology and Mediterranean Warfare at Sea in the Sixteenth Century*, Cambridge, 1974, 165. Even detailed data compiled in the early seventeenth century to measure the maximum range of cannons did little to assist gunners because of the wide variety of factors that interfere in the firing of mortar and the achievable range of the ballistic. *Idem.*, 277–281.

So, facing some criticism of his theory, Galileo must have been pleased to find an ally in one of his young and promising students, Torricelli. At this stage, near the end of Galileo's life, Torricelli adopted Appollonius' surviving work on conic sections and Eudoxian proportion theory, and came to Galileo's defence by supporting the notion that geometrically and mathematically the theory was in fact correct. The arguments made by Torricelli and Galileo in support of their claims now extended beyond the utility of their measurements to include the theoretical value of their work. But, as we shall see later, it is important to note that Galileo and Torricelli maintained in their presentation to the Tuscan Court that their measurements were still useful to gunners – this, of course, interested their Medici patrons and supported the efficacy of their work.

According to Torricelli, the arguments against Galileo's theory were riddled with numerous distortions caused by the firing of a gun, such as friction, the quantity and quality of the gunpowder, the tilt of the gun when fired, and so on. He claimed that they could not possibly effectively invalidate Galileo's theory explained under perfect hypothetical circumstances. In other words, regardless of the experiments made by opponents, he claimed that he and Galileo were simply pursuing mathematical and geometrical demonstrations in support of their theory. In a letter to Michelangelo Ricci in 1646, Torricelli wrote in defence of his and Galileo's theory that their speculations were 'purely geometrical'.[36] This may have been an easy way to avoid at least the criticism regarding the physical accuracy of their claims, even if, as we shall soon see, Torricelli still had to continue to defend Galileo from the criticism that their work was of no practical benefit to gunners. In any case, there is no doubt that Galileo and Torricelli were using mathematical and geometrical principles as foundations for their work on the motion of projectiles. At some point they may have even believed in their practical application – certainly Galileo's early work would seem to suggest so, and as we shall see below, Torricelli had also seemingly appealed to the utility of his work, if only to preserve his courtly status and reputation. Meanwhile, it is clear that, as we have seen with Viviani's early education, experiments were barely on the periphery of their concerns.[37]

We cannot be sure how worried Galileo was about the criticisms he faced on projectile motion, yet he surely would have felt a concern to defend his theory. It was still of vital importance to the Medici, and as Rose argues, there could have been a strong desire 'to seem to have discovered the long sought after general solution and yet still retain his integrity'.[38] This means that it was not only important for Galileo to maintain the mathematical and theoretical value of his work, but also that he had to argue for its practical application and thus preserve his status and reputation within the Court. In any case, Galileo was nearing the end of his life and we may believe that for the young Torricelli there was so much more at stake in defending his work on projectile motion. He was just beginning to

[36] P. Galluzzi and M. Torrini (ed.), *Le opere dei discepoli di Galileo Galilei*, 2 vols., Florence, 1975–1984, i, 276. As cited by Segre, *In the Wake*, 93.

[37] Segre, *In the Wake*, 98–99.

[38] Rose, 158.

establish his career in the Medici Court and would have seen this debate as an opportunity to strengthen his status as Tuscany's leading mathematician and successor to Galileo. He not only supported Galileo's theory, but in *Opera Geometrica*, he used the theory to follow in his teacher's footsteps by printing an illustration of an instrument that could measure the range of projectiles. As Segre points out, it was even described in Italian so that it could be understood and applied by gunners.[39] Therefore, as they embarked on their careers inside the Tuscan Court, the central issue that was at stake for Galileo's successors in Tuscany, including both Viviani and Torricelli, was the validity of their natural philosophical interests. That is, how they constructed their knowledge claims according to their natural philosophical concerns, and how they presented those claims in order to increase their status and reputation inside the Court. As Segre argues with regard to this case of projectile motion, 'in the particular case of Galileo, Torricelli, Viviani, and other Galilean followers in the service of princes, it was an example of both the kind of theoretical and mathematical knowledge they were expected to teach and the practical knowledge of direct benefit to their employers'.[40]

Galileo fell into his last illness in December, and died in January 1642. This only allowed a few weeks for Torricelli and Galileo to collaborate. According to Drake, during this time Galileo simply dictated to Torricelli what was intended to become the great mathematician's next publication: a sequel to the dialogue between his three interlocutors, Simplicio, Sagredo, and Salviati.[41] On this occasion, the dialogue was an attempt to demonstrate Euclid's definitions of same ratios (Book V, Definition 5), greater ratios (Book V, Definition 7), and compound ratios (Book VI, Definition 5), a task apparently not completed to Galileo's satisfaction by Euclid, especially regarding the need to lay down a clear definition of equal multiples. So Galileo was further developing the geometrical and mathematical tools that had previously enabled him and Viviani to demonstrate the acceleration of falling bodies and the postulate regarding the final speeds of bodies falling along inclined planes.

These tools were amongst the necessary foundations of Galileo's mathematical natural philosophy and they also soon became important parts of the basis for the natural philosophical concerns his students carried into the construction of knowledge claims. Galileo did not have many students, but in those who did become part of his school, he instilled this same mathematical dedication to natural philosophy. More specifically, as we are now seeing in the life of Vincenzio Viviani, and as we shall continue to see in the biographies of most of the other Cimento members, the Galilean tradition that many seventeenth-century Tuscan natural philosophers were pursuing entailed an interest in physico-mathematics,

[39] Segre, *In the Wake of Galileo*, New Jersey, 1991, 91–92.
[40] *Ibid.*, 88.
[41] Drake, *Galileo at Work*, 421; Favaro, *Le Opere*, viii, 349–362; See also 'Euclid Book V from Eudoxus to Dedekind', in *History in Mathematics Education* (ed. I Grattan-Guiness), n.s.21, 1987, 52–64. Reprinted: S. Drake, *Essays on Galileo and the History and Philosophy of Science*, 3 vols., Toronto, Toronto University Press, 1999, iii, 61–75.

with a strong anti-Aristotelian and pro-mechanist agenda. Experiments of various kinds were undoubtedly part of their work, but they played a minor role next to the mathematical and natural philosophical skills and commitments pursued by Galileo and his followers.

In the meantime, there is no evidence at all of an atheoretical and inductivist experimental method in Viviani's early career and interests in Euclidean geometry, or in Torricelli's short career. As we shall see in Part Two, the mathematical interests of Tuscany's leading natural philosophers continued to play a part in Italian early modern thought as the Accademia del Cimento got under way. Its most active members continued to discuss and publish improvements and comments on such ancient authors as Euclid, Archimedes, and Apollonius. Furthermore, in Part Three we shall see the presence of these mathematical and natural philosophical concerns in the cultural and political aims and interests of courtly life.

Through the analysis of letters and manuscripts we have gained an insight into the natural philosophical and disciplinary concerns that dominated Galileo's, Torricelli's, and young Viviani's work and it is through these same sources that we shall continue to see these issues playing through the academicians' work inside the Cimento – despite the non-committal rhetoric of the *Saggi*. So for now, it is important to continue our analysis of Viviani's career, as the Medici Court built up its interests in natural philosophy and edged closer towards the foundation of the Cimento, only ten years after Torricelli's death.

4. HOW TORRICELLI'S DEATH BROUGHT VIVIANI'S CAREER INTO THE SPOTLIGHT OF TUSCANY'S INTELLECTUAL COMMUNITY

After Galileo's death, Viviani and Torricelli pursued their interests in mathematics within the context of their new duties at the Tuscan Court. In 1644 Viviani was called upon by the Grand Duke to perform the job of assistant to the Court's engineer, Baccio del Bianco. As tension mounted between Tuscany and the Papal States over territorial disputes in Umbria, Viviani was required to help fortify the region's borders.[42] By 1653, he took over del Bianco's job and was appointed first engineer to Ferdinando II.[43] In the meantime, in 1642, Ferdinando II nominated Torricelli to take over the role of First Mathematician to the Medici Court, lecturer in mathematics in the Florentine Academy, and lecturer on military fortifications at the Accademia del Disegno.[44] Although Torricelli did not publish any of his many manuscripts during his lifetime, aside from *Opera Geometrica*, he still occupied the remainder of his career with the study of such topics as the cycloid,

[42] BNCF, Ms. Gal. 155, f. 4r; E. Cochrane, *Florence in the Forgotten Centuries. 1527–1800*, Chicago, 1973, 199.
[43] Favaro, *Amici e Corrispondenti*, 1038; Bonelli, 670.
[44] Segre, *In the Wake*, 61–62.

the volume of solids, and improving Cavalieri's notion of indivisibles. During this time, he and Viviani developed quite a close friendship, discussing various topics they had worked on with Galileo. They even collaborated on the construction of the barometer in 1643.[45]

It is beyond our scope here to explore the details of Torricelli's unpublished works. For the purposes of the present argument, it is sufficient to have discussed his early interest in projectile motion, an example of his natural philosophical concerns, and to have recognised how those concerns remained during the next few years of his life. Therefore, Torricelli's early interests in mathematics continued during his career inside the Tuscan Court, as he not only pursued his own physico-mathematical and natural philosophical aims, but also applied their benefits for his patron, the ruling Medici family. His career, as well as Viviani's, was built upon the pursuit of the mathematical principles that Galileo instilled in his students, within the setting of the Tuscan Court.

After Torricelli's death in 1647, Cavalieri died within only one month and mathematician Vincenzio Renieri later that same year. In 1649, Galileo's son, Vincenzio Galilei, also died at a young age. These sudden deaths of Tuscany's most influential and industrious natural philosophers, including the losses of Castelli and Gasparo Berti years earlier (both died in 1643), left Viviani, at the age of 26, as the only Tuscan natural philosopher to be in a position to carry on with Galileo's work. Viviani indeed must have seen himself in this light, since he came to give himself the title of Galileo's 'last disciple'. In 1674, as a preface to his *Quinto libro degli elementi d'Euclide*, and in response to those who, so he believed, were envious of him, he wrote:

> The fact is that through my good fortune, I am his last disciple, because he was my teacher continually during the last three years of his life, and from all of us who were present while he took his last breath (who apart from two priests, included Torricelli, his son Vincenzio Galilei, and others from his home), I alone have survived them all.[46]

So began the next phase of Viviani's life with the responsibility of filling much of the void in Tuscan natural philosophy that was created by the deaths of so many illustrious thinkers within a short period of time. Viviani was to take on greater responsibilities for his Medici patrons, but throughout he maintained his dedication to preserving Galilean thought and to exercising his mathematical skills. From 1647 to 1649, Viviani took over the role of mathematics lecturer at the Accademia del Disegno, and the rather prestigious roles within the Court of tutoring mathematics to the pages of the ruling Medici family and contributing to the very

[45] The communications about the barometer between the two were published by Fabroni, *Lettere inedite*, ii, 24–25. The story is also retold by Favaro in *Amici e Corrispondenti*, 1032–1033. These authors narrated how Torricelli became interested in replacing water with mercury in the water pump and then devising an instrument to measure the pressure on the liquid and the relation with the vacuum. Viviani, by this time good friends with Torricelli, was excited by this project and assisted in the production of such an instrument. This will be discussed in more detail in Chapter Five.

[46] BNCF, Ms. Gal. 243, f. 119r. See also V. Viviani, *Quinto libro degli elementi d'Euclide*, Florence, 1674. As cited by Bonelli, 658 n.8. Viviani's account of this scene in his autobiographical letter in 1697 is similar. There he mentioned living with Galileo for three years, 'and in the last three months with Torricelli, and we were present with three priests, his own son and all of his family taking part in the blissful passing of his great soul to his Maker'. BNCF, Ms. Gal. 155, f. 9r–9v.

informal natural philosophical academy under the supervision of Ferdinando. While his responsibilities as a court mathematician often varied according to the desires of the Grand Duke, the travels of the Tuscan Court, the arrival on the scene of other thinkers, all his life he was kept busy by his duties to the Medici Court. Particularly exhausting was the role of engineer, which required substantial travel on horseback and led to illness on more than one occasion.[47]

Also during the late 1640s, Viviani was asked to take on some archival work. Before his death, Torricelli made it known to his executor that he wished Cavalieri and Ricci to organise his unpublished works. However, given Cavalieri's untimely death and Ricci's reluctance to carry out the task because of his duties to the Roman Court, the responsibility fell on Viviani's shoulders.[48] While Torricelli failed to publish anything other than his *Opera Geometrica* during his lifetime, he still accumulated quite a bit of work that remained unorganised in manuscript form, leaving Viviani with a difficult task.

Furthermore, after the death of Galilei, Viviani was left with the responsibility of reordering Galileo's papers,[49] a task that he pursued with much vigour and eventually led to his first major manuscript, *Racconto Istorico*, completed in 1654. Although this work remained unpublished during Viviani's lifetime, he continued to collect, translate, and restore Galileo's unpublished manuscripts and letters by maintaining a close relationship with Galileo's heirs.[50] Viviani's intention was eventually to publish his own collection of Galileo's works, and a biography of his late teacher more extensive than the *Racconto Istorico*. However, the project was uncompleted, since Viviani was never satisfied that he had collected all of Galileo's papers and because he was annoyed at the unshakeable stance of the Catholic Church regarding the prohibition of the *Dialogue*. Even Leopoldo, once appointed Cardinal in 1667, was unsuccessful in his efforts to facilitate Viviani's wish to publish a complete collection of Galileo's works.[51]

Although he never published his collection of Galilean papers, his efforts eventually resulted in the mass of Galilean manuscripts now found in the Biblioteca Nazionale Centrale in Florence. This has contributed immensely to the preservation of Galileo's memory and it is also the basis upon which many may remember Viviani himself.[52] In fact, he emphasised his dedication to Galileo by erecting two huge plaques lauding Galileo's career on the façade of his home and a bust of the mathematician over the doorway. Viviani was also largely responsible for erecting Galileo's tomb and sepulchre in Santa Croce.[53]

47 Borsetto, 'Introduction', in *idem*. (ed.), *Vita*, 64.
48 Bonelli, 665–666.
49 Viviani, *Vita* (ed. Borsetto), 66.
50 Viviani's search spanned from 1656 until 1677, enlisting the help of Lorenzo Magalotti, Michelangelo Ricci, as well as friends and colleagues in Paris, Rome, Bologna, Venice, and various other cities. Favaro, *Amici e Corrispondenti*, 1112. Favaro goes into a detailed description of Viviani's searches in *Documenti Inediti per la storia dei manoscritti galileiani nella Biblioteca Nazionale di Firenze*, Rome, 1886, 211, n.14.
51 Favaro, *Amici e Corrispondenti*, 1120.
52 Favaro, *Documenti Inediti*, 41; A. Procissi, 'I manoscritti superstiti dell'Accademia del Cimento' in *Celebrazione della Accademia del Cimento nel Tricentenario della Fondazione*, Pisa, 1957.
53 Bonelli, 683.

Here, then, we have Viviani in the lead up to the Cimento's foundation in 1657: he was a natural philosopher, mathematician and geometer; life-long servant of the Medici Court as a mathematician and an engineer; and archivist, through his long search for Galilean and Torricellian papers. This, it may be fair to say, is hardly the picture of an experimental philosopher prepared to lead his colleagues into the dawning of 'modern experimental science'. In fact, when we take a glimpse at Viviani's interests in the lead up to the Cimento's foundation, we see that, rather than simply producing experimental matters of fact, he and his colleagues were interested in much more complex natural philosophical issues.

5. THE SPEED AND PROPAGATION OF SOUND

One document written by Viviani, now held in the Galilean collection of manuscripts, and first published by Giovanni Batista Clemente Nelli in 1754, lists Viviani's exploits inside the Cimento. Perhaps concerned with the anonymity of the *Saggi*, Viviani may have felt a need to make his own record of his contribution to the group. One of the issues mentioned in this list, 'the concept of the equability of sound, and its uses',[54] exemplifies Viviani's physico-mathematical and natural philosophical commitments that he undoubtedly took into the Cimento. In a moment we shall examine what is meant by the 'equability of sound' and how Viviani's natural philosophical concerns can be clearly seen in his experiments on this topic. But before doing so, it is important to remember that despite these experiments being reported in the *Saggi*, Viviani worked on this topic in October 1656, several months before the Cimento began. So we are catching a glimpse of Viviani's construction of knowledge claims just before the formal foundation of the Cimento. What this means is that we shall see how Viviani, who was soon to become a prominent member of the supposedly strictly atheoretical experimentalist academy, actually speculated upon natural phenomena and constructed his claims in a mechanistic natural philosophical style.

During the years leading up to the Cimento's first recorded meeting in June 1657, Viviani was a leading member of the previously mentioned informal academy that met under the protection and supervision of the Grand Duke Ferdinando II. Not a great deal is known about the organisation and movements of this group, although surviving manuscripts suggest that its members undertook quite a bit of experimental activity on a wide range of topics of interest to the Grand Duke. Ferdinando brought one such topic, the speed and propagation of sound, to Viviani's attention, probably in October 1656. Viviani narrated this entire episode in a letter to an unnamed correspondent, most likely written some months later.[55] He commented how one day the Grand Duke had called him for

[54] Nelli, *Saggio*, 110–111.

[55] BNCF, Ms. Gal. 268, ff. 155r–158v; Abetti and Pagnini (eds.), 449–452. The letter is undated so we cannot be entirely sure of when it may have been written. Judging from the references to Gassendi's 'Philosophy', meaning, presumably, the 'Syntagma philosophicum', published in 1658, it is possible that Viviani wrote the letter that same year.

an interview and had asked him to give his opinion on the following questions: (1) Do sounds of different magnitudes and projected in different directions travel at equal speeds? (2) Is the speed of sound distorted by wind? Viviani gave the Grand Duke his opinion, that the time it takes for sound to travel a certain distance is not altered by the magnitude of the sound or the direction in which it is projected, or even the strength of the wind. Viviani explained his familiarity with the topic from his discussions with Borelli and from what had been written by Gassendi.[56] Viviani's response was judged compatible with the conclusions that Ferdinando had reached after performing the experiments himself only days earlier.[57] At this point in this discussion, Viviani referred to a related issue that, so he stated, could be of 'great use' and that for 'some time he had been curious to clarify'[58]: whether sound travels at a uniform speed. Ferdinando became so interested in this issue that, on 10 October 1656, he sent Viviani out with Borelli and other court mathematicians to test it. This was the third and last of the experiments on the speed of sound later mentioned in the *Saggi* and it is the experiment that Viviani laid claim to having invented in his list by referring to the 'equability of sound'.

The experiment involved the use of the pendulum to show that the sound reached twice the original distance in exactly double the time. In other words, that the speed of sound is uniform. The 'uses', that is, practical applications that Viviani referred to in his letter and in his statement declaring his intellectual ownership of this concept, may be gathered from the later account in the *Saggi*. They were 'to get the exact distance of places, especially at sea', a practice that would also be useful in cartography, where it was sometimes necessary to calculate the distances between towns. This type of knowledge, it was believed, could also be used 'to know how far away the clouds are and at what distance from the earth thunder is made, measuring the time from when the lightning is seen until we hear the thunder'.[59] On 12 October, Viviani performed another experiment regarding the questions Ferdinando had put to him in their meeting. The Grand Duke was curious about the exact distance between his palace in Florence, and Petraia, the Medici palace on the outskirts of town. This was the distance that the sounds created in his experiment had to travel. So Viviani repeated the experiment reportedly

[56] 'Da ciò che ne dice detto Gassendi'. BNCF, Ms. Gal. 268, ff. 156v; Abetti and Pagnini (eds.), 450. If Viviani did indeed refer to Gassendi's opinion on the topic in this conversation in 1656, we could only assume that this reference was from private correspondence with Gassendi or from a manuscript of Gassendi's 'Syntagma philosophicum', in which he discussed the movement of sound.

[57] As Viviani revealed in this letter, Ferdinando had measured the time it takes for sound to travel between the nearby Medici villa of Petraia, and the Grand Ducal palace in Florence. BNCF, Ms. Gal. 268, ff. 156v; Abetti and Pagnini (eds.), 450.

[58] BNCF, Ms. Gal. 268, ff. 156v; Abetti and Pagnini, 450.

[59] 'Le conseguenze poi che si pretendono di cavare da questa equabilità sono, fra l'altre, che per via di lampi e di suoni di diversi tiri potremo aver l'esatta misura delle distanze de'luoghi, e particolarmente in mare ... Sarà ancor facile e curioso a sapersi quanto da noi siano lontane le nuvole, e in che distanza da terra si creino i tuoni, misurando i tempi da che si vede il baleno a che quegli si sentono. ... Con questo stesso mezzo del suono potremo raggustar le carte de'luoghi particolari, e formar piante di diversi paesi, pigliando prima gli angoli di posizione delle città, castelli e villaggi per situarli acconciamente a' lor luoghi.' Magalotti, 243–244.

performed earlier by Ferdinando, by firing mortar from Petraia and measuring the time taken for the sound to reach the palace in Florence. That time was recorded as 41 'vibrations' of the pendulum.

There is no doubt then that Viviani had an interest in this topic and saw practical benefits that could be gained from experiments concerned with the movement of sound. More importantly, we can already see that these were traditional experiments quantitatively confirming theories that had been arrived at previously. This was certainly not an exercise in inductive 'fact-gathering'. With this in mind, we can now take a slightly closer look at the natural philosophical questions at stake for Viviani, and even make an early judgement regarding how this episode may guide us towards a new understanding of post-Galilean thought. In particular, since, by Viviani's own admission, his opinion on the movement of sound depended largely on Gassendi's writings on the topic, it is important that we identify what Viviani understood from Gassendi's work.

In his *Syntagma philosophicum* in the *Opera Omnia*, published posthumously in 1658, Gassendi made an analogy between the propagation of sound and the ripples that a pebble causes when dropped into a still body of water. Regardless of the weight of the pebble or the force with which it hits the water, the ripples always travel at the same speed. Gassendi claimed that the same could be said of the unseen ripples through the air propagated by sound.[60] This was also mentioned in the *Saggi*:

> Just as we see circular ripples made when a pebble is thrown into still water, these ripples being propagated on and on in ever larger circles until they reach the bank exhausted and die there, or, striking it with force, are reflected back; precisely in this way ... the subtle air around sonorous bodies travels over immense distances in fine ripples, and meeting our organ of hearing in the form of such waves, and finding it soft and pliant, imprints on it a certain trembling which we call sound.[61]

Gassendi's analogy, as the academicians themselves pointed out, was problematic. In the same paragraph that Magalotti mentioned the waves, he also noted the experiments performed in January 1662 showing that the size and speed of the ripples do in fact depend on the force with which the rock is made to hit the water.[62] Additionally, Gassendi's theory was actually more complex and pertained more to his natural philosophical beliefs than the wave analogy indicates.

If we are to understand Viviani's theoretical concerns in the construction of his sound experiments in 1656, the phrase to note in this passage is how 'the subtle air around sonorous bodies travels over immense distances'. Gassendi believed in the atomistic structure of all matter, including air and water, and insisted that minuscule pockets of vacuous space exist between the atoms. These actually allow

[60] P. Gassendi, *Opera Omnia*, 6 vols., Lyons, 1658–1675, i, 420–422.

[61] 'Si come veggiamo l'acqua stagnante increspársi in giro per una pietruzza che in lei si getti, e tali increspamenti andarsi via via propogando in cerchi successivemente maggiori, tanto ch'è giungono stracchi alla riva e vi muoiono, o che percuotandola con impeto, da essa per all'in là si reflettono, così per appunto asseriscono la sottilissim'aria dintorno al corpo sonoro andarsi minutamente increspando per immenso tratto, onde incontrandosi con tali ondeggiamenti nell'organo del nostro udito, e quello trovando molle e arrendevole, gl'imprime un certo tremore che noi suono appelliamo.' Magalotti, 242.

[62] *Ibid.* These experiments were recorded in the official diary: BNCF, Ms. Gal. 262, f. 132r.

for the movement and collision of sound atoms between the atoms of air, and thus the travel of sound. It is unclear how Gassendi believed the sound atoms interact with the atoms of air and how the flow of 'ripples' is supposedly created. In any case, he argued that the movements of all atoms are uniform and perpetual; in other words, that the speed of sound is uniform and does not vary. Once motion is imparted without constraints, it continues equably over 'immense distances'.[63] So, while scholastics considered sound to be a qualitative form, the academicians, led by Viviani, constructed an explanation based on Gassendian atomistic matter and mechanical motion.

Therefore, these experiments regarding sound, performed only months before the Cimento's foundation, show how Viviani exercised his natural philosophical skills and commitments. Furthermore, although not officially part of the Cimento's work, Viviani's experiments on the speed and propagation of sound found their way into the *Saggi* still with these subtle yet unmistakably corpuscularian references. These hints of a corpuscularian–mechanical natural philosophy were still present because the experiments themselves were carried out in confirmation of a natural philosophical claim. They were certainly not part of a Baconian 'fact-gathering' programme or a more systematic inductive procedure. Rather, natural phenomena were interpreted and experiments constructed according to mechanistic and corpuscularian natural philosophical beliefs. We may also add that Viviani was framing the efficacy of these experiments by mentioning their utility in navigation and meteorology. As we have seen with Torricelli's work on projectiles, this was a plea to the Tuscan Court for its support and favourable consideration of the work carried out by Viviani and his colleagues who assisted him. This case study, therefore, strengthens our understanding of Viviani's theoretical and disciplinary aims and interests and his contributions to the Cimento.

6. 1659–1703

Returning now to Viviani's activities outside the Cimento, it will be evident that during the rest of his life, a great deal of his work was of a humanist nature, including commentary upon and restoration of ancient texts. The first of these works, *De maximus et minimus*, was a controversial attempt to restore Apollonius' Book V of *The Conics* in 1659. This was during the height of the Accademia del Cimento's activities, meaning that another academician making his presence known inside the Tuscan Court, Borelli, was not far from the action. Later we will look closely at Viviani's first encounter of substance with Borelli – how Borelli came across an Arabic translation of Apollonius' missing books and how this episode characterises some of the intellectual concerns dominating the pursuits of these natural philosophers. For now, however, it is important simply

[63] Brundell, *Pierre Gassendi*, 54–59; M.J. Osler, *Divine Will and the Mechanical Philosophy*, Cambridge, 1994, 182–194.

to recognise that, from 1659 until his death in 1703, Viviani continued to pursue his interests in mathematics, albeit in the rather humanist manner of restoring and commenting upon ancient material.

In 1674, eight years after being confirmed as First Mathematician to the Grand Duke and 11 years after being assigned a pension by Louis XIV of France, Viviani published *Quinto Libro degli Elementi d'Euclide*. This was a restoration of Euclid's Book V following in the vein of Torricelli's and Galileo's combined efforts in 1641 to clarify and improve the famous ancient text. This publication was followed by *Diporto Geometrico* in 1676, a text consisting of answers to 12 geometrical problems posed for all Italian mathematicians. According to Favaro, Viviani was not very interested in publishing this work since he had found it too easy. In fact, he took on the task only in response to the persistent requests from Leopoldo and colleagues, and reportedly took no more than six days to arrive at solutions using basic geometry, including the work of Euclid and Apollonius. As the title, translated as 'Geometrical Recreation', suggests, this was nothing more than entertainment for Viviani, and the little importance he attached to the text is reflected in its dedication to beginners in geometry.[64]

For the remainder of his life, Viviani occupied his time with publications regarding architectural and engineering matters in Florence,[65] a translation of Euclid's *Elements* (1690), a restoration of Aristeo's five last books,[66] and of course, his never-ending search for material related to Galileo's life and works. The restoration of Aristeo's books in 1701 was the first and only publication Viviani dedicated to Louis XIV, almost 40 years after being awarded a pension by the French King, and only two years after finally being named as a foreign associate of the Académie Royale des Sciences in Paris. In 1696, Viviani was also appointed a fellow of the Royal Society of London. Viviani died in 1703 at the age of 81 and was laid to rest by Galileo's side in Santa Croce in Florence.

7. CONCLUSION

Historian Luigi Tenca describes Viviani as a very calm individual who 'would have wished to live in agreement with everyone, not because of a weakness of character, but because of a natural gentlemanly spirit. A true gentleman, in the purest sense of the word.'[67] This description would fit perfectly with the notion that a gentlemanly environment with trustworthy individuals was the ideal setting in mid to late seventeenth-century Europe for the birth of experimental science. However, Tenca's description, along with the traditional stories we have seen in

[64] Favaro, *Amici e Corrispondenti*, 1071.

[65] V. Viviani, *Discorso al Serenissimo Cosimo III Granduca di Toscana intorno al difendersi da' riempimenti e dalle corrasioni de' fiumi applicato ad Arno*, Florence, 1688.

[66] V. Viviani, *De locis solidis secunda divinatio geometica in cinque libros iniuria temporum amissos Aristaei Senionis Geometrae*, Florence, 1701. Aristeo's work has been lost since antiquity, but Viviani's attempt to restore this classical study of conic sections, is consistent with his earlier interest in restoring Apollonius' *Conics*.

[67] L. Tenca, 'Le relazioni fra Giovanni Alfonso Borelli e Vincenzio Viviani', *Rendiconti* (1956), **90**, 109.

Chapter One, fails to discuss the natural philosophical and disciplinary concerns that shaped Viviani's and the academicians' careers. By contrast, by exploiting the work by Nelli, Targioni Tozzetti, Antonio Favaro, Luciana Borsetto, and Maria Luisa Bonelli, as well as the publications, letters, and manuscripts by Viviani himself, we may come to appreciate the natural philosophical complexities behind Viviani's career, including a glimpse of his concerns in the months leading to the foundation of the Accademia del Cimento. The new picture that we can form from these sources shows Viviani as an archivist, a humanist scholar, a keen mathematician, geometer, and natural philosopher of mechanist/Gassendist leanings. Therefore, through a close look at the career of one of Galileo's most respected students, and one of the Cimento's most prominent members, we have begun our investigation of the group's activities, based not on finding clues about a supposed experimentalist programme, inductivist method, and gentlemanly behaviour, but rather on an understanding of the natural philosophical skills, commitments, and agendas that dominated the academicians' activities. With this in mind, we can now prepare ourselves for a look at one of Viviani's fellow academicians, Borelli, to investigate whether Viviani had the same natural philosophical concerns as some of his colleagues inside the Cimento. Then, we shall see in Parts Two and Three, how those concerns contributed to the academicians' construction, interpretation, and presentation of knowledge.

CHAPTER THREE

GIOVANNI ALFONSO BORELLI (1608–1679)

Borelli's career can be divided into four different stages: his education in Naples and Rome; his tenure as mathematics professor at Messina; his time in Florence; and finally, his last years in southern Italy. Consequently the following account of Borelli's life and works follows those four phases. We shall see that although several details of Borelli's life remain ambiguous, such as his reasons for relocating at different points in his career, the lack of documentation regarding his personal motivations does not prevent us from carefully examining his life and works, or from gathering clues regarding his intellectual, social, and political concerns. In particular, through the available primary and secondary sources regarding Borelli's career, we shall gain an understanding of the natural philosophical skills and commitments he brought to the construction, interpretation, and presentation of experiments inside the Accademia del Cimento.

1. BORELLI IN ROME: HIS EDUCATION UNDER CASTELLI AND HIS INITIATION INTO THE GALILEAN SCHOOL

Very little is known as certain about the early years of Borelli's life and education in Naples, but it is believed that he may have been a student of the Neoplatonist Galilean supporter and anti-Spanish activist, Tommaso Campanella.[1] Because of his political activities against the Spanish rulers in the Kingdom of Naples, Campanella was imprisoned there from 1599 to 1626.[2] According to Thomas Settle, Miguel Alonso, Borelli's father, was actually a sympathiser of Campanella and as a consequence was also imprisoned in 1615. Soon afterwards, Alonso was exonerated, released, and allowed to return to his duties as a guard at Castel Nuovo where Campanella was being held.[3] The circumstances here are not

[1] T.B. Settle, 'Giovanni Alfonso Borelli', in *Dictionary of Scientific Biography* (ed. C.C. Gillispie), 18 vols., New York, 1991, ii, 306.

[2] According to Yates, Campanella believed that Calabrians needed to overthrow the Spanish rulers and establish a republic based on new religious ethics. For this reason, Campanella was tortured during his first years in prison and narrowly escaped the same fate as Giordano Bruno. Frances Yates, *Giordano Bruno and the Hermetic Tradition*, London, 1964, 364–366.

[3] Settle, 'Giovanni Alfonso Borelli', 307.

entirely clear, but it is thought that during his imprisonment, Campanella was still allowed to have visitors and students. So it was through his father's position at Castel Nuovo and his association with the rebellious Campanella, that Borelli is believed to have met and studied under one of Galileo's and Copernicus' most ardent supporters.[4] During this time, it is also believed that Borelli may have been studying medicine at the University of Naples.[5] Although there is little evidence in support of this suggestion, it would seem likely in light of the interests in medicine and physiology that Borelli was to show later in his career.

It is not known exactly when Borelli moved to Rome or if he went with Campanella, who was taken to Rome still under arrest in 1626. In any case, it is around this time that Borelli found himself in the papal city. Once there, he became a student of Benedetto Castelli, one of Galileo's leading students and certainly amongst the best of Italy's mathematicians during the seventeenth century. Now that Borelli's education had taken on a slightly more formal tone than his presumed previous tutelage under Campanella, we may gain some valuable clues regarding Borelli's early intellectual interests by analysing what Castelli might have taught his students in Rome.[6]

In 1613, following his education under Galileo at Padua, Castelli was appointed professor of mathematics at the University of Pisa. For the rest of his life he remained a close correspondent and collaborator to Galileo, particularly in the field of hydrostatics, where Castelli made some significant contributions to the Galilean theory regarding floating bodies.[7] In 1626, Castelli was appointed to the chair of mathematics at the University of Rome, as well as papal consultant on hydraulics. Importantly, in addition to his responsibilities to the university and the Papal Court, Castelli set up a network of young Galilean scholars in Rome, including Borelli. This was the Galilean school he referred to on more than one occasion in his letters to Galileo.[8] There have been very few analyses and biographies of Castelli and the work he carried out in Rome, so we can be even less certain of precisely what kind of knowledge he passed on to his students. In any case, in order to understand the role he may have played in shaping Borelli's early intellectual interests, we may draw on our knowledge of some of the other characters and concerns present in Rome at that time.

[4] *Ibid.* Campanella's presence in this early stage of Borelli's education is mentioned in quite a few sources, but there would appear to be few details available regarding what natural philosophical lessons, including atomism and natural magic, he may have given Borelli.

[5] *Ibid.*; G. Barbensi, *Borelli*, Trieste, 1947, 19.

[6] 'mio primario maestro'. G. Borelli, *Discorso del Signor Gio: Alfonso Borelli, accademicio della Fucina e professore delle scienze matematiche nello Studio della nobile città di Messina, nel quale si manifestano le falsità, e gli errori, contenuti nella difesa del Problema Geometrico, risoluto dal R.D. Pietro Emmanuele*, Messina, 1647, 15. As cited by U. Baldini, 'Giovanni Alfonso Borelli e la rivoluzione scientifica', *Physis* (1974), **xvi**, 107, n.29. Borelli is not recorded as having graduated from the University of Rome where Castelli occupied the chair of mathematics, but we can be sure that he was indeed Castelli's student. This is evident in the recommendations Castelli gave for Borelli's employment and Borelli's reference to Castelli as 'my first teacher' in his first publication in 1646.

[7] According to Segre, Castelli's major scientific work was an extension of the study on hydraulics that he and Galileo had performed years earlier. Castelli, *Della misura dell'acque correnti*, 1628. Segre, *In the Wake*, 52.

[8] Favaro (ed.), *Le Opere*, xviii, 303–304.

To begin with, we should note that this was during the height of Galileo's controversy with the Catholic Church, and of course in the city where he was tried by the Inquisition. Castelli, like all of Galileo's students, was greatly interested in the outcome of the trial and very familiar with the contents of the *Dialogue*. With this in mind, we may recall that Torricelli was also a student of mathematics at the Sapienza during this period and a member of the 'Galilean school'. In a letter to Galileo in September 1632, just months after the Holy Office placed Galileo under house arrest, Torricelli referred to his mathematical education under Castelli, including the geometry of the ancients extending to the contemporary studies in astronomy: 'having practised so well all the geometry of Apollonius, Archimedes, Teodosio, and having studied Ptolemy and seen nearly everything by Tycho, Kepler and Longomontano, finally accepting Copernicus'.[9]

So despite the lack of historical studies pertaining to Castelli's own work whilst in Rome, we can be sure that he would have had a significant impact on his students in the fields of mathematics and natural philosophy, particularly Galilean astronomy and physics. Judging from Torricelli's letter, there is no doubt that Castelli kept his students well aware of Galileo's mathematical, geometrical, and mechanistic natural philosophical opinions, including the arguments in Galileo's *Dialogue*. Furthermore, we may assume that Castelli did not actually teach Copernicanism as a true description of the cosmos because of the condemnation he would have received from the Catholic Church. But Torricelli claimed that he was still being encouraged to accept Copernicus and Kepler, at least privately, while perhaps discussing their works only as hypotheses in public.

Since Borelli was the same age as Torricelli, and was also being tutored by Castelli during this time, it would not be unfair to assume that he also would have familiarised himself with Galileo's physics and astronomy, including the revision of ancient geometrical writings. Therefore, Borelli's presence in Rome during Galileo's trial, as well as his education under one of Galileo's leading disciples in mathematics, provides us with our first clues to the physico-mathematical and natural philosophical commitments Borelli was to pursue throughout his career.

2. 1635–1656: POLITICS, MATHEMATICS, AND MEDICINE IN BORELLI'S SICILY

While Torricelli joined Galileo in Arcetri, resulting in his succession to the position of First Mathematician to the Grand Duke of Tuscany in 1642, Borelli moved to Sicily. In 1635 under Castelli's recommendation, Borelli was called before the Senate of Messina to give a speech demonstrating his knowledge and ability in the field of mathematics.[10] He was subsequently appointed to the

[9] *Ibid.*, xiv, 387.

[10] Barbensi, 19; T. Derenzini, 'Giovanni Alfonso Borelli, Fisico' in *Celebrazione della Accademia del Cimento nel Tricentenario della Fondazione*, Pisa, 1957, 38, n.8.

position of lecturer of mathematics at the University of Messina.[11] In a moment we shall see some evidence of Borelli's natural philosophical concerns while in Sicily, but it is important first to understand the volatile political environment which he entered through his employment by the Messinese Senate. This will help us to form an idea of how Borelli began his career involved in political, social, and intellectual issues, and why he may have abandoned Tuscany in 1657 in favour of a return to Sicily.

Since the mid-thirteenth century, following the death of Frederick II, the last of the great Hohenstaufen rulers, the Sicilian and the Neopolitan Kingdoms had been under the rule of the great Spanish houses such as the Aragon and then the Hapsburgs. During the first half of the sixteenth century, the Hapsburg Holy Roman Emperor Charles V (who was also King of Spain) wished to dismantle the ancient feudal system of administration and government in Naples. He was successful in his attempt to consolidate the power of his own kingdom in southern Italy by establishing a more centralised form of government and avoiding any revolts. Yet while Naples was governed closely according to the directions and interests of Spain, Sicily was quite a different case. The Hapsburgs were very interested in maintaining their power in Sicily because of the lucrative production and exportation of silk, wheat, and other products from the island. The position of Sicily at the centre of the Mediterranean Sea also provided a strong incentive for dominating the trade networks established by the Sicilian cities. However, they found it hard to break down the existing power and influence of the local aristocracy, and the Sicilian cities held on to a great deal of autonomy.[12] This was especially the case in Messina, where Borelli first made his mark in natural philosophy and where our attention momentarily lies. Here we shall see the strong link between the political and intellectual environments in Borelli's new place of employment.

A Spanish representative guarded the Hapsburgs' interests in Messina and presided over the movements of Spanish troops and the fortifications of the city, but it was the elected members of the Senate who essentially controlled the city's administration, particularly the valuable exportation of silk. In fact, according to Meli, it was the funds from the silk industry that sustained the resurgence of intellectual activity in Messina at the beginning of the seventeenth century, and the

[11] According to Meli, that appointment was not made until 1639, 4 years after Borelli's arrival in Sicily. Neither Meli or any other biographer suggests exactly how Borelli may have made a living in Sicily between 1635 and 1639, but it is understood that after his arrival in Messina and speech to the Senate he was immediately in favour with some of the local powerful families. D.B. Meli, 'The Neoterics and Political Power in Spanish Italy: Giovanni Alfonso Borelli and his Circle'. *History of Science* (1996), **xxxiv**, 60.

[12] As Helmut Koenigsberger points out, the Sicilian towns 'maintained their local laws and privileges against all foreigners'. H. Koenigsberger, *The Government of Sicily under Philip II of Spain*, London, 1951, 47. According to Meli, this was because Sicily was never actually conquered by 'Spain' (that is, the Kingdom of Aragon, as it was known at the time), but only offered to Aragon after the departure of the French in 1282. Therefore, while accepting Aragonese, and later united 'Spanish' rule, they were not prepared to allow the Spanish Emperor complete control of the island. Meli, 'The Neoterics', 66. For a comprehensive coverage of Sicily's political history, see M.I. Finley, D.M. Smith, and C. Duggan (eds.), *A History of Sicily*, London, 1986.

outcomes of the Senate elections often determined if some of those funds were to be distributed to the University of Messina and other academic institutions such as the politically powerful Accademia della Fucina.[13]

The Fucina was established in 1639, when the university fell under full control of the Senate and Borelli was appointed to the chair of mathematics.[14] According to Settle, this institution was founded by Messina's aristocracy. The Senate was willing to pay philosophers such as Borelli well for their contributions to this community and to the revival of the city's intellectual activities. The interests of the Fucina's members not only included the pursuit of knowledge and the production of literary art, but also discussion regarding politics and Spanish rule.[15] The Accademia della Fucina was therefore an aristocratic group harbouring some potentially rebellious attitudes towards the Spanish rulers. Borelli's association with this community is evident not only in his close friendship with some of Messina's aristocratic rebels, but also in the accusations made against him by the Spanish governor of even leading the intellectual anti-Spanish movements.

So Borelli's new circle of friends in Sicily included mathematicians, physicians, and astronomers all involved in political movements aimed at destabilising the controversial presence of a Spanish representative, and possibly selecting a local noble as the king of Sicily.[16] This was the volatile political environment in which Borelli was working. By 1639 he had become a very well-respected and leading member of this community and its intellectual and political activities. When he arrived in Pisa in 1656 to work under Medici patronage, Borelli found that his natural philosophical skills were once again used by his patrons to raise the status of their Court. Inversely, it was also in these environments, where artists and thinkers were highly valued, that Borelli used the Court to raise his own status. So the mix of political and intellectual aims and interests continued for Borelli throughout his career. We shall return to these issues in a little while when we examine Borelli's, and the Cimento's, work. Questions regarding the social and political foundations and purpose of the Cimento will be examined in detail in Part Three. In the meantime, we can now turn our attention once again to Borelli's intellectual ambitions and disciplinary interests while in Messina.

In 1640, Borelli must have shown some interest in occupying the vacant chair of mathematics at the University of Pisa and becoming a member of the Tuscan Court. He was recommended for the position, once again by Castelli in a letter to Galileo in May that year.[17] Borelli was overlooked for the position in favour of Vincenzio Renieri. This is not surprising since Borelli did not yet have any publications to his name, and despite his standing in Sicily, he only had one year of experience in the chair of mathematics at Messina. Nevertheless, we may imagine

13 Meli, 'The Neoterics', 66.

14 Barbensi, 19.

15 Settle, 'Giovanni Alfonso Borelli', 307.

16 The figures Meli points out include the poet Simone Rao, a student and friend of Borelli, physician Domenico Catalano, amateur mathematician Daniele Spinola, and Iacopo Ruffo, a Messinese noble and patron of Borelli in Sicily. Meli, 'The Neoterics', 62.

17 Favaro (ed.), *Le Opere*, xviii, 188–189.

that Borelli, as a student of Galilean natural philosophy, with friends and colleagues such as Torricelli and Castelli closely connected to Galileo and the Tuscan Court, had considerable interest in visiting or working in Tuscany. This opportunity finally came at the end of 1642, when he was sent on an errand by the Senate of Messina to tour the main centres of learning in northern Italy and recruit, if possible, new members for the university's growing community of elite intellectuals. Although he missed out on meeting Galileo himself,[18] Borelli took the opportunity to meet Viviani as well as his fellow student from Rome, Torricelli, and probably even the Grand Duke Ferdinando II,[19] who eventually offered Borelli the chair of mathematics in Pisa in 1656.[20] Additionally, he met another of Galileo's famous disciples in Bologna, Bonaventura Cavalieri. Bologna's authorities were also so impressed by the young neopolitan that they even considered him to take over the chair of mathematics after Cavalieri's death in 1647.[21] Evidently, by the beginning of the 1640s, Borelli was starting to make himself known in Italy as one of the region's leading mathematicians and followers of Galileo. During the following years, he continued to enhance his reputation through his first publications. This is also where he began to demonstrate the mathematical and natural philosophical concerns that were to dominate his work in physics, astronomy, and physiology throughout his career.

The best-known publication to come from Borelli while he was in Messina was not a mathematical or geometrical treatise. Instead, it was his first attempt to publish some of his ideas in the field of medicine, including a study in anatomy and physiology. In 1647 an epidemic of fevers swept through the entire island of Sicily, and authorities wished to find the cause of the disease and how such epidemics could be avoided. In pursuit of such knowledge, Borelli was asked by the Senate of Messina to produce a fully funded investigation of the disease and its spread. This resulted in the 1649 publication with the following title: *Delle cagioni delle febbri maligne di Sicilia negli anni 1647 e 1648*.

This work, Borelli's first published effort in the field of medicine, is also widely considered as his most important writing whilst in Messina. This opinion is usually grounded by linking his study of the epidemic to his very last publication, *De motu animalium* (1680), a mechanical explanation of the movements of animals based on the motions of pulleys and levers. But for traditional historiographies searching for the so-called origins of modern science, the interest in these texts lies mostly in Borelli's supposed use of an 'experimental method' through dissections and experiments, rather than his use of a mathematical and mechanical natural philosophy. For example, in his investigations regarding the cause and the spread of the disease, Borelli travelled to several other Sicilian cities and accumulated quite a few notes from autopsies. As a result of this accumulation of data,

[18] It is possible that Borelli may have been in Tuscany while Galileo was still alive, but there is no evidence to confirm that they met.
[19] Targioni Tozzetti, i, 205.
[20] According to Barbensi, that offer came as a direct result of Borelli's visit to Tuscany. But once again, we cannot be entirely sure that this is true. Barbensi, 21.
[21] Settle, 'Giovanni Alfonso Borelli', 308.

authors such as Gustavo Barbensi consider Borelli as 'the person who introduced the experimental method in the study of living matter, in particular physiology'.[22] Tullio Derenzini also claims that after studying Galilean physics under Castelli, Borelli acquired a passion for 'the new philosophy of nature and the Galilean experimental method that he then used with resoluteness and rigour in all his research'.[23] So these traditional historiographies, like those we have already encountered in Chapter One associated with the experimental origins of Italian science, have preferred to focus most of their attention on the notion of experimental method and associated origin stories for modern science. Derenzini, in particular, maintains the traditional story centred around Galileo by suggesting that Borelli was applying an experimental method actually established by Galileo.

Yet there is so much more that can be discerned from looking carefully at Borelli's arguments. His best-known publication from the early stages of his career, *Delle cagioni*, actually reflected some of the natural philosophical principles that he had probably already developed in his education and that he was to demonstrate throughout his career. Borelli carefully set out an argument aimed not only at providing a solution to the problem of how disease was spread, but also refuting scholastic physiological beliefs. Moreover, it was an argument that could not be derived by applying some supposedly universal method with no theoretical presuppositions and aiming to reach 'scientifically objective' conclusions.

Borelli divided the text into three parts; The first argued the impossibility of the disease being caused by the disequilibrium of the four Galenic elements as a result of meteorological conditions. The second claimed to establish that astrological explanations could not adequately account for the epidemic. Finally, the third section offered an alternative explanation that was strongly based on the work by Santorio di Capodistria (1561–1636), a member of the anatomical school at the University of Padua, and eventually also a colleague of Galileo there. While eager to listen to new theories and methods, di Capodistria still supported the Galenic theory regarding the transpiring of the skin through tiny pores. There is no doubt that in *Delle Cagioni* Borelli took di Capodistria's work to devise his own medical theory. He first concluded that the disease must get into the body from the outside, and proceeded to explain how chemical pestilent particles enter the body through the tiny pores in the skin.[24]

We may certainly take this to be the first hint of Borelli's belief in a micro-mechanical and corpuscularian philosophy, but as Ugo Baldini notes, this text can still provide us with more clues regarding Borelli's natural philosophical interests. These clues can be found not in his corpuscularian suggestions, but in his approach to criticising Galenic and astrological belief systems. In his examination of the epidemic in the first two sections of the text, Borelli refused to resort simply to presenting his arguments purely by reporting his experiments and observations. Instead, he attempted to use classical sources in order to deconstruct the

[22] Barbensi, 38.
[23] Derenzini, 37.
[24] P. Galluzzi, 'G.A. Borelli dal Cimento agli Investiganti', in *Galileo e Napoli* (eds. F. Lomonaco and M. Torrini), Napoli, 1987, 343.

Galenic and astrological arguments made by his contemporaries, and criticise their conclusions. According to Baldini: 'The result of such a form of argument ... consists of the possibility of extracting from an analysis of the texts referred to by traditional medics, conclusions that could negate them.'[25] As Baldini concludes, Borelli attempted to dismantle the old natural philosophical knowledge systems, so that he would be left only with some supposedly concrete fundamentals, from which he could build a new solution to the problem.[26] We may add that Borelli was resorting to his familiarity not only with the classical texts, but also with Italian sixteenth- and seventeenth-century arguments in physiology. This is a strong clue suggesting that Borelli was not a revolutionary experimentalist detached from natural philosophical concerns, and applying some type of atheoretical, inductivist method. Instead, at the opening stages of his career he was already beginning to construct knowledge claims according to his familiarity with classical sources and the natural philosophical commitments of his contemporaries and predecessors. In fact, before he had even written *Delle cagioni*, Borelli had been working on the restoration of classical mathematical sources such as Euclid. Moreover, Borelli's refinement of his knowledge of mathematics and geometry assisted him to construct some knowledge claims characteristic of the emerging physico-mathematical culture discussed in Chapter One.

Before the opportunity came to write his treatise on medicine, Borelli was interested in improving the geometrical definitions and propositions in Euclid's *Elements*. The aim of this exercise was to establish solid and reliable geometrical principles as a foundation for studying natural phenomena. Although Borelli did not complete and publish his *Euclides restitutus* until many years later in Pisa in 1658, he recalled having conceived some of the ideas in this text as early as in 1642, while he was visiting old acquaintances and establishing new connections in Tuscany.[27] Borelli's introduction to *Euclides restitutus* explains how he discussed the restoration of the most relevant parts of Euclid's *Elements* with some of the members of the Tuscan Court. He was especially eager to find out whether the Tuscan mathematicians could provide demonstrations and definitions clarifying Book V of the *Elements*, the very book on proportionalities that had occupied Galileo's and Viviani's time during their work on the scholium supporting Galileo's key postulate in his theory of accelerating falling bodies.[28] But Borelli was not satisfied with the work being carried out by the Tuscan thinkers on the subject, and he set about producing his own restoration of the *Elements*. Borelli's *Euclides restitutus* was therefore intended to be a clarification of Euclid's propositions with the particular aim of providing a clearer explanation of the notion of proportionalities.[29]

[25] U. Baldini, 'Giovanni Alfonso Borelli biologo e fisico negli studi recenti', *Physis* (1974), **xv**, 247.
[26] *Ibid.*, 248.
[27] Ugo Baldini even suggests that Borelli finished the manuscript in Sicily. Baldini, 'Giovanni Alfonso Borelli e la rivoluzione scientifica', 123.
[28] C. Vasoli, 'Fondamento e metodo logico della geometria dell'Euclides Restitutus del Borelli', *Physis* (1969), **xi**, 582.
[29] Settle, 'Giovanni Alfonso Borelli', 310.

It is important to note here for our understanding of Borelli's skills as a mathematician that these were not modest aims that he was expressing. He was not only attempting to clarify an ancient geometrical proposition, but as he expressed in the Proem, he was also looking to establish clearer and more rigorous demonstrations and definitions of the principles that shaped the foundations of seventeenth-century geometry. He achieved this aim by simplifying the ancient text to the point of reducing Euclid's 473 propositions to just 230.[30] In the process, according to Cesare Vasoli, he strengthened the use of logic that is traditionally associated with geometry by providing more rigorously demonstrated propositions from which further geometrical claims could be easily deduced.

This analysis of Euclid was valuable, both for possibly facilitating advances in geometry as well as in the mixed (or physico-)mathematical fields, indeed, anywhere that geometry was applied. Borelli quite arguably had these aims. In fact, Vasoli quite enthusiastically grasps this point, arguing that such improvement of the logical method or procedure embedded in geometry, also provided Borelli with a solid basis for the application of geometry to his study of natural phenomena. For example, in the following passage it is evident that Vasoli recognises the extent to which Borelli's work on Euclid's theory of proportionalities reflected his aims and interests in physiology:

> The physiologist who searches ... for the inevitable connection between the big geometrical 'machine' of the 'macrocosm' and the very fine structure of the 'minute mechanisms' which make up the 'wonderful organs' with various vital functions, knows, in fact, that only the definition of the determining concept of 'proportionality', with all its consequences in the field of geometry and mechanism, can constitute the secure acid test of a mathematical process free from any obscurities or conceptual uncertainties.[31]

In other words, Borelli's devotion to the mathematical sciences was such that he believed that Euclid could provide him with the help he required to understand the micro- and macrostructures and movements of the universe, including the properties of the human body, as well as the matter and motion of the entire cosmos. This emerging commitment to physico-mathematics was part of the culture of natural philosophising in the seventeenth century that was discussed in Chapter One. Furthermore, in Part Two, when we examine the case studies regarding the Cimento's work, we shall see that these were precisely the type of cognitive concerns debated by the academicians in the Tuscan Court. Therefore, as we continue to analyse Borelli's career, it will become clearer how his work on classical mathematics was helping to establish the foundations of a physico-mathematical practice and an anti-scholastic and mechanistic natural philosophy inside the Accademia del Cimento.

Borelli's work on Euclid was certainly not his last show of interest in ancient texts, since he soon became involved in restoring and using other ancient authors. Indeed, it is believed that as early as 1654 Borelli participated in the posthumous publication of a work by Francesco Maurolico, a sixteenth-century Messinese

[30] Vasoli, 582.
[31] *Ibid.*, 575.

scholar of ancient Greek geometrical texts. Maurolico had attempted to reconstruct Books V and VI of Apollonius' lost sections of the *Conics*, but never published this work before his death in 1575. According to Baldini, this unpublished work fell into the hands of Borelli, who made annotations along the margins of the manuscript and was in all probability responsible for its publication in 1654.[32] This surely was to help Borelli tremendously just four years later when he actually found a twelfth-century manuscript of Apollonius' lost books in the library of the Grand Duke of Tuscany.[33]

Soon after his travels around Italy, Borelli saw his first opportunity to publish some of his own geometrical findings. This opportunity came in 1644 and 1645 through two works by a mathematician from Palermo by the name of Pietro Emmanuele. Emmanuele proposed a solution to a geometrical problem being considered by Italy's mathematicians at that time.[34] Detecting some errors in this work based on his knowledge of ancient texts, Borelli responded with a critical account of Emmanuele's solution.[35] This small contribution to a little-known geometrical debate, was Borelli's first attempt to present his mathematical and geometrical thoughts to the public. More importantly, according to Pietro Nastasi, this was a work that reflected Borelli's skills in mathematics and that he had used to complete his restoration of Euclid's *Elements* and that he even referred to in his later writings containing geometrical principles.[36]

In conclusion to this analysis of Borelli's first years and earliest treatises in Rome and Messina, we may state that he was clearly devoted to mathematics and natural philosophy. This is made obvious when we recap what has been examined of his life and works so far in this chapter. First, Borelli's education in Rome was grounded in mathematical and natural philosophical concerns. Second, Borelli demonstrated his mathematical and natural philosophical interests from as early as 1642 when he developed his ideas for *Euclides restitutus*. Third, *Delle cagioni*, his first writing on medicine and physiology also contained clues about how he was developing his anti-scholastic commitments. This is all to say that Borelli did not actually develop an experimentalist method, let alone some species of the modern scientific method, as the traditional historiographies have claimed.[37]

[32] F. Maurolico, *Emendatio et restituio conicorum Apollonii Pergaei*, Messina, 1654. See Baldini, 'Giovanni Alfonso Borelli e la rivoluzione scientifica', 120, n. 65.

[33] L. Guerrini, 'Matematica ed erudizione. Giovanni Alfonso Borelli e l'edizione fiorentina dei libri v, vi e vii delle *Coniche* di Apollonio di Perga', *Nuncius* (1999), **xiv**, 510.

[34] P. Emmanuele, *Lettera intorno alla soluzione di un problema geometrico*, Palermo, 1644. The problem regarded the symmetrical composition of a triangle with unequal sides.

[35] This publication was titled: *Discorso del Signor Giovanni Alfonso Borelli, accademico della Fucina e professore delle scienze matematiche nello Studio della nobile città di Messina, nel quale si manifestano le falsità, e gli errori, contenuti nella difesa del Problema Geometrico, risoluto dal R.D. Pietro Emmanuele*, Messina, 1646. Borelli's criticism did not focus on the problem as much as Emmanuele's methodology, arguing that the mathematician from Palermo failed to apply ancient mathematical principles before reaching a conclusion. P. Nasatasi, 'Una polemica giovanile di Giovanni Alfonso Borelli', *Physis* (1984), **26**, 217.

[36] *Ibid.*, 216.

[37] This is quite apart from the issue of whether such a modern method – unique, transferable, and efficacious – exists except in the legitimising discourses of modernity. See Schuster and Yeo, ix–xxxvii.

Instead, even while undergoing his education, he had begun to construct knowledge claims according to a deep interest in exploring ancient geometrical sources and the accumulation of mathematical and natural philosophical knowledge from his contemporaries and predecessors. What will become more apparent as we continue our study of Borelli's career, is that he was using established natural philosophical principles to construct physico-mathematical knowledge claims. Furthermore, while Borelli used experiments throughout his career in the fields of astronomy and physiology, these experiments were still constructed according to wider natural philosophical concerns and performed in very particular cultural and political circumstances. As we now leave this phase of Borelli's career behind and move on to his achievements and conflicts in Pisa, we shall see that these natural philosophical concerns largely motivated and left their imprint in all of Borelli's works, including his contributions to the Accademia del Cimento.

3. BORELLI AND VIVIANI

In 1656, Borelli finally succeeded Vincenzio Renieri in the chair of mathematics at the University of Pisa. This turned out to be the most important move of Borelli's career. In the decade he spent in Tuscany he not only made contributions to the Accademia del Cimento, but he was also involved in several episodes of intellectual innovation and controversy, as well as the foundation of his school of iatrophysicians at Pisa. For the moment, we shall examine some more of the collaborative work carried out by Borelli and Viviani while they were both in Tuscany.

We have already seen how Borelli and Viviani were interested in a corpuscularian and mechanical natural philosophical explanation regarding the speed of sound in 1656. Despite their clear collaboration on this issue, on three other occasions between 1657 and 1667, they seemingly opposed each other and even insulted one another behind each other's back. Importantly, these controversies were not based on opposing natural philosophical opinions, since Borelli and Viviani actually agreed on their theoretical speculations. Instead, the disputation concerned the courtly recognition that each thought he deserved for his efforts, although they were still sharing and collaborating on some strong mathematical and mechanical natural philosophical aims.

From one of these episodes, the restoration of Apollonius' the *Conics*, we shall see that in 1658, when, according to traditional historiographies, the academicians were supposedly committed to performing only experimental science, Borelli and Viviani were actually working on understanding and improving classical mathematical sources. For both Borelli and Viviani, this humanist practice was a vital part of their natural philosophical activity, including their contributions to the Accademia del Cimento. Furthermore, the involvement of the Medici princes in this activity provides us with a vital clue regarding how the academicians struggled for status inside the Court, and how the Court itself looked to capitalise on the work produced by its subjects. Historians have traditionally described this episode as the defining moment of the relationship between Borelli and Viviani. Just as Viviani was publishing a restoration of Apollonius' missing books, Borelli was undertaking the task of translating a long-lost Arabic version of the famous text. Despite the lack of evidence to suggest that this created any friction between Borelli and Viviani, M.L. Bonelli and Targioni Tozzetti suggest that the Apollonius episode sparked an intense rivalry between these two that led directly to Borelli's eventual departure from Tuscany in 1667.[38]

In addition to this supposed clash of egos, some historians have also discussed the outcome of the entire situation and have come up with similar conclusions regarding the relationship between the two Court members. Both Giovanni Giovannozzi and Modestino del Gaizo quite rightly deny, on the lack of evidence, that this issue sparked a bitter rivalry between Borelli and Viviani. However, Giovannozzi claims that Viviani's publication still became obsolete in the face of Borelli's seemingly more relevant work on Apollonius: '[O]nce published, the world could say that Viviani's divination was like a post eventum prophecy, and deny its value.'[39] Segre contributes to this view by suggesting that Viviani's restoration of Apollonius' Book V, *De maximis et minimis*, 'proved to have little scientific value' following Borelli's discovery of the Arabic translation of the original text. Middleton adds that Viviani would have been bitterly disappointed with the prospect of being upstaged by Borelli on this topic.[40] In summary, the suggestion from all these authors is that controversy was bound to arise once these two brilliant members of a Galilean tradition, Borelli and Viviani, were made to work on the same topic. Furthermore, it is supposed that since Borelli discovered and made public a manuscript descended from Apollonius' original treatise, Viviani's restoration of Book V had become obsolete.

However, there is no manuscript evidence suggesting that a rivalry had been born between Borelli and Viviani on the basis of their individual work on Apollonius. What is certain is that the aims of both writers to restore Apollonius' work on conic sections reflected their common interests in mathematics, geometry, and natural philosophy. In fact, I suggest that they may have even felt that they were collaborating on the topic. After all, Borelli's work did not upstage Viviani's, but rather assisted it in becoming a triumphant display of Viviani's ability in mathematics and the geometrical foundations of his natural philosophy. Viviani's was certainly not a work of 'little scientific value'. Indeed, the work by both Viviani and Borelli on the *Conics*, contained technical material that they were later both perfectly willing to try to deploy in physics experiments and in the articulation of natural philosophical claims. In other words, at the height of the existence of the Accademia del Cimento, when it was supposed that together these academicians were demonstrating a devotion to atheoretical experimental science, they were actually far more interested in making natural philosophical claims that were constructed on the basis of their mathematical skills.

[38] Targioni Tozzetti, i, 212–214.
[39] G. Giovannozzi, 'La Versione Borelliana dei Conici di Apollonio', in *Memorie della Pontificia Accademia Romana dei Nuovi Lincei*, 2 vols, Rome, 1916, ii, 11.
[40] Segre, *In the Wake*, 102; Middleton, *The Experimenters*, 313.

4. APOLLONIUS' LOST BOOKS

While Apollonius' first four books of the *Conics* were available to medieval and Renaissance scholars, the last four books had been missing since antiquity. In fact, the entire second half of Apollonius' work on conics might still be missing today had it not been for the ancient Arabic translators of classical Greek geometrical texts, who preserved seven of the eight books in their translations. Credit here for the survival of Apollonius' work should also be given to the mathematical interests of Italy's early modern thinkers, in particular, Borelli and Viviani. Before examining the efforts of these two members of the Tuscan Court in 1658 to restore Apollonius' work and the important natural philosophical implications of those efforts, we will quickly review Apollonius' writing on conic sections. We will pay particular attention to the all-important Book V that so strongly captivated the interests of our protagonists even before the Arabic translations were rediscovered and published by Borelli.

Apollonius of Perga was a student of Euclidean geometry in the third century BC and before his death c.190 BC, he managed to complete his writing on the geometrical properties of cones with circular sections, that is, conic sections. Books I and II are simple studies of the cone; its geometrical construction and the description of the fundamental properties of the conic sections: the parabola and the hyperbola, as well as the tangents. Meanwhile, Books III and IV are attempts to improve on Euclid's work on the locus and provide some new theorems regarding how the conic sections can meet.[41] Apart from some original propositions put forward in Book IV, the first half of Apollonius' treatise seems to have been aimed at revising and clarifying several geometrical problems mentioned or discussed by earlier writers, in particular Euclid, Aristaeus, and Menaechmus. As Apollonius himself stated, the theorems put forward in this section were 'worked out more fully and generally than in the works of other authors'.[42]

While the first half of Apollonius' text is based on the work carried out by his predecessors, Books V, VI, and VII, do seem to consist largely of his original ideas. In these books Apollonius dealt with normals to conics, and in Book V in particular he discussed extremal normals, their maximum and minimum lengths. Apollonius was interested in the lines of maximum and minimum length that can be produced from any point to a curve.[43]

With this in mind, it would now be to our benefit to look a bit more closely at Borelli's and Viviani's works regarding the translation and restoration of Apollonius' *Conics*. In our study of the natural philosophical construction of the Accademia del Cimento's experiments, it will be of great assistance for us to

[41] This refers to Book III of Euclid's *Elements*. This text on the theory of circles deals with the geometrical properties of circles, especially how to construct a circle from three points. Heath (ed.), *Euclid's Elements*.

[42] T.L. Heath (ed.), *Apollonius of Perga*, Cambridge, (1961 printing), lxx.

[43] It is beyond our scope here to discuss the details of Apollonius' work regarding the notion of extremal normals. Maximum and minimum lengths to conics are, however, explained in Propositions 82 and 83 of Book V of *The Conics*. *Ibid.*, 140.

observe how these types of mathematical and geometrical concerns translated into the work carried out by these two academicians.

Italian mathematicians and philosophers were already familiar with the first four books of Apollonius' work by the sixteenth century when they were translated into Latin. During this time, as has already been mentioned, the Sicilian mathematician Maurolico attempted a reconstruction of the Books V and VI based on what was said in the first four books. This was revised and published by Borelli in 1654. Meanwhile, Viviani published his own restoration of Book V in 1659. Furthermore, Apollonius' work on conic sections was also used by Galileo in the Fourth Day of *Two New Sciences* when providing a geometrical demonstration of the motion of projectiles. There is no doubt then that by the mid-seventeenth century, Italy's leading mathematicians not only used Apollonius' work, but were also eager to restore and comment on the missing books.

Viviani's restoration of Book V was a project that he had long been working on and it coincided with his later commentaries on Euclid as well as Aristaeus' text on conics.[44] Yet in June 1658, while he was completing his Apollonian treatise, Borelli discovered an Arabic manuscript actually containing a translation of the *Conics*.[45] In 1590, Cardinal Ferdinando de' Medici, who was soon to take over the Grand Ducal crown from his brother Francesco I, visited Rome and became acquainted with a learned Arab by the name of Ignazio Naheme. Naheme handed Ferdinando a collection of Arabic manuscripts that the future Tuscan Grand Duke took back to Florence. Perhaps not completely aware of the value of these papers, the members of the Tuscan Court failed to have them translated.[46] They remained untouched in the Grand Duke's archives until Borelli recognised their importance in 1658 and showed the enthusiasm to work on a translation of the text. Almost immediately upon his discovery, Borelli received confirmation that the manuscript did indeed contain all eight books of Apollonius' *Conics* from a local archbishop learned in Arabic.[47] Once he received

[44] According to Pappus of Alexandria, Aristaeus' treatise on conics, *Solid Loci*, a copy of which has never been found, was written before Apollonius began working on his *Conics*. M. Clagett, *Archimedes in the Middle Ages*, 4 vols., Philadelphia, 1980, iv, 74.

[45] The following account of the discussions and negotiations that resulted between Borelli, Viviani and the Grand Duke because of Borelli's discovery, have been documented in G. Giovannozzi (ed.), *Lettere inedite di Gio. Alfonso Borelli al P. Angelo di S. Domenico sulla versione di Apollonio*, Florence, 1916. See also Giovannozzi, 'La Versione Borelliana'; Barbensi, 26–32. More recently, Luigi Guerrini has expanded upon Giovannozzi's account of the publication of the Apollonian translation by referring to a collection of unpublished letters between Borelli and one of his collaborators on this project, Carlo Dati. These letters elucidate some of the concerns and frustrations Borelli had with the process of translating and publishing Books V, VI and VII. L. Guerrini, 'Matematica ed erudizione. Giovanni Alfonso Borelli e l'edizione fiorentina dei libri v, vi e vii delle *Coniche* di Apollonio di Perga', *Nuncius* (1999), **xiv**, 215–247.

[46] Ferdinando I handed the manuscripts to orientalist G.B. Raimondi who showed some interest in their translation, but died before he could even begin. By 1645, Michelangelo Ricci revived the interest of several other Tuscan court members in the content and translation of the Arabic papers, including Torricelli, Antonio Nardi and Raffaello Magiotti, but once again this enthusiasm was short-lived and for reasons unknown the work was never done. Giovannozzi, 'La Versione Borelliana', 4–5.

[47] *Ibid.*, 6. While Borelli originally believed that the Arabic manuscript contained all eight books, in actuality, it was still missing Book VIII.

this confirmation, the Grand Duke gave Borelli permission to travel to Rome and seek some assistance in the translation of the papers. What followed was an exchange of letters between Borelli and Leopoldo discussing the difficulties Borelli faced in carrying out the translation. Additionally, Borelli and Viviani also corresponded during this time and negotiated the efficacy of Viviani's forthcoming publication in light of Borelli's translation of the original text.

Once Borelli arrived in Rome in July 1658, he found an oriental scholar by the name of Abramo Echellense to begin the task of translating the text. During the following months, Borelli attempted to facilitate Echellense's job by explaining the mathematical propositions to him. However, from July to September, Borelli wrote to Leopoldo about the frustratingly slow progress of the translation and claimed to find many errors in the Arabic manuscripts. In fact, on 14 August, as he was approaching the end of this task, he wrote to Leopoldo: 'Thanks to God, I am almost at the end of this very difficult translation of Apollonius, which I could in good conscience call my own composition, because it has been necessary firstly to discover the demonstrations to be able to construct them from this very unsound and faulty manuscript.'[48] So difficult did this task become that Borelli returned to Tuscany in December 1658, and from there he corresponded with Father Angelo Morelli from the Pious Schools on the editing of the text. It would seem that Morelli continued to work with Echellense towards understanding the mathematical propositions in the Arabic manuscript in order to facilitate the translation. Meanwhile, each section that was completed was immediately sent to Borelli for editing. The process was a long and laborious one, and although they completed a translation of all seven volumes available, only Books V, VI, and VII were published in 1661.[49]

During his stay in Rome while undertaking this task, Borelli also maintained a correspondence with Viviani. On 29 June 1658, Borelli wrote to Viviani stating his find of the missing books and the plans that were being carried out to translate them.[50] According to Modestino del Gaizo, by the beginning of July the Grand Duke granted Viviani permission to publish his *De maximis et minimis*. In fact, del Gaizo notes that upon hearing of this news, Borelli wrote as follows to Viviani on 20 July, perhaps abiding by the Grand Duke's decision: '[Y]ou should go ahead and enjoy this advantage, which I doubt whether I can enjoy, although I have by me a very compendious treatise on the conics, all written out in my own

[48] 'Sono già per gratia d'Iddio quasi alla fine di questa stentatissima traduttione d'Apolloniio, la quale io con buona coscienza potrei chiamare mia compositione propria, perchè è stato necessario ritrovar prima le dimostrattioni per potere cavar costutto da questo manuscritto tanto difettoso, e scorretto'. As cited by Giovannozzi. *Ibid.*, 7.According to Guerrini, Giovannozzi fails to acknowledge the role Carlo Dati played in negotiating the illustration and printing of the translated text, further issues which frustrated Borelli in his eagerness to complete the project. Guerrini, 511–512.

[49] Giovannozzi, (ed.), *Lettere inedite*, i-ii. In light of the news that Dutch orientalist and mathematician, Jacob Golius (1596–1667), was also working on a reconstruction of Apollonius' work, Dati and Leopoldo agreed that only the 'new' books merited publication. Guerrini, 514.

[50] BNCF, Ms. Gal.254, ff. 105r-106v.; Barbensi, 30.

hand.'[51] Meanwhile, according to Barbensi, Borelli must have been ordered to delay his own translation.[52]

This story seems quite plausible. As Middleton notes in his brief review of these events, a letter written by Leopoldo on September 1661 suggests that the Grand Duke gave the command for the publication of Borelli's translation in that year, although there are no other hints here regarding the degree of involvement of the Medici brothers in Borelli's and Viviani's publications.[53] In addition to this piece of evidence, we may be willing to believe that surely the Grand Duke and the Prince would have wished both authors under their patronage to be successful. For this to occur they would have preferred that there was plenty of time for Viviani's work to be digested and applauded by the European Courts before the release of Borelli's translation of the original text. This type of interest by the Grand Duke and the Prince in the publication of works by their natural philosophers, reflects the social and political aims and ambitions of the Court and its members.

However, despite the impression given by Barbensi and del Gaizo that so much depended on the judgement of the Grand Duke, we may perhaps see things differently. There is no solid evidence that the Grand Duke and the Prince dictated the pace and direction of Borelli's and Viviani's works on this topic. In the correspondence between Borelli and Viviani on the one hand, and between Borelli and Leopoldo on the other, there is no definite indication that it was only up to the Grand Duke whether these texts were to be published separately. As has already been noted, Borelli experienced a great deal of difficulty in completing the translation; he could not have hoped to have his work fully prepared for publication before 1659, when Viviani was having his restoration of Book V printed. Also, Borelli gave no indication that he actually wished to publish his translation virtually simultaneously with Viviani's. Indeed, far from being competitive, jealous, or bitter about the entire situation, both men seemed to be happily applauding and even assisting each other's efforts.[54]

In support of this argument, we may look more closely at Borelli's letter to Viviani on 20 July 1658, particularly at an earlier section of the letter that del Gaizo failed to cite. Borelli wrote:

[51] As translated by Middleton, *The Experimenters,* 313. 'Tiri pure avanti, e goda di questo benefizio, del quale io dubito di non poter godere, ancorchè abbia presso di me un compeniosissimo trattato dei Conici disteso tutto di mia mano ...' BNCF, Ms. Gal.254, f. 107r; M. del Gaizo, *Contributo allo studio della vita e delle opere di Giovanni Alfonso Borelli*, Memoria letta nella tornata del 2 febbraio 1890, 14.

[52] Barbensi, 30.

[53] Middleton, *The Experimenters*, 314. This letter was to Melchisadec Thevenot. BNCF, Ms. Gal. 282, f. 57r.

[54] This point was also made by Middleton, *The Experimenters*, 314. The only competition with which Borelli and his collaborators seemed concerned, was that of Dutch mathematician, Jacob Golius. See note 49 above.

> I also concur in and approve your decision and that of all your friends to send to the printer your discoveries about the conic sections; and I shall be able to testify along with the others that you have had no knowledge of these last books.[55]

There is a great deal to be understood from this passage. First, Borelli referred to Viviani's 'friends', who persuaded him to publish *De maximus et minimus*. 'Friends' is hardly a term commonly used at the time in reference to a Medici Grand Duke or Prince. This would indicate that despite hearing of Borelli's venture to publish a translation of the original *Conics*, Viviani still decided on his own accord, and perhaps in consultation with other Court members, to publish his restoration of Book V. Second, Viviani not only wished to publish his work, but also wanted it to be made clear that he had never seen any part whatever of the Arabic manuscript and therefore could not possibly have plagiarised from the original. His intention to make this point clear is further evident in the preface to *De maximis et minimis*, where Viviani even cited the sentence in this letter pertaining to his having no knowledge of the contents of the Arabic manuscript. Furthermore, by making this point in his letter, Borelli seemed to be showing wholehearted support for the efficacy and success of Viviani's publication. We may add that Borelli and Viviani corresponded on a number of issues during these months of Borelli's sojourn in Rome, and at all times both thinkers showed exemplary conduct and constantly expressed their support for the other's work.[56]

Viviani's *De maximus et minimus* turned out to be an excellent prediction of Apollonius' propositions and geometrical demonstrations of normals to conics. Natucci, the only historian to note the accuracy of Viviani's restoration, even suggests that once the translation of Apollonius' work was published under Borelli's editorship, '[I]t then became possible to ascertain the substantial similarity between the two works.'[57] Indeed, not only did Viviani manage to provide propositions similar to those in Apollonius' original treatise, but the logical construction of the geometrical demonstrations also closely resembled the original text.

We therefore may not be able to determine the precise role the Medici brothers had in setting the publication dates of Borelli's and Viviani's works concerning Apollonius' *Conics*. The most that can be said on this is that the status of both the Court members and the Court itself would have benefited from these publications. What we may learn with some certainty from this encounter regarding the *Conics*, is that Viviani and Borelli were quite supportive of each other's work and shared a humanist type of concern for the restoration of ancient mathematical treatises.

However, we shall see in the remaining sections of this chapter and in our analysis of the Cimento's work in Part Two, that Viviani, and particularly Borelli, were looking well beyond merely refining their skills in pure mathematics. For example, while he was working in Pisa between 1657 and 1667, Borelli began to

[55] As translated by Middleton. *Ibid.*, 313. 'Io similmente concorro, e approvo la soluzione di V.S. e di tutti i suoi amici di mandare alle stampe le sue invenzioni intorno ai conici, et io potrò testificare fra gli altri, che ella non ha avuto notizie di questi ultimi libri'. BNCF, Ms. Gal. 254, f. 107r.; Bonelli, 676.

[56] Borelli's correspondence to Viviani during this period can be read in BNCF, Ms. 254, ff. 103r–114r.

[57] A. Natucci, 'Vincenzio Viviani', in Gillispie (ed.), x, 49.

use his knowledge of Euclid and Apollonius in order to demonstrate the mechanics that he believed were applicable to both terrestrial and celestial motion, including the structure and movements of the human body. This means that Borelli was using his knowledge of the classical, mathematical, and geometrical propositions that he had begun to strengthen since his time in Messina, to help develop a physico-mathematical domain of great relevance to natural philosophy: the application of mathematics to his observational work in 'celestial physics' and experimental and observational work in physiology. This, as we have seen in Chapter One, was part of the culture of anti-Aristotelian and mechanical natural philosophising that had begun to emerge during the seventeenth century. Furthermore, this shows that traditional and cultural historiographies that focus on the misleading notion that the academicians were practicing an atheoretical experimental science within the gentlemanly interactions and etiquette of the Medici Court, reveal nothing about the intellectual skills and commitments of each academician.

5. *THEORICAE MEDICEORUM PLANETARUM EX CAUSIS PHYSICIS DEDUCTAE*

Besides Borelli's intense interest in the restoration of Apollonius and Euclid, he still made many other contributions to Tuscan intellectual and courtly life through the Accademia del Cimento, particularly in the fields of kinematics, percussion, and pneumatics. His activities in these areas between 1657 and 1667 eventually led to three important publications completed after his tenure at the University of Pisa.[58] These works, which reveal many of the natural philosophical concerns Borelli carried into the construction of the Cimento's experiments, will be examined in some detail later in this chapter. But for the time being, we shall look at one of Borelli's most important works published while he was still in Florence in 1666. This was his treatise on the movements of Jupiter's satellites, titled: *Theoricae mediceorum planetarum ex causis physicis deductae*. This text reveals the mathematical and natural philosophical commitments that Borelli had continued to pursue since his education and his earlier work in mathematics and mathematical humanism. Historians of Italian science have often ignored this publication, preferring instead to highlight Borelli's treatises on medicine and mechanistic physiology, which are often believed to represent his observational and experimentalist tendencies.[59] However, astronomy was a mixed mathematical

[58] These works included: *De vi percussionis liber*, Bologna, 1667; *De motionibus naturalibus a gravitate pendentibus, liber*, Bologna, 1670, and *De motu animalium*, Rome, 1680.

[59] We have already discussed Borelli's early work regarding the epidemic that swept through Sicily in 1647–1648. His last publication, *De motu animalium*, published posthumously in 1680–1681, is an excellent demonstration of his mechanical philosophy and will be discussed in more detail shortly. Meanwhile, another reason for the relative unimportance assigned to *Theoricae* is because it came only a few years before Newton's brilliant work on universal gravitation and has since been largely forgotten. An example of this type of historiographical position can be seen in Middleton who does not even refer to Borelli's astronomy as amongst 'the best known of his works'. Middleton, *The Experimenters*, 28.

discipline where many claims were not obvious to the senses for Copernicans such as Borelli, and where a speculative 'celestial mechanics' was required, linking one's preferred natural philosophy to an account of structure and causation in a Copernican cosmos. Therefore, it was an immensely important field for Galileo's followers and represented many of their natural philosophical skills and commitments. Borelli was able to draw on nearly all the cognitive tools he had acquired throughout his education and career, including Galilean terrestrial mechanics, Keplerian celestial physics, and Euclidean and Apollonian geometry.

The story behind the composition of the *Theoricae* begins in 1664, while Borelli was engaged in a discussion with several of Europe's astronomers about the supposed parabolic trajectory of a comet seen in December of that year.[60] It was during Borelli's observations of the comet from the Medicean fortress of San Miniato near Florence that he first made some calculations regarding the movement of Jupiter's satellites. According to Domenico Bertoloni Meli, these calculations were requested by the Grand Duke and Prince Leopoldo, who wished to expand on Galileo's discovery of the Medicean stars. More specifically, the Medici were involving themselves in a discussion initiated by the telescope maker, Giuseppe Campani and Bolognese astronomer Domenico Cassini in July 1664.[61] Campani and Cassini claimed to observe the shadows made by Jupiter's satellites. After news of these claims reached Tuscany, Borelli was asked to make his own calculations in August of that year. They were not only reported to agree with Campani's and Cassini's observations, but apparently also fed Borelli's curiosity about the movement of the planets.[62]

So following Borelli's calculations of the movements of Jupiter's moons, as well as his observations of the comet, he was permitted to establish an astronomical observatory in San Miniato, containing Campani's latest telescope, the same instrument he and Cassini used to observe the shadows of Jupiter's satellites.[63] During the following months, Borelli often stayed at the fortress, making his observations and calculations regarding the trajectory of the Medicean planets. His efforts at San Miniato soon led to the publication of *Theoricae*: a theory of the path of Jupiter's moons and the causes for their movement. This theory called upon a Keplerian model of the Copernican heliocentric system. Additionally, it was based on a commitment to Galileo's terrestrial mechanics and its application to the heavens, considerable mathematical knowledge, including statics and the conics, and a broadly mechanistic approach to the problem of planetary motion. With these tools Borelli was intent, as is revealed in the introduction to the

60 After sighting the comet and monitoring its trajectory, Borelli claimed that its parabolic path beyond the sublunary region gave definitive proof that the planets did not move on solid celestial spheres. This led to the publication of his *Lettera del movimento della cometa apparsa il mese di Dicembre 1664* (Pisa, 1665) and to several heated discussions with Adrien Auzout. This episode is discussed in more detail in Chapter Eight.

61 D.B. Meli, 'Shadows and Deception: from Borelli's *Theoricae* to the *Saggi* of the Cimento', *British Journal for the History of Science* (1998), **31**, 386; Middleton, *The Experimenters*, 256.

62 The correspondence on this issue can be found in Targioni Tozzetti, ii, 749–750. This episode is also discussed by Meli, 'Shadows and Deception', 386. And Middleton, *The Experimenters*, 256.

63 Settle, 'Giovanni Alfonso Borelli', 310; Meli, 'Shadows and Deception', 386.

Theoricae, on providing a mechanistic explanation concerned with the movements of planets 'without', as he put it, 'intelligences or angelic faculties'.[64]

Borelli came up with the following three points to explain the movements of Jupiter's satellites, based on the notion that a centripetal tendency balanced a centrifugal force to keep the planets or satellites in equilibrium and on a regular path around its central body. First of all, in his publication he claimed that all celestial bodies have a 'tendency' to move directly towards the centre of their orbit, 'just as we observe that all heavy bodies have a natural instinct for approaching our Earth, doubtless impelled by their inborn force of gravity'. This statement may immediately cast some doubt for the reader over the mechanistic aims that Borelli expressed earlier – his intention to avoid 'intelligences or angelic faculties' – since rather than discuss mechanical forces of nature he began by describing a seemingly Aristotelian, or even animistic, 'natural instinct' of planets. In fact, as if to emphasise this scholastic aspect of his work, Borelli illustrated this 'instinct' by comparing it to a magnetic faculty: 'Therefore it will not be impossible that the body of a planet should have a certain faculty similar to the magnetic faculty by means of which it approaches the solar globe itself.'[65]

Borelli then had to explain why he believed planets remain in their orbits and do not move rapidly towards their central body because of the centripetal tendency. This part of his theory was based largely on his knowledge of statics, but it began with a simple explanation of the circular motion of planets, which was borrowed largely from Kepler. Borelli asserted that the rays of light, emanating from the central body as it revolves, push the satellite around. This was another theory obviously based on Kepler's celestial physics. But Borelli came up with an important difference that was to demonstrate his strong belief in Galilean terrestrial mechanics and a mechanistic, rather than a Neoplatonic, Keplerian, notion of inertia. He purported that instead of the planet requiring the constant push from the rays to keep it moving, as Kepler suggested, each ray actually imparts an increment of motion upon the orbiting body. In other words, the satellite receives an impetus from each ray, an increment of motion that remains in the body. However, as Alexandre Koyré's and Westfall's expositions make clear, the planets

[64] R.S. Westfall, *Force in Newton's Physics: the Science of Dynamics in the Seventeenth Century*, London and New York, 1971, 214.

[65] As cited by A. Armitage, 'Borelli's Hypothesis and the rise of celestial mechanics', *Annals of Science* (1950), **vi**, 273. As both Armitage and Koyré note, despite this reference to the magnetic faculty of the sun, Borelli was not actually suggesting that this is some sort of attractive or gravitational force acting at a distance from the central body to the orbital body, but only implied that it is a 'tendency' in the orbiting body. That is, much like Galileo's explanation of accelerating falling bodies in free fall, there is no force attracting the body, but rather simply a natural tendency for the body to move towards its natural place. See also Koyré, *The Astronomical Revolution*, 519, n.19. This, of course, still also borrowed from scholastic natural philosophy, based on evaluating the qualities that determine a body's existence and movement. Borelli's natural philosophical position, however, in the emerging tradition of physico-mathematics, should attempt to describe the dynamics of celestial motion on the basis of mathematical rather than qualitative reasoning, with no such vague references to tendencies and 'natural instinct'. Nevertheless, as we review the following two points of Borelli's theory of planetary motion, we shall see that he certainly did not neglect to apply his skills in mixed mathematics and Galilean terrestrial mechanics.

moving in this solar whirlpool do not accelerate indefinitely, but reach a final velocity set by the overall speed of the whirlpool.[66]

Returning to the issue of centrifugal tendency, which will counterbalance the centripetal tendency, Borelli went on to claim that the impetus provided by the light rays would push the planet outward, for the following reason. As we have seen, the repeated impulse from each ray increases the speed of the orbital body, up to a terminal rotational velocity. This causes, from moment to moment, a tendency to move along the tangent to the circle at the point under consideration. Borelli understands such tangential tendency to manifest itself at each instant as a centrifugal tendency whereby the orbiting body has a natural impulse to move in a radial, outward direction from the central body. Finally, according to Borelli, what restrains the planet from moving in this outward direction is the previously described centripetal magnetic tendency pushing the planet inwards. These two tendencies are therefore keeping the planet in equilibrium and on its orbit. Or as Borelli himself put it: '[T]he virtue of the planet to approach cannot overcome the contrary repelling virtue, nor can it be conquered by the other.'[67]

This notion of tendencies leaves us with Borelli's final point in which he attempted to demonstrate mathematically these contrasting centrifugal and centripetal tendencies. To arrive at an understanding of how Borelli constructed this notion of contrasting forces in equilibrium, we briefly need to examine the cognitive tools that he used in the construction of his claim, particularly his knowledge of statics. It is beyond the scope of this analysis of Borelli's career to launch into a thorough investigation of the details of the attempt by some seventeenth-century natural philosophers to apply statics, or statical analogies, to dynamical explanations of celestial and terrestrial motion. However, we are fortunate enough to be able to draw from Richard Westfall's thorough and erudite analysis of Borelli's forces in celestial physics in order to grasp Borelli's skills and commitments in mixed mathematics and mechanical natural philosophy.[68]

According to Westfall, during the 1660s, Borelli was preoccupied initially with understanding Galilean kinematics and then finding a quantitative dynamical explanation for the motion of bodies. In particular, Borelli drew from his knowledge of static forces in order to make the argument that a body will be in motion only if the forces, or tendencies, acting upon it are not in equilibrium. In effect, Borelli was using his skills in mixed mathematics, especially measuring the statical weights needed to hold bodies in equilibrium, in order to recognise, as Westfall put it, 'the relation of dynamics to statics'.[69]

Borelli did not apply these skills only to his work in celestial mechanics. As we shall soon see, in his works dealing with the force of percussion and the movements of animals, he was also heavily committed to applying statical analogies to

[66] Koyré, *The Astronomical Revolution,* 492; Westfall, *Force in Newton's Physics*, 219.

[67] G. Borelli, *Theoricae mediceorum*, Florence, 1666, 77. As cited by Westfall, *Force in Newton's Physics*, 219.

[68] Westfall, *Force in Newton's Physics,* 213–221.

[69] *Ibid.*, 215.

dynamical problems. Furthermore, we shall also see in Chapter Six that these skills in working from statics toward a more general mechanics, were also part of the work carried out by some of the members of the Accademia del Cimento regarding the effects of heat and cold. This will demonstrate how the mathematical and mechanical commitments that we are examining in this chapter, are therefore crucial for our understanding of the experiments carried out by the Cimento.

Writing two generations ago, Angus Armitage, a noted historian of astronomy, also described these fundamental propositions in Borelli's *Theoricae*, concluding that from Ptolemy to Borelli, the Sun had passed from being a 'symbolic geometrical centre' to 'a source of physical force'. From what we have seen so far, we may be willing to accept this statement, since this brief look at Borelli's model seems to show him to be following Kepler in no longer adopting the purely mathematical and geometrical value that was assigned to the position and power of the Sun in Ptolemaic astronomy and even in Copernicus' own system. But such a conclusion provides us with little insight into the details of Borelli's proposals. Indeed, we need to look further into Borelli's theory in order to understand just how important it was for him also to provide mathematical demonstrations of planetary motion. In other words, Borelli's work was an attempt to construct a plausible celestial mechanics. We have already seen how he used his skills in mixed mathematics to illustrate the dynamics of the solar whirlpool. Now we are about to see that he continued to speculate upon mechanical celestial motion by constructing some strong links between the movements of planets in elliptical orbits, including geometrical demonstrations involving the conics, and Galilean natural philosophical principles regarding terrestrial mechanics.

Following the description of his theory regarding the solar whirlpool and its opposing tendencies, the first major mathematical point Borelli was willing to make in his work, was his acceptance of the elliptical planetary orbits proposed by Kepler. Aside from Descartes' ideas about the dynamics of celestial motion, including the role of the Sun and the structure of the vortex, nobody else since Kepler and the rise of Copernicanism in the early seventeenth century had attempted to give dynamic explanations of the physical causes of planetary motion.[70] Even Kepler's work had dealt with this problem only to a limited extent with his use of magnetic attraction and the sun's rays. But attractive forces were not favoured by Borelli as solid and suitable explanations (despite his use of the term 'magnetic faculty' to illustrate a planet's centripetal motion), and as we have just seen, he added the notion of impetus to the push provided by the sun's rays. So, as Koyrè suggests, Borelli felt content with Kepler's unity of terrestrial and celestial physics, but may have believed that he needed to improve on the explanations of elliptical orbits and planetary motion.[71] How he undertook this task is of much interest to us since it reflects some of the mathematical and mechanical commitments we have already encountered in his career.

[70] For an analysis of Descartes' physico-mathematical and natural philosophical positioning in the early seventeenth century with regard to his celestial mechanics, see J.A. Schuster, ' "Waterworld": Descartes' Vortical Celestial Mechanics', in J.A. Schuster and P. Anstey (eds.), *The Science of Nature in the Seventeenth Century: Patterns of Change in Early Modern Natural Philosophy*, Dordrecht, 2005, 35–79.

[71] Koyré, *The Astronomical Revolution*, 472.

It takes a lengthy study of the type produced by Koyrè to gain a comprehensive understanding of the geometrical details behind Borelli's astronomy, but here, in the interest of brevity, we simply may attempt to grasp how important Apollonius was to Borelli's belief in elliptical orbits. His first reference to ellipses was not through a comparison of Kepler's theory to his own, but interestingly, he aimed 'to give an exact explanation' of how ellipses can be formed from conic sections and thus how these orbits of planets can be deduced from geometrical principles. Drawing from the work of Ismael Bullialdus and Kepler, Borelli imagined a scalene cone in the heavens, across which would exist a plane, ERK (Figure 2).[72] That plane served to construct a conic section, creating an elliptical shape inside the cone. Borelli then drew the transverse axis of the ellipse, EK, intersecting with a series of lines parallel to the circular base of the cone. These intersections created two foci equal distances from the ellipses' centre, M and H. One focus, H, is supposedly where the luminary (that may be the Sun, Earth, Jupiter, or any body with satellites) is positioned. The other focus, M, is on the axis of the cone, AI, which Borelli regards as 'the centre of uniform motion'. To understand the meaning and significance of this term, we must imagine that when the celestial body is at aphelion E, it is actually travelling briefly on a point of the circumference of the circle with the diameter ED, and its centre at S. Similarly, at perihelion K, it is traversing a point in the circle PK. Borelli's reasoning here is that circular motion is uniform, thus giving the planets moving along the ellipse ERK, a 'centre of uniform motion' at the axis of the cone, AI. Furthermore, the circles are of different sizes, but bodies that travel along their circumferences move in equal times, meaning that the moment when the planet is at K, it will be travelling faster than when it is at E, since PK is a much larger circle than ED. In sum, the planet will traverse parts of these innumerable circles, and as it does so, its speed will increase or decrease according to the size of each circle it traverses. Borelli himself acknowledged that it is difficult to imagine planetary motion being based on such perfect geometrical relationships and a fictitious cone in the heavens, but this does not sway him from reaching the following conclusion: '[E]ven though no real cone be supposed [to exist] in the Universe, it is nevertheless possible for elliptical motion to take place in exactly the same way as it would if we assumed the existence of a solid cone of that kind.'[73] Here Borelli denies that he is merely describing a hypothetical geometrical relationship between the celestial bodies – this is not just an exercise in mixed mathematics. Instead, he insists that he is speculating about the dynamics of celestial motion in a way that should be expected of a physico-mathematician.

Now, we must remember that the plausibility of this geometrical demonstration as evidence in favour of elliptical orbits, depended upon Borelli's natural philosophical view that natural phenomena can be uncovered through the application of mathematics and geometry. But this was still not enough for Borelli, who wished to provide mechanical terrestrial evidence in favour of elliptical

[72] I.Buolliau, *Astronomia Philolaica*, Paris, 1645.

[73] G. Borelli, *Theoricae mediceorum*, Florence, 1666, 66. As cited by Koyré, *The Astronomical Revolution*, 478.

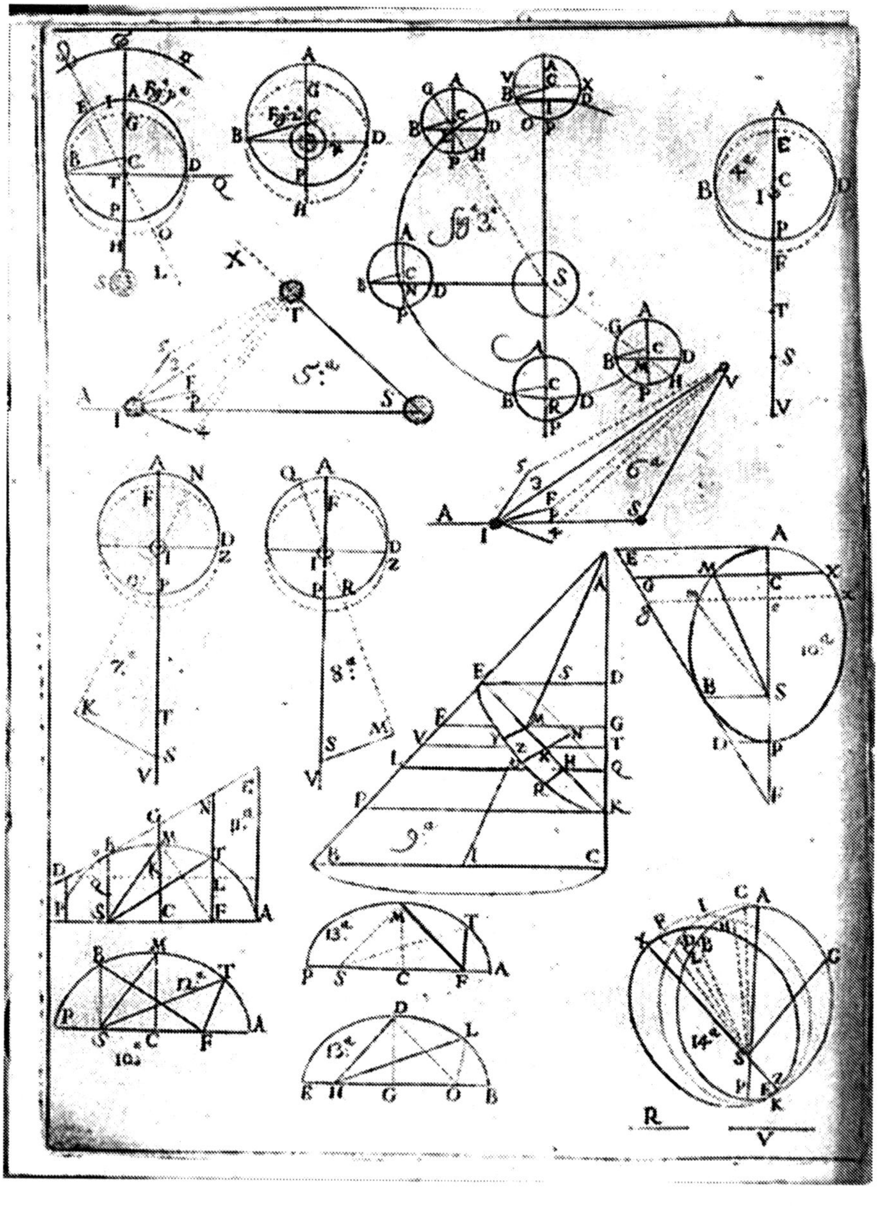

Figure 2. Borelli's geometrical construction of an ellipse within a scalene cone, used as a geometrical demonstration of the elliptical orbits of planets. Diagram labelled '9:a'. Borelli, *Theoricae mediceorum planetarum ex causis physicis deductae,* Florence, 1666. (Courtesy of the IMSS Biblioteca Digitale.)

orbits. He did this quite cleverly by suggesting that even though there is an equilibrium between the outward and inward tendencies of a planet or satellite, that equilibrium is not perfect and the planet always swings, at times in favour of one tendency, and at other times in favour of the other. As an example, Borelli argued that the position of a pendulum in equilibrium is vertical, but when released from any height, rather than return to equilibrium, it gains enough velocity to push through the resisting tendency and reach the same height on the opposite side of the swing. As it descends again, the pendulum goes through the same motion to return to its initial position of release. On earth, resistance eventually reduces the pendulum's velocity and therefore forces it to return to its natural place, but 'in the supremely fluid ether', Borelli suggested, the swing of the pendulum would be perpetual.[74] Furthermore, according to Koyré, this implied that the planet's velocity changes just as the velocity of the pendulum always varies during every oscillation.[75] So Borelli was returning once again to the skills in mixed mathematics that he used to begin his speculation on planetary motion, including his knowledge of statics as explained by Westfall. Only now, as Koyré showed, Borelli was using these skills to demonstrate that there is no distinction between terrestrial and celestial motion and that a simple physics experiment could be used to argue about the movements and trajectories of planets.

Besides the use of Apollonius and the pendulum for the geometrical and terrestrial demonstrations of elliptical orbits, Borelli also employed a mathematical approach to help explain the variations in the velocity of the planets at different distances. He relied on the characteristics of a balance to describe the resistance of planets to the impulse from the rays according to their distance from the luminary. In other words, the further the planet is from its central body, the less resistance it needs to balance the impetus from the sun's rays. Here, Euclid's definitions regarding proportions were crucial to the formulation of Borelli's proposition that the resistance of the planet is proportional to its distance from the sun. Or as Borelli claimed: '[T]he velocity acquired by the planet ... will increase in the same ratio as the distances decrease.'[76] This is a perfect example of Borelli's intention to use terrestrial mechanics to explain celestial movement, but also the application of his skills in mixed mathematics to physical problems.

In concluding our analysis of Borelli's astronomical interests as expressed in his *Theoricae*, we may note that his aim was to clothe certain Keplerian astronomical concepts with more plausible mathematical principles and a mechanical natural philosophy. That is to say, that he was using Apollonian and Euclidean demonstrations, in order to construct heuristic support for a mechanistic explanation for terrestrial and celestial motion. In fact, by applying terrestrial physics based on solid mathematical demonstrations, much like Galileo, to the construction of a celestial mechanics, Borelli improved on Kepler's theories of planetary

[74] Borelli, *Theoricae*, 66. As cited by Koyré, *The Astronomical Revolution*, 505.
[75] Koyré, *The Astronomical Revolution*, 505.
[76] Borelli, *Theoricae*, 74. As cited by Koyré, *The Astronomical Revolution*, 507–508. The use of 'resistance' may seem odd here in light of Borelli's previous claim that the resistance of celestial bodies was zero, but Koyrè suggested that this may have been due to a confusion on Borelli's part of the term used in celestial and terrestrial ontology.

motion. As Koyré notes: 'In his view, motion in a straight line, and linear velocity, persist in the skies exactly as they do on Earth. Being a better Galilean than Galileo himself, he could apply all the progress achieved by the Galilean revolution to a profitable study of the Keplerian problem.'[77]

Therefore, the physico-mathematical and mechanistic natural philosophical tools that Borelli had been sharpening during his education and career were on display in his most important astronomical treatise, published as his career inside the Cimento was coming to a close. Despite Borelli's interests in astronomy and the observations he performed while working for the Accademia del Cimento, no astronomical observations were published in the *Saggi*. As we shall see in Part Three, this may well have been due to the Prince's concern that the publication be free of the type of mechanical and anti-Aristotelian controversial arguments that Borelli was formulating. In any case, we can now take a closer look at the other three publications produced by Borelli from 1667 onwards. It will be important to remember that while these were written after the Cimento's collapse, Borelli was still relying on the natural philosophical skills and commitments he had developed during his ten years in Tuscany.

6. BORELLI'S LIFE BEYOND THE CIMENTO: 1667–1679

Borelli's three most important writings after the *Theoricae* that display his natural philosophical commitments, were *De vi percussionis* (1667), *De motionibus naturalibus* (1670), and *De motu animalium* (published posthumously, 1680). The first two works contained Borelli's analysis of such physical questions as the force of percussion and the weight of air, the latter a favourite topic for the Cimento. In the meantime, *De motu animalium* reflected the second dimension of his work – how he used his mechanics to explain the structure and movement of the human body. Borelli's physiological treatise even recalled various propositions put forward in *De vi percussionis* and *De motionibus naturalibus*. For this reason it is important that we explore what Borelli worked on in the first two physics treatises before we examine his interests in physiology in the concluding section to this chapter. So in this section we will be looking first at his interests in the force of percussion, and second at his work regarding positive levity, as examples of the anti-scholastic and mechanistic principles that Borelli had developed during his career and published following his work on planetary motion.

On 5 May 1665, Michelangelo Ricci wrote to Leopoldo about his impressions of Borelli's 'profound speculations' in physics and astronomy and insisted that '[I]t would perhaps be good that Sig. Borelli apply himself to putting to light a treatise on the composition of motion, and its increases and decreases.'[78] Leopoldo relayed this suggestion to Borelli, and then responded to Ricci on Borelli's behalf,

[77] Koyré, *The Astronomical Revolution*, 473.

[78] 'La Scrittura del Sig. Dottor Borelli è sì piena di profonde speculazioni [...], sarìa forse bene che s'applicasse il Sig. Borelli, a dar in luce un Trattato della Composizione de' Moti, e dell'aumento e diminuzione loro'. Targioni Tozzetti, i, 424.

suggesting that he would attempt to compose such a treatise after completing his work on physiology. But perhaps Ricci's suggestion reinvigorated Borelli's interests in terrestrial mechanics, since he put aside his physiological studies in favour of writing a book on motion. According to Ricci, motion was a particularly important topic since so many others, including Torricelli, Roberval, Descartes, and Kepler, had dedicated so much time to it, 'speculating on geometrical, astronomical, and physical things'.[79] Ricci's call for Borelli's participation in the investigation of natural motion was reinforced in another letter, just 20 days later, on 25 May 1665. Ricci insisted that Borelli should dedicate himself to investigating the force of the percussion, and indeed, this is exactly what Borelli did in *De vi percussionis*, published just before his return to Messina in 1667.[80] While this work covered a variety of topics, its central aim was to explain the force of percussion according to a mathematical and mechanical natural philosophy. What this means is that Borelli employed the same skills in mixed mathematics that we saw from his work in astronomy, in order to try to resolve another physics problem.

The issue in question, as Aristotle had put it, was to explain why a heavy axe, as an example, has virtually no effect when rested on a piece of wood, but has a much greater impact when it is made to fall from a great height. Scholastics believed that the increased force is a result of the velocity of the moveable; the velocity supposedly artificially increases the weight of the object.[81] In contrast to this traditional opinion, Borelli attempted to construct an explanation for the force of percussion based on the same mathematical notion mentioned earlier regarding an equilibrium of opposing forces – this time the forces in question were those of impact, and the resistance of the body being impacted upon. This was another case where motive forces were analysed and quantified according to a mechanics of motion grounded in statics. So once again Borelli was seeking to establish an anti-Aristotelian and mechanistic dynamics, and just as we have seen with his astronomical speculations, this was to include his abilities as a mathematician, especially his knowledge of statics. Furthermore, this was based largely on the work carried out by Galileo concerned with percussion, and this is where our brief analysis begins.

In his *Mechanics* (c.1590), Galileo claimed that to study percussion, one must consider

> [T]hat which has been seen to happen in all other mechanical operations, which is that the force, the resistance, and the space through which the motion is made respectively

[79] 'perchè quivi pescano molti che oggidì vanno speculando per le cose Geometriche, Astronomiche, e Fisiche. V.A. si ricorderà quanto capitale ne faceva il Torricelli, e quanto se ne sia prevalso il Robervallio, ed altri Matemateci famosi e Descartes in Filosofia, e Keplero nell'Astronomia'. *Ibid.* It is interesting to note how Ricci justified the importance of the study of motion. He believed that it had been a relevant topic for natural philosophers specialising in a variety of disciplines, including mathematics, philosophy, and astronomy. This displays a widening of the scholastic sense of the boundaries of natural philosophy, seeing the reach the articulation of a natural philosophical topic, motion, spread across a field of endeavours.

[80] Ricci wrote: 'Si fece gran perdita con la morte del Sig. Galileo, especialmente della dimostrazione tanto stimata da lui, e da tutti gli intendenti, della Forza della Percossa, per la quale ha ingegno molto proporzionato il Sig. Borelli'. *Ibid.*, 425.

[81] Barbensi, 70.

> follow that proportion and obey those laws by which a resistance equal to the force will be moved by this force through an equal space and with equal velocity to that of the mover.[82]

That was to say, that it is not only the weight of the body in motion that determines the force of percussion, but the distance it travels and its velocity before impact that is required to overcome the resistance of the body being impacted upon. Galileo elaborated on his argument in *Two New Sciences*. There he presented several experiments where the force of percussion was tested and examined. More specifically, on one occasion, Galileo used weights placed on one side of a balance to counter the impact created by water and therefore to come up with a measurement for percussion.[83] Galileo was using his skills as a mathematician to compile a quantitative demonstration of the dynamics of the force of percussion.

In *De vi percussionis*, Borelli agreed with the Galilean proposition that this force is not measurable through the weight alone of the body that is being moved to create an impact. Yet according to Westfall, Borelli was not at all interested in performing the same type of experiments with static weights as Galileo, and argued more strongly that the critical factor in determining the force of impact is simply the height from which the body is dropped. For this reason, Borelli placed greater emphasis on explaining the impetus created by impact, which according to Westfall, became an analysis of the 'energy of percussion'.[84] This was beginning to appear like the quantitative dynamics that Borelli may have been looking to achieve, but in Westfall's exposition of this theory, Borelli was still relying on measuring the dimensions related to the force of a moving body, rather than creating a dynamical model for describing the changes in motion produced by the impetus.[85] For example, Borelli contended that the energy of percussion was only measurable in terms of both bodies involved in impact – the height from which the mover is dropped and the distance the moved is displaced.

This recalls the type of mixed mathematical skills that we saw him apply in his work on astronomy, where dynamical forces were measured according to a type of statical mechanics involving the notion of equilibrium between two opposing forces. Significantly, we shall see in Part Two that these are also the type of mechanistic tools most of the Cimento academicians applied between 1657 and 1662 while they experimented on air pressure, the vacuum, and the effects of heat and cold. This will show how crucial Borelli's contribution was to the Cimento and how that contribution consisted of far more than following mere rules of courtly etiquette. Furthermore, it will continue to show how the academicians were concerned with constructing and interpreting their experiments according to their natural philosophical commitments, rather than according to some putative belief in an atheoretical and inductive experimental method. But before reaching these case studies, we will investigate some of the other work Borelli completed before his death (as well as the work carried out by the other academicians, examined in

[82] Galilei, *On Motion and On Mechanics*, 180.
[83] Favaro (ed.), *Le Opere*, xiii, 323–325.
[84] Westfall, *Force in Newton's Physics*, 222.
[85] *Ibid.*

Chapter Three). In order to emphasise the natural philosophical commitments that Borelli developed during his career inside the Tuscan Court, we shall now see how he handled the topic of positive levity in his publication,

7. *DE MOTIONIBUS NATURALIBUS*

As was mentioned, one of the topics of interest in Part Two will be the academicians' work on air pressure. Borelli's contribution to this case study was quite substantial, since he saw it as an opportunity to put forward a strong argument against the Aristotelian notion of positive levity. Additionally, as can be seen from his 1670 publication, *De motionibus naturalibus*, he was still eager to apply his skills in mathematics and mechanical natural philosophy to problems in natural motion. Much of the material in this text came from the work carried out by Borelli for the Cimento, but our analysis here will not focus on the presentation of these experiments in the *Saggi*. Rather, we will examine how Borelli approached the question of positive levity in *De motionibus naturalibus* and other manuscripts. These sources will directly reveal the intellectual agenda that Borelli brought to the Cimento.

One of the most critical experiments cited by Borelli against positive levity was an attempt to measure the force of the supposed levity of a wooden cylinder inside a container full of water or any other liquid. Borelli wished to show that, contrary to scholastic belief, the force known as positive levity is not what causes the cylinder to float above the surface of the liquid.[86] The cylinder was placed on top of a metal plate and a small amount of water was contained around its base to prevent air from getting in underneath the wood. This would mean that for the cylinder to be lifted, the force lifting it would have to be equal to or greater than the resistance caused by its own weight as well as the repugnance to the vacuum created by the contact between the wood and the metal. Using a balance to lift the cylinder with a counterweight, Borelli was therefore able to record the statical weight that he believed signified the force of the cylinder's resistance to being lifted.[87]

Next, Borelli placed the cylinder inside a container full of mercury where it floated freely. This time he attempted to measure the force levitating the cylinder by placing weights on the wood until it was prevented from floating.[88] Borelli, as well as the academicians who carried out this experiment inside the Cimento, recorded that the weight needed to stop the cylinder from levitating was greater

[86] See Chapter IV of G. Borelli, *De motionibus naturalibus*, Bologna, 1670. This experiment is also described in a manuscript published by Abetti and Pagnini entitled: 'Against positive levity. By Sig. Dott. Gio. Alfonso Borelli. Hypothesis revealed to the senses'. BNCF, Ms. 268, f. 177r.; Abetti and Pagnini (eds.), 419.

[87] In the *Saggi*, the academicians measured this statical resistance to be about three libbra (about one kilogram). Magalotti, 218.

[88] The academicians measured this weight to be five libbra (about 1.7 kg), 2 libbra greater than the weight needed to lift the cylinder of wood.

than the weight used to overcome the cylinder's resistance to being lifted. With these procedures, Borelli was once again using dead weights to measure opposing forces. That is, to describe quantitatively the dynamics of a body being levitated, Borelli relied on his skills as a mixed mathematician and used his knowledge of statics to conclude that recording the amount of the dead weights used in the experiment was the same as measuring the opposing forces.

Once the cylinder, weighed down by additional weights, touched the bottom of the container full of mercury, the resistance described by the weight on the balance in the first half of the experiment, would be recreated. Now, if the levitating force is indeed greater than the resistance, as Borelli found to be the case through the use of a mechanics based on statics, once the weight was taken off and the cylinder released to these natural forces, the lesser force, the resistance to being levitated, should be overcome by the greater force and the cylinder should rise. The problem for Borelli was that this simply did not occur. With the statical measurements of opposing forces Borelli was hoping to compile a case against the scholastic belief in positive levity. His experiment, however, was clumsy and did not achieve his desired aim. Nevertheless, desperate to reach a conclusion against levity, he argued that a body 'that moves upwards in any liquid, is not pushed spontaneously by an internal principle commonly called levity'.[89]

Borelli's conclusion seems rather unconvincing, since his experiment was both clumsily constructed and interpreted. He did not provide an alternative explanation for the failure of the cylinder in the container full of mercury to rise after the weight was taken off, when the forces that he measured through statical weights, theoretically demanded that it should do so. We may even be tempted to suggest that perhaps Borelli missed this opportunity to put forward a stronger mechanistic argument based on the properties and movements of atoms affecting the levity (positive or otherwise) of the wooden cylinder. After all, as we shall see in Part Two, it was not unusual for Borelli and his mechanist colleagues inside the Cimento to discuss the movements of corpuscles with regard to pneumatics and hydrostatics. So it is difficult to imagine that any reader of this experiment in the seventeenth century could have been convinced that Borelli had proven the non-existence of positive levity, a natural philosophical principle that was a cornerstone of Aristotelianism and that could not be easily dismissed.

Nevertheless, despite the questions surrounding the efficacy of this experiment, evidently Borelli was still quite intent on making a claim against Aristotelianism, demonstrating the natural philosophical framing of his work.[90] Furthermore, as we have also seen from his work concerned with the force of

[89] BNCF, Ms. 268, f. 117v.; Abetti and Pagnini (eds.), 420.

[90] In fact, in *De motionibus naturalibus* Borelli cited one more experiment against the Aristotelian notion of positive levity. This included the movement of smoke inside a void. Since, as Borelli observed in an experiment also suggested to the Cimento, the smoke was seen to fall inside the vacuous bulb of the barometer, instead of rising as in a normal environment, Borelli concluded the following: 'It should be the opposite when the said smoke has an internal principle called levity of moving upwards.' BNCF, Ms. Gal. 268, f. 116r; Abetti and Pagnini (eds.), 421. Clearly then, Borelli constructed this experiment with the aim of disproving scholastic beliefs that there was a natural positive levity in all bodies consisting of air.

percussion, he was clearly attempting to apply his knowledge of mixed mathematics to problems in physics, a practice that was characteristic of the rise of mechanical natural philosophising in the seventeenth century.

8. CONCLUSION: *DE MOTU ANIMALIUM*

Through the works mentioned above, Borelli claimed that no terrestrial and celestial motions are beyond the scope of mechanical and mathematical laws. We saw to a certain extent that this was also what he was dedicated to proving whilst participating in the Cimento's meetings. But this was still not the full extent of Borelli's contribution to early modern Italian natural philosophical endeavour, nor does it cover all of his studies while working for the Tuscan Court.

While teaching at the University of Pisa, Borelli was also leading a school of physicians dedicated to a corpuscularian and mechanical natural philosophy. This group included some famous names, such as Marcello Malpighi (1628–1694), Lorenzo Bellini (1643–1704), and Carlo Fracasatti (c.1630–1672). It also involved a close correspondence and collaboration with English anatomists John Finch (1626–1682) and Thomas Baines (1622–1680). Anatomy and physiology had long been a favourite topic for Borelli, since his early work in Messina where he studied how the body transmits illnesses. Now in Pisa, he was gaining the opportunity to explore these disciplines in detail, and for many years, he had been working on the publication of a book on the macro and micro mechanical movements of the muscles in animals. He intended to show that the muscles move just as pulleys and levers do, and that the microstructures inside the body cause the contraction of muscles, including the heart. By the end of the 1670s, following his previous writings on the force of the percussion and positive levity, Borelli dedicated his time to completing his anatomical treatise. According to Barbensi, this was the second part of Borelli's ultimate plan: once he had set out what he believed to be the mathematical and mechanical laws of terrestrial and celestial motion, he was then in a position to write about the mechanical movements of the body.[91] In other words, this treatise was the last piece in Borelli's mechanistic agenda, expanding from celestial motion, through terrestrial physics and finally the human body. Considering that Borelli had long been working on the publication of an anatomical treatise, it is possible that he may not have been as systematic as Barbensi suggests in the compilation of his major works. In any case, his collaboration with some talented anatomists in Pisa, eventually resulted in his posthumously published *De motu animalium*.

Borelli began this text by dismissing the Aristotelian notion that animal spirits flow to the nerve ends and cause the movement of body parts. Instead he proposed that the muscles constitute the machine, or the motor, of the body's movements. He classified the mechanics of the muscles according to the different

[91] Barbensi, 82.

geometrical shapes of the fibres that contract and expand. This allowed Borelli to calculate the movements of the fibres on the basis of Euclidean proportion theory. Euclid's second book on geometrical algebra was particularly relevant for Borelli in this study, since, for example, he could articulate geometrically his argument about how tendons and fibres contract according to their shapes.[92]

It is beyond our interests here to discuss the details of Borelli's studies in anatomy and physiology, since this work was not part of the Cimento's agenda.[93] Nevertheless, Borelli's *De motu* is a fine representation of the seventeenth-century accumulation of work in mathematics and mechanics. Borelli's efforts in his last publication resembled Cartesian physiology somewhat, although he never publicly or privately admitted to subscribing to Descartes' beliefs.[94] In any case, he borrowed heavily from Galileo, Torricelli, as well as contemporary mathematicians, in order to construct a mechanical natural philosophy that explained all celestial, terrestrial, and physiological matter and motion. Furthermore, rather than rely on any type of atheoretical, universally applicable experimental method, Borelli constructed arguments informed by, and expressive of, his natural philosophical agenda.

So in summary, we have noted Borelli's early training in mathematics and mechanics, we have then seen how his interests in these fields began to expand into the natural philosophical framing of such disciplines as physiology. Finally, when he moved to Pisa, he enriched his association with one of the most supportive Courts and some of the most enlightening minds of seventeenth-century Italy. This is where he collaborated with Viviani on the value of Euclidean geometry and its application to natural philosophy and where, as we shall also see later in more detail, he performed experiments for the Cimento framed by natural philosophical opinions and contention. Finally, if we consider his publications during the last 14 years of his life, we can conclude that, while participating in the Cimento's meetings, he never abandoned his natural philosophical beliefs in favour of a supposed atheoretical experimental method. In fact, quite contrary to the academicians' supposed inductivist experimental program, Borelli, like his natural philosophising predecessors in the Tuscan Court, was performing 'a priori' experiments. That is, he was constructing and interpreting experiments for the Cimento in order to support claims derived from his mathematical and mechanical natural philosophy.

The same can be said for Viviani. Despite following very different early career paths, both Viviani and Borelli dedicated their careers to reviving ancient mathematical theories, and applying them to their studies in physics as part of their mechanistic natural philosophical beliefs. When both Viviani and Borelli found

[92] See G.A. Borelli, *On the Movement of Animals* (tr. P. Marquet), Berlin, 1989, 12.

[93] It is certainly strange that a group such as the Cimento, that also included Redi, who was such an accomplished naturalist, did not record any experiment in anatomy, apart from only a couple of zoological experiments.

[94] In addition, Ugo Baldini makes the point that by the time Borelli began developing his iatrophysics in Messina, Descartes was as yet practically unheard of in Italy. So since early in his career, Borelli became attached only to Gassendian and Galilean natural philosophy. Baldini, 'Giovanni Alfonso Borelli biologo e fisico negli studi recenti', Physis (1974), **16**, 237.

themselves working inside the Cimento and for the Tuscan Court, they demonstrated the mathematical and mechanical interests that they had adopted and perfected from their Galilean educations.

So we now find ourselves in a good position to understand the mathematical and mechanical views shaping the Cimento's work. The grasp that we now have of the intellectual aims and interests of the two most prominent mechanist members of the Cimento, will provide us with an excellent insight into the construction, interpretation, and presentation of the Cimento's experiments. Once we explore the lives of some of the other academicians, we will then be in an even stronger position, being able to examine the conflicts that existed within the Cimento on natural philosophical grounds. These will be the inner workings revealed in their studies on air pressure, the vacuum, the freezing process, and the properties and effects of heat and cold.

CHAPTER FOUR

WHAT IT MEANT TO BE A CIMENTO ACADEMICIAN

In a letter to Tozzetti in July 1759, Giovanni Batista Clemente Nelli indicated that from the names mentioned in the official Cimento diary, the only members of the Accademia, in addition to Borelli and Viviani, appeared to be the following: Alessandro Segni, the group's secretary at the time of the Cimento's foundation in 1657; Lorenzo Magalotti, Segni's successor to the secretarial position in 1660; Paolo del Buono and Candido del Buono, two Tuscan mathematicians and brothers; Antonio Uliva, the mathematician and Calabrian activist against Spanish rule in southern Italy; and finally, the only two Aristotelian sympathisers in the Cimento, Alessandro Marsili and Carlo Rinaldini.[1] To this list, Tozzetti added Francesco Redi, a former student of medicine at the University of Pisa, whose achievements in natural philosophy certainly did not go unnoticed in Tuscany, but who appeared to make very few contributions to the Cimento's meetings.[2] Tozzetti also mentioned Carlo Dati, a Florentine disciple of Galileo, a loyal Court member, and a natural philosopher also deeply interested in the disciplines of mathematics and astronomy.[3]

All published and unpublished manuscripts pertaining to the Cimento, indicate that these were the only members of the Accademia. For this reason, references here to the 'academicians' will mean the collection of the 11 names mentioned above, from Borelli to Dati. But this is not to suggest that membership was based on regular attendance at the Cimento's meetings. Paolo del Buono was never even in Tuscany at any stage of the Accademia's history, Segni seemed never to be present after 1660, Redi was never mentioned in the diaries, and Borelli, Rinaldini, and Uliva often could not attend meetings because of their teaching commitments in Pisa. So instead, the qualification for membership that they all had in common was that by the time Leopoldo opened his academy on 19 June 1657, all 11 natural philosophers were already employed by the Tuscan Court, or had at least maintained strong ties to the Medici family.

[1] Nelli, *Saggio*, 82.
[2] Targioni Tozzetti, i, 418–419.
[3] *Ibid.*, 443–447.

Additionally, the last person worth considering as an academician is Leopoldo himself. After all, the Prince was educated in natural philosophy, he supervised everything that was being done by his courtiers, and maintained ties with most of the Cimento's correspondents, including Michelangelo Ricci, Nicholas Heinsius, Ismael Boulliau, Honorè Fabri, and Domenico Cassini. Magalotti even wrote that Leopoldo was 'satisfied to act as an academician, and not as a Prince'.[4] In fact, the Prince was often as active in the performance of the experiments as any regular member of the Cimento. He suggested several experiments that were intended to bring about resolutions to certain natural philosophical problems. However, on most occasions his role in the group's history will still be considered here as separate from the academicians themselves. Although he was a student of Galilean natural philosophy, his primary concerns when founding the Cimento and supervising its activities were political. Leopoldo and his brother, Grand Duke Ferdinando II, were intent on founding an academy in order to advertise its exploits and elevate the status and reputation of the Tuscan Court. For this reason, the Prince's role in the Cimento will be examined, instead, in the case studies, and mostly in Part Three, with regard to the writing and editing of the *Saggi*, over which Leopoldo took so much control. In the meantime, the biographical details of the remainder of the academicians will continue to prepare us for an understanding of the natural philosophical skills and commitments of the period.

1. CARLO RINALDINI AND ALESSANDRO MARSILI: DEFENDING SCHOLASTICISM

The only two academicians to express their sympathy for Aristotelian beliefs were Carlo Rinaldini and Alessandro Marsili. Because of Rinaldini's and Marsili's scholastic commitments, traditionally historians have had difficulty describing their role in the Cimento. For example, according to Tozzetti, Rinaldini was 'one of the more active and useful Cimento academicians'.[5] Middleton agrees with this description, on the basis of Rinaldini's illustrious career as a professor in Pisa, and his role in the editing of the *Saggi*.[6] But beyond Middleton's mention of Rinaldini's experiment on the convection of the air, neither Tozzetti nor Middleton provides an account of how Rinaldini may have contributed to the Cimento's experiments. Similarly, Marsili has often been described as an unimportant or marginal character in the Accademia's history.[7] Indeed, we can imagine that as an ardent Aristotelian he would have often felt isolated in an academy that consisted of such vocal anti-scholastics as Borelli and Viviani, and also contained several other mechanists and experimentalists.

[4] 'si contenta di far da Accademico, e non da Principe'. Fabroni (ed.), *Delle lettere familiari*, i, 86.
[5] Tozzetti, i, 430.
[6] Middleton, *The Experimenters*, 34.
[7] *Ibid.*; P. Galluzzi (ed.), *Scienziati a Corte: l'arte della sperimentazione nell'Accademia Galileiana del Cimento (1657–1667)*, Livorno, 2001, 22.

But this does not mean that Rinaldini and Marsili should go unnoticed in our analysis of the Cimento's natural philosophical interests. On the contrary, the presence of both these court philosophers is crucial to our understanding of the Accademia's workings since they created a striking contrast to Borelli's and Viviani's mechanist commitments, and helped to create a situation in which the academicians' individual natural philosophical concerns were regularly debated. In particular, Rinaldini still provided extensive reading lists for the Prince; he continually argued with Borelli and Viviani about the natural philosophical significance of the group's experiments; and he was heavily involved in the editing of the *Saggi*. All these details contributed to the foundations and workings of the Accademia, as well as the refinement of the presentational techniques used in the *Saggi* that will be analysed in Part Three.

The details of Rinaldini's early life and career are not entirely clear. Aside from possessing skills in practical mathematics that he would have used early in his career as an engineer for the papal military forces, Rinaldini was also a promising philosopher. In 1649, the Grand Duke of Tuscany appointed Rinaldini to the position of senior professor of philosophy at the University of Pisa.[8] As an indication of how highly Ferdinando and Leopoldo de'Medici valued Rinaldini, he was also appointed mathematics tutor to Prince Cosimo, the future Grand Duke Cosimo III. Furthermore, as Tozzetti points out, Rinaldini was not only fortunate enough to be approved by the Tuscan princes for these two highly prestigious positions, but he was also frequently asked to participate in the Court's philosophical discussions.[9] In November 1656, he was also assigned the task of providing Leopoldo with two extensive reading lists of contemporary publications in natural philosophy.[10]

It is no surprise, therefore, that after gaining the approval of the princes, Rinaldini became a valuable member of the Accademia del Cimento in 1657. But what exactly were Rinaldini's natural philosophical aims and interests? And how did he gain so much trust inside the princely court? To begin with, we should note his professional background as an engineer. He displayed his expertise in mixed mathematics by publishing several pamphlets dealing with mathematical problems. Furthermore, in his lectures at Pisa, Rinaldini was seemingly not afraid to teach the condemned works of Galileo and the controversial atomistic philosophy of Pierre Gassendi.[11] All this would seem to suggest that he was perhaps a part of the culture of Galilean philosophising in seventeenth-century Italy that appealed to the development of mathematical skills in natural philosophy. But this type of speculation still does not reveal a great deal of insight into Rinaldini's natural philosophical skills and commitments. After all, it was certainly not unusual, even for seventeenth-century scholastics, to practice mathematics and engineering, and we cannot be sure that he was teaching Galilean astronomy and

[8] Biographical details of Rinaldini's career can be found in: Tozzetti, i, 346.
[9] *Ibid.*
[10] BNCF, Ms. Gal. 275, ff. 44r–49v. It is unclear what these lists meant for the foundation and organisation of the Accademia del Cimento. They will be discussed again in the introduction to Part Two.
[11] Middleton, *The Experimenters*, 35.

Gassendian atomism as anything other than hypothetical, as indeed the Catholic Church permitted.

So instead of judging Rinaldini to be a mechanist on the basis of the few biographical details that can be found about him, we may find that more is revealed about his natural philosophical agenda in his actual contributions to the Cimento. As an example, when the academicians examined the properties and effects of heat and cold, Rinaldini provided enthusiastic opposition to the notion that heat could be transmitted through the air in the form of corpuscles. Instead he insisted that the quality of heat that some substances possessed or were given, interacts with the surrounding air to provide the effects that could be seen on thermometers and other measuring instruments. These were the principles of the qualitative structure and movements of nature that scholastics had long used to explain natural phenomena.[12] Furthermore, they were the principles that Rinaldini maintained were true while he was working in the Tuscan Court, from the time of his appointment until his departure in 1667.[13]

In fact, despite Rinaldini's obvious admiration for Galileo, when inside the Tuscan Court he regularly opposed the Galilean and mechanistic claims made by Borelli and Viviani on a number of topics, and often proposed alternative explanations of natural phenomena based on Aristotelian beliefs. These disputes were also particularly clear when Rinaldini, Borelli, and Viviani collaborated on a lengthy editing process of a draft of the *Saggi* written by Magalotti in early 1662. The comments that they made about the draft were kept amongst the Galilean manuscripts and were published by early twentieth-century editors of the Cimento's work, Abetti and Pagnini.[14] During the following chapters we will often refer to these comments to gather clues about the natural philosophical opinions of the academicians, which Borelli and Rinaldini in particular, were not afraid to express. For the moment, we may simply note that Rinaldini provided a counterweight inside the Cimento to the mechanistic opinions of Viviani, and in particular, Borelli. This alone made him an invaluable member of the Cimento since it allows historians to witness the natural philosophical concerns and contentions that often determined how the academicians constructed, interpreted, and presented their experiments.

While Carlo Rinaldini was quite vocal about his support for Aristotelian natural philosophy, another Cimento academician expressed his scholastic opinions more quietly. Marsili was another Cimento academician who admired Galileo, yet seemed determined to defend Aristotelianism against the claims being made by the mechanists in the Cimento. Marsili spent his early life and career in Siena. He was born there in 1601 and he graduated from the University of Siena in law and in philosophy. In 1627, at this same institution, he was appointed lecturer of

[12] This case will be analysed in more detail in Chapter Six.

[13] Rinaldini left Pisa citing health problems and the unfavourable climate. Yet it is more likely that he simply left to take up the chair of philosophy at Padua where he was offered a significant salary increase. He stayed in Padua, until he retired and returned to his place of birth, Ancona. There he died in 1698. Tozzetti, i, 346.

[14] Abetti and Pagnini (eds.), 324–348.

logic and philosophy. It was also in Siena where Marsili had the opportunity to meet Galileo in 1633, just after the Inquisition had condemned Galileo for teaching and supporting Copernicanism as the truth.[15]

In a letter to Galileo in October 1636, Marsili admitted that from the 'few months' he had spent in Galileo's company, he had learnt a great deal. In fact, he confessed that this period of his life had been the most educational of his entire career. So by the time Marsili was appointed to the Chair of Philosophy at Pisa in 1638, he would have certainly been familiar with controversial Galilean natural philosophical beliefs, including arguments regarding atomism and the void. Indeed, Marsili admitted that he even admired Galileo's natural philosophical opinions. The admiration was evidently mutual, since in a letter to Leopoldo in March 1640, Galileo praised Marsili's intellect and even recommended Marsili for the Chair of Philosophy in Pisa, to which he was eventually appointed.[16]

Considering his background, his association with Galileo, and the reputation he probably carried to Pisa as a worthy natural philosopher, it is not surprising that Marsili became a member of the Accademia del Cimento. However, much like Rinaldini, throughout the construction and interpretation of the Cimento's experiments, Marsili surprisingly maintained Aristotelian rather than mechanist commitments. As an example, in an attempt to prove the abhorrence of a vacuum in the Torricellian tube, Marsili suggested an experiment designed to prove that a vaporous substance was always present in this barometric instrument. As we shall see in Chapter Five, Marsili's colleagues in the Cimento, believed this experiment to be faulty and he was unsuccessful in attempting to prove his pro-Aristotelian argument. For this reason, this experiment was omitted from publication in the *Saggi*.[17] Nevertheless, Marsili was clearly attempting to demonstrate his Aristotelian convictions and to organise the downfall of mechanistic and corpuscularian beliefs that maintained the anti-scholastic position held by Galileo and his followers that a vacuum could be created.

It would appear that this was Marsili's biggest contribution to the Accademia's activities. Although Middleton claims that Marsili might have been responsible for the suggestion of other pro-scholastic experiments,[18] it is likely that Rinaldini was in fact behind most of the claims being made against the mechanists in the group. Therefore, Marsili was not as heavily involved in the Accademia's activities as Rinaldini, and according to Middleton, because of the obvious allegiance Marsili showed to 'peripatetic prejudices', his 'contribution to the Academy was unimportant'.[19] That is to suggest, that had it not been for his devotion to Aristotelianism, Marsili could have been a more useful member of the Cimento. However, Michael Segre and Paolo Galluzzi provide an alternative

15 During the first five months of his imprisonment, Galileo was consigned to the archbishop of Siena, allowing Marsili the opportunity to visit him often during that period.

16 Favaro (ed.), *Le Opere*, viii, 496, 502, 542.

17 We shall see in Chapter Five that the omission of this experiment from the final publication may have been partly due to the concern shown by Borelli and Viviani that the experiment was a blatant attempt to support an Aristotelian concept.

18 Middleton, *The Experimenters*, 34.

19 *Ibid.*

theory to that of Middleton's that may help us to attribute a greater role to both Rinaldini and Marsili in the Accademia's experiments.

Segre and Galluzzi suggest that the Prince may have invited Rinaldini and Marsili to join the Cimento so as to structure the group in a way similar to Galileo's famous dialogues.[20] In a bid to make the Cimento seem like a living and continuous homage to Galileo and to give the impression to the Catholic Church that this was not an academy fostering only anti-scholastic opinions, Leopoldo supposedly intended Rinaldini and Marsili to play the role of Simplicio, the Aristotelian philosopher being convinced by the arguments of the other more reasonable modern interlocutors in Galileo's *Dialogue* and *Two New Sciences*.

Considering the tension that existed between the moderns in the group, namely Borelli and Viviani, and the Aristotelians, Rinaldini and Marsili, Segre's and Galluzzi's suggestions do not seem at all far-fetched. At the very least, it seems a more reasonable description of the role of Rinaldini and Marsili inside the Cimento, than Middleton's claim that only Rinaldini was of possible service to the academy, while Marsili was of no use at all because of his natural philosophical prejudices.[21]

So we may deduce from Rinaldini's and Marsili's presence in the Accademia and their defence of scholastic principles, that they were indeed crucial to the dynamics of the Cimento. Whether their presence there was intentionally arranged to create contention or not, their arguments in defence of scholastic natural philosophy provided a perfect counterweight to Borelli's and Viviani's mechanistic skills and commitments. This demonstrates how important natural philosophical concern and contention was to these academicians and how little we can detect of the practice of the so-called atheoretical experimental science that traditional and some 'cultural' historians have claimed was the centrepiece of the Cimento's work.

2. THE CONTRIBUTIONS OF ANTONIO ULIVA, CARLO DATI, CANDIDO DEL BUONO, AND PAOLO DEL BUONO

Just as few details can be found about the lives and careers of Rinaldini and Marsili, the situation is similar for historians wishing to piece together the lives of some of the other lesser-known academicians: Uliva, Dati, and the del Buono

[20] Segre, *In the Wake*, 138; P.Galluzzi, 'L'Accademia del Cimento: 'gusti' del principe, filosofia e ideologia dell'esperimento', *Quaderni Storici* (1981), **xlviii**, 801.

[21] On the other hand, it could be argued that this scenario described by Segre and Galluzzi is unlikely considering that when compiling the *Saggi*, Leopoldo insisted that Magalotti present the academicians' work without a trace of the natural philosophical debates that Rinaldini and Marsili helped to instigate through their Aristotelian arguments. In fact, as we shall see in Part Three, this was a policy in the Cimento's writing and editing process that even Borelli, Viviani, and Rinaldini were required to maintain, despite their heavy involvement in the group's natural philosophical concerns. So the doubt over Segre's and Galluzzi's theory lies in why Leopoldo would purposefully set up an academy with individuals maintaining contrasting natural philosophical beliefs if he was discouraging his academicians from airing those contrasting beliefs in public.

brothers. Uliva and Dati in particular had ambiguous links to the 'Galilean school', apart from their participation in the Cimento, and so it is difficult to determine why they were even invited to join the Medici Court and the Cimento. In the meantime, considering that Paolo del Buono never attended a single meeting of the Accademia, it is also difficult to find why the Cimento carried out several of his suggested experiments. These questions will be partially answered here as we see the natural philosophical skills and commitments of these academicians unfold and become intertwined with their social and political aims and interests. The most ambiguous of these figures was Uliva.

Uliva's family origins and date of birth are unknown, what is known about his educational background is unclear, and his career before arriving in Tuscany seemingly did not include natural philosophical training of any sort.[22] What is known mostly about Uliva's life before his involvement with the Medici Court, is that during the late 1640s he had participated in a plot to overthrow the Spanish rulers of the Kingdom of Naples. For this reason, from 1649 until 1652, he was imprisoned in Reggio Calabria, in all probability also his city of birth.[23] So, possibly, more could be said about Uliva's career as a political activist in the south of Italy, than his accomplishments as a natural philosopher.

According to historian Domenico Bertoloni Meli, the attempted social revolt Uliva became involved in was not a popular rebellion, but rather sought greater power for aristocrats, restoring privileges to southern Italy's nobility.[24] With these ideals in favour of preserving the power of aristocratic and noble families in Italy, it is not hard to imagine that Uliva would have relished the opportunity to work in a major Italian princely court. Indeed, before his involvement with the anti-Spanish movement, he had served as theologian in Rome to Cardinal Francesco Barberini, nephew of Pope Urban VIII. In fact, according to Meli, Barberini may have recommended Uliva to a position at the Medici Court.[25]

Very little is known about Uliva's movements after he was released from prison in 1652. He left Calabria soon after his release and reappeared in Tuscany in 1657. Under the Cimento diary entry for 21 June 1657, his name was introduced as one of three academicians, along with Borelli and Rinaldini, responsible for setting up the experiments before the meetings.[26] At this stage he seemed to be just a visitor to the Court with no official position, but in 1663 he was finally appointed professor of medicine at the University of Pisa.[27] Also at this time, he evinced some interest in natural philosophy when he began composing a treatise on the nature of fluids and another on Euclid's Book V. Although these were never published, Leopoldo certainly deemed Uliva's work worthy of the respect of other court members, including Viviani, who was asked to suspend work on his

[22] See U. Baldini, 'Un libertino accademico del Cimento. Antonio Oliva', *Annali dell'Istituto e Museo di Storia della Scienza*, Monografia 1, 1977.

[23] Meli, 'The Neoterics', 65.

[24] *Ibid.*

[25] *Ibid.*

[26] BNCF, Ms. Gal. 262, f. 3v. Uliva was never named in the performance of any specific experiment.

[27] Middleton, *The Experimenters*, 36.

own treatise on fluids in 1660 so that Uliva could complete and publish his work on the same topic.[28]

Uliva never published this treatise, and all that has been found of his writings about the movements of fluids have been some pages containing a list of chapter titles, but these headings still reveal something about his work on hydrostatics and his natural philosophical interests.[29] According to Meli, as part of Uliva's analysis of the properties of water, he would have used oak galls as colour indicators. The surviving manuscript with the chapter titles suggests that this led him into an analysis of the generation of galls in oak trees that was aimed against traditional beliefs regarding the generation of animals. In his *De motu animalium*, Aristotle claimed that all plants exclusively possessed vegetative souls with nutritive and reproductive powers, while animals possessed sensitive souls providing them with the power of movement and perception. This means that plants and animals have different forms and generate separately.[30]

Yet in a direct contradiction of Aristotle, Uliva seemingly rejected the notion that the generation of animals depended upon the existence and functions of sensitive souls. He appeared to argue that an oak tree, a plant with a vegetative soul, according to Aristotelians, could be responsible for the generation of insects, animal beings with sensitive souls. If Uliva had managed to compile a convincing presentation of such a claim, it would have been a blow to the credibility of Aristotle's metaphysics. In fact, this type of argument against traditional beliefs about the properties of vegetable and animal souls, would have also assisted physiologists, including Borelli, to compile an account of the movements of animals and humans based on a strict mechanical philosophy consisting of pulleys and levers, and the indivisible microstructures that cause the movements of the muscles, rather than mystical qualities such as forms and the functions of the separate souls.[31] So, as Meli put it, Uliva was proposing a theory that did not comply with Aristotle's natural order, and that revealed somewhat of an 'intellectual agenda'.[32]

Uliva left Pisa in 1667 to return to Rome, but immediately found himself in trouble with ecclesiastical authorities who accused him of following a libertine Francophile movement.[33] Possibly in order to avoid a sentence from the Inquisition, he threw himself from a window of the palace of the Holy Office. According to Middleton, there is so little to show for Uliva's time with the Tuscan Court, that we can only conclude that he 'accomplished nothing at the

[28] Tozzetti, i, 434.
[29] BNCF, Ms. Gal. 268, ff. 173r–174r.
[30] Aristotle, *Generation of Animals* (tr. A.L. Peck), London, 1963, lviii.
[31] J. Henry, *The Scientific Revolution and the Origins of Modern Science*, London, 1997, 69.
[32] Meli, 'The Neoterics', 65.
[33] We could have reason to speculate that Uliva's imprisonment might have also been due to his pursuits in biology. His interests in 'medical novelties', as Meli describes such studies as the generation of vegetative and sensitive souls, indicate that Uliva could have believed in a theologically heterodox movement, such as Paracelsian or Helmontian philosophy. Such religious and political movements, considered to be radical, were not uncommon throughout seventeenth-century Europe. See W.R. Newman and L.M. Principe, *Alchemy Tried in the Fire: Starkey, Boyle and the Fate of Helmontian Chymistry*, Chicago and London, 2002.

Accademia del Cimento'.[34] While there is some truth to this argument, we should not be too dismissive of Uliva's presence inside the Accademia. From the little evidence that exists of Uliva's work whilst in Tuscany, it would appear that he was committed to an anti-Aristotelian agenda, much like another member of the Cimento from southern Italy who had also escaped Spanish rule to work in the Medici Court, Giovanni Borelli.

It is not known if there was any connection between Uliva and Borelli before they both came to the Tuscan Court, or if the two were in alliance when they both left their positions in Pisa in 1667 to return to the south. The only definite connection between them is in the fact that they were both members of the Cimento, and, as a letter from Borelli to Leopoldo in December 1664 indicates, that they collaborated on some projects while in Pisa.[35] Therefore, although Uliva may not have featured heavily in the academicians' experiments, he was nonetheless a 'neoteric', as Meli describes those in the seventeenth century interested in philosophical and medical novelties.[36] In an environment such as the Cimento where natural philosophical issues were repeatedly contested, Uliva was one more academician who probably would have argued his support for Borelli and Viviani and against the Aristotelian beliefs of Rinaldini and Marsili. In this sense he was another very important member of the Cimento, probably contributing to the natural philosophical debates within the group.[37]

In the meantime, a similar summation may be made for Dati, whose contributions to the Cimento were also ambiguous and who is also described by Middleton as having had only a 'slight influence' on the Accademia.[38] Dati was certainly not a political activist like Uliva, but before participating in the Prince's academy he had forged a career that had largely nothing to do with natural philosophy. Dati instead specialised in language and literature. He was a Florentine gentleman who, as secretary of the prestigious literary Accademia della Crusca, studied the Tuscan language and even claimed its supremacy in his 1657 publication, *Discorso dell'obbligo di ben parlare la propria lingua*.

According to Tozzetti, despite this apparent lack of interest in experimental studies of nature's structure and movements, Dati was still highly respected by Leopoldo for his knowledge of mathematics and philosophy.[39] Indeed, Dati proved to be a useful member of the Medici Court because of his ardent support for the physico-mathematical beliefs of the 'Galilean school'. He may not have made too many contributions to the experiments performed inside the Cimento, but in 1663 he published an important text under the pseudonym of Timauro Antiate.[40] This consisted of Torricelli's previously unpublished letters

34 Middleton, *The Experimenters*, 36.

35 BNCF, Ms. Gal. 277, ff. 61r–63v.

36 Meli, 'The Neoterics', 57.

37 Considering the possibility that Uliva could have also been either a Neoplatonist, Paracelsian, or Helmontian philosopher, he might have been included in the Cimento as representative of a third alternative in natural philosophy: a non-mechanical anti-Aristotelian.

38 Middleton, *The Experimenters*, 31.

39 Tozzetti, i, 443.

40 T. Antiate, *Lettera ai Filaleti della vera storia della cicloide e della famosissima esperienza dell'argento vivo*, Florence, 1663.

and manuscripts about his barometric experiment, and defended the academicians' conclusions regarding their own experiments on air pressure. Therefore, Dati showed his support for some of the mechanistic interpretations the Cimento academicians had made during their barometric experiments in 1657.

Furthermore, Dati left behind a manuscript that may tell us even more about his natural philosophical agenda. Published by Tozzetti under the title *Dissertazione di Carlo Dati sull'utilità, e diletto che reca la geometria*, this text reveals the value Dati placed on geometry as a tool for investigating nature.[41] Here he claimed that neither philosophy nor experiments can adequately help us obtain the truth, but that geometry does instead lead us to a reliable knowledge of nature.[42] We cannot be sure of exactly what this meant for the Accademia's experiments or in what ways Dati involved himself in the group's discussions. Indeed, since Dati's name was not mentioned in the academicians' diaries, it may still be true that despite his evident interests in natural philosophy, he had only a 'slight influence' over the Cimento, as Middleton states. In any case, as Segre puts it, Dati's support for a physico-mathematical and mechanical natural philosophy indicates 'that empiricism was not the only tendency in the Accademia del Cimento'.[43] Just like Uliva, despite an ambiguous background, Dati's presence in the Tuscan Court from 1657 to 1667 would have contributed to the natural philosophical concern and contention that dominated the construction and interpretation of the group's experiments.

Slightly more is known about the del Buono brothers than the academicians so far mentioned in this chapter. However, their involvement with the Cimento was still seemingly unclear. According to Tozzetti, Candido, a priest and the elder of the two, had studied mechanics under the guidance of Galileo, hinting at his possible natural philosophical commitments.[44] But once inside the Cimento, he only suggested a few experiments in 1657 measuring the weight and pressure of various liquids. These were not published in the *Saggi* and they seem to have contributed little to the academicians' achievements. Meanwhile, Tozzetti argues that Candido may have invented the 'arcicanna', a mounting for large telescopes, but even this has been attributed in other sources to another of his brothers, Antonio Maria.[45]

Paolo, in the meantime, developed a greater talent in mathematics and natural philosophy than his elder brother. Middleton suggests that Paolo too had briefly studied under Galileo in 1641 before studying mathematics at the University of Pisa under the guidance of Famiano Michelini, one of Galileo's pupils and later a highly valued engineer for the Tuscan Court.[46] Following his graduation at Pisa in 1649, he stayed in Tuscany and was in contact with Prince Leopoldo until 1655. During this time, in 1652 he carried out several observations of a comet. Contrary to Aristotelian beliefs in the sublunary existence of comets,

[41] Tozzetti, ii, 314–327.
[42] See also Segre, *In the Wake*, 137.
[43] *Ibid.*
[44] Tozzetti, i, 435.
[45] *Ibid.*, 436; Middleton, *The Experimenters*, 30.
[46] Middleton, *The Experimenters*, 30.

Paolo reported on its apparent lack of parallax, implying that comets must actually travel far beyond the distance of the moon.[47] These observations would have been particularly useful to Borelli when he too observed the lack of parallax of a comet in 1664.

One can image that Paolo, with this background and with a talent for observations and natural philosophy, would have become one of the Cimento's biggest contributors had he stayed in Florence. But instead, he entered the service of Emperor Ferdinand III in Germany where he was appointed master of the Imperial Mint and also worked as a mining engineer. In the meantime, his name occasionally appeared in the Cimento diary, when he suggested experiments for the Prince's academy through correspondence. In particular, Paolo contributed to the academicians' work regarding the compression of air. In fact, in an attempt to find some consistency in their corpuscularian explanations of the properties and movements of air and liquids, Paolo and the group's other mechanists tested the compressibility of water. On 10 September 1657, the academicians carried out an experiment suggested by Paolo for this purpose.[48] Although not completely satisfied with their observations, the experiment was still published in the *Saggi*.[49]

So, although he was never in Tuscany during the Cimento's history, Paolo continued to work on the practical application of mathematics in Germany, while he also corresponded with the academicians about the natural philosophical interpretations of their experiments. Obviously from such a distance he could not have contributed a great deal to the Cimento, but he was nonetheless another voice inside the Accademia more interested in natural philosophical contention than the application of a supposedly unbiased and atheoretical experimental method. These were the intellectual interests that Paolo shared with each of his fellow academicians.

3. FRANCESCO REDI AND THE EXPERIMENTAL METHOD

The details of Redi's career are far less ambiguous than those of the other academicians examined in this chapter. He was born in Arezzo, Tuscany in 1626, attended the Jesuit College in Florence in his youth, and then graduated in medicine from the University of Pisa in 1647. Following his graduation, Redi travelled throughout Italy before settling in Florence again as a member of the Tuscan Court. In 1655 he entered the Accademia della Crusca, where he was eventually elected president in 1678. In 1666 he was appointed Chief Physician to the Grand Duke Ferdinando II, a position he retained under Cosimo III. And finally he became a founding member of the Accademia del Cimento and was also appointed head of the Medicean pharmacy. With so many duties involved in his service to the Grand Duke and with such varied expertise, Redi is described by

[47] Tozzetti, i, 439.
[48] BNCF, Ms. Gal. 262, ff. 32r–v.
[49] Magalotti, 215. This will be examined in greater detail in Chapter Five.

Paula Findlen as 'the perfect example of a natural philosopher whose career was made at court'.[50] Indeed, Redi was probably considered as an integral part of the Tuscan Court by his Medici employers, especially because of his skills in the field of medicine and natural history. These were disciplines that were evidently encouraged by the Grand Duke, who also employed Marcello Malpighi and Nicolaus Steen, among other contemporary physicians, and of course encouraged Borelli's school of physiologists at Pisa. With this in mind, Tozzetti even believes that Redi was Ferdinando's 'most favoured' court philosopher.[51]

However, despite the clarity of Redi's biographical details and his position inside the Tuscan Court, some uncertainty remains about his role in the Cimento. Redi's name was never recorded in any of the Accademia's diaries or manuscripts. In fact, the only indication that he was an academician exists in three letters he wrote in 1659, 1660, and 1686.[52] In the first two Redi told his correspondents about how he was heavily engaged in work with the Accademia, while in the last letter he boasted to one of his friends about 'having been one of the first founders of the famous Tuscan Accademia del Cimento'.[53] Yet while he confirmed his status as an academician, Redi did not provide any indication in these letters of what experiments he may have helped to carry out or discuss. What is more, Redi was a naturalist and a man of medicine enrolled in an academy seemingly interested mostly in physico-mathematical and mechanical topics in natural philosophy. So what exactly was his role in the Cimento?

According to Findlen, Redi was certainly not a regular natural philosopher such as Borelli or Viviani. His interests were only in natural history, excluding him from pursuing the topics that seemed to attract the attention of his fellow academicians, such as air pressure, the vacuum, the effects of heat and cold, the speed of sound, or the path of comets. Instead, Findlen describes Redi as the philosophical voice inside the Cimento, commentating on the importance of relying purely on experiments for investigating nature constructed within the virtuous and gentlemanly community of the Medici Court. That is, according to Findlen, Redi's 'scientific method', or more specifically his 'experimental method', used to construct efficacious and unbiased knowledge claims, helped to earn him a reputation inside the Tuscan Court as a reliable philosopher and knowledge maker, and therefore made him a valuable contributor to the Accademia.[54] In other words, the most outstanding characteristics of Redi's career were not, supposedly, his natural philosophical skills and commitments, but rather that he was a classical loyal courtier who relied on the status and reputation of his royal patrons in order to carry out an efficacious and unbiased experimental method. Findlen, therefore, concludes that Redi and his fellow court members were

[50] Findlen, 'Controlling the Experiment', 39.

[51] Tozzetti, *Atti e memorie inedite dell'Accademia del Cimento e notizie aneddote dei progressi delle scienze in Toscana*, 3 vols., Florence, 1780, i, 251.

[52] See Middleton, *The Experimenters*, 34, 52.

[53] Tozzetti, *Notizie*, i, 450. Curiously, the first two letters were written on 25 April 1659 and 9 May 1660, while the Cimento was actually in recess. Middleton suggests that Redi might have therefore been referring to the informal academy being run by the Grand Duke. Middleton, *The Experimenters*, 52.

[54] Findlen, 'Controlling the Experiment', 43–45.

interested in 'the social circumstances of their philosophical endeavours more than their specific intellectual goals'.[55] This is an example of the type of 'cultural' historiography also found in the writings of Biagioli and Tribby, discussed in Chapter One.

While Findlen provides us with an interesting account of how Redi made a living inside the Tuscan Court, we are left wondering what this means for our understanding of the natural philosophical significance behind Redi's work. Despite Findlen's analysis, we remain unaware of how Redi fitted into an institution consisting of several thinkers intent on contesting each other's natural philosophical beliefs, and as we have seen, showing little interest in the pursuit and application of any type of inductive, atheoretical 'experimental method'. To address these issues fully we need to understand that although the academicians, especially Redi, did indeed perform many experiments of various types, this does not mean that they developed some type of experimental method with which they were capable of producing matters of fact, free from any theoretical commitments. This leaves us then to examine the work that Redi carried out inside the Court, and to assess what his work means for our understanding of the Cimento's experiments.

It would seem that Redi could have only possibly been responsible for two of the experiments narrated in the *Saggi* and recorded in the Cimento diary. They were the academicians' only zoological experiments, one on the digestive system of some animals,[56] which Redi also examined in his 1671 publication, *Esperienze intorno a diverse cose naturali*, and one on the effects of snakebites,[57] also analysed by Redi in his own publication, *Osservazioni intorno alle vipere* (Florence, 1664). But the published text that best reveals Redi's natural philosophical interests while participating in the Cimento is his *Esperienze intorno alla generazione degl'insetti* (Florence, 1668), an epistle dedicated to Redi's fellow Crusca and Cimento academician, Dati.

Here Redi expanded on the work mentioned earlier by Uliva regarding the generation of insects in the galls of oak trees. He described several experiments and literally thousands of observations that tested the Aristotelian notion that insects are generated spontaneously. This was a traditional scholastic theory based on the belief that the soul is the inseparable form of each living body of the animal and vegetable kingdom. In opposition to this belief and in a move to show the need to discard any reliance on ancient sources for accumulating knowledge of nature, Redi claimed that the galls acted as a site for insects to lay their eggs and to regenerate. As Uliva had argued against Aristotelianism, if the generation of insects from plants depended on the existence of sensitive souls, then there appeared to be a transcendence from the vegetative soul of the plant to the sensitive soul of the insects. What this meant for seventeenth-century naturalists such as Redi was that Aristotle's metaphysical basis for his theory on the generation of animal beings was problematic and that the only way of ascertaining the truth was through first-hand observations of nature.

[55] *Ibid.*, 43.
[56] Magalotti, 247.
[57] *Ibid.*, 276.

Evidently Redi was preoccupied with an anti-Aristotelian agenda. He wished to question the validity of traditional beliefs in studies of natural history and to establish the efficacy and authority of his experiments and observations over the ancient writings. But it would be a giant historiographical leap for us to assume that just because he used experiments, Redi was applying some type of 'experimental science', or 'method', implying that he was compiling unbiased and theory-neutral matters of fact. At the same time, it cannot be maintained that Redi and his fellow court members valued an experimental method 'more than their specific intellectual goals'.[58] As we have seen through the lives and careers of the academicians, including Redi, those 'intellectual goals' cannot be dismissed so easily in our studies of the experimental life in seventeenth-century Tuscany.

4. THE CIMENTO'S SECRETARIES AND THE LAST WORD ON COURTLY CULTURE AND EXPERIMENTAL SCIENCE

Redi was not the only academician dedicated to courtly life. Both of the Cimento's secretaries, Segni and Magalotti were also courtiers during most of their careers. Like Redi, they gained great status and reputation from their long association with the princely Medici family. However, in contrast to Findlen's account of the virtuosity of the Tuscan Court and the birth of experimental science, the courtly lives of these secretaries, especially Magalotti's, did not mean that they were exempt from natural philosophising or that they were instead more interested in simply performing or talking about experiments to boost their status within the court.

After serving Leopoldo as the Cimento secretary from 1657 to 1660, Segni was retained in the service of the Grand Duke in 1662 as a librarian, and in 1674 he again served Leopoldo as the superintendent of the Prince's secretariat. He remained in the Tuscan Court until his death in Florence in 1697. He thus spent his entire life serving the Medici family. But with an interest in literature rather than natural philosophy, Segni made no contributions to the Cimento's experiments.[59] Additionally, after he was replaced by Magalotti, it is not likely that his association with the Cimento continued. Indeed, there is no indication anywhere that he contributed in any way after 1660. For this reason, the life of Segni is of little interest to us.

On the other hand, Magalotti's life is certainly worthy of careful consideration.[60] He entered into the Tuscan Court at the age of 22, and spent a long career serving Leopoldo, as well as the Grand Dukes Ferdinando II and Cosimo III. Furthermore, he was trained and was interested in a variety of disciplines in natural philosophy, and he was also skilled in law, literature, and language. This background would suggest that Magalotti was just as much the classical courtier

[58] Findlen, 'Controlling the Experiment', 43.
[59] Tozzetti, *Notizie*, i, 447; Middleton, *The Experimenters*, 35.
[60] Sources on Magalotti's life include: Middleton, *The Experimenters*, 31–33; Fabroni (ed.), *Delle lettere familiari*.

as Redi, and judging from his rhetoric in the *Saggi*, we could even have reason to believe that he too was dedicated to establishing the use of an experimental method in Italy. But a deeper insight into the intellectual interests that Magalotti declared in his letters and manuscripts, as well as the details of his career, reveal instead his natural philosophical beliefs and the subsidiary role of experiments in his and the Cimento's work.

Magalotti was born in Rome to a noble Florentine family in 1637, and there attended a seminary run by Jesuits where he was taught philosophy by none other than Uliva, until he was 18. This was also where Magalotti became acquainted with the scholastic philosopher and mathematician, Honorè Fabri. Although Magalotti excelled in literature and in philosophy during these early years of his education, in 1656 he travelled to Pisa to study law, and according to Fabroni, during that same year he also became interested in anatomy under the guidance of Marcello Malpighi and may have even attended some of Borelli's classes on physiology.[61] Throughout his life, Magalotti seemed to draw from his education under each of these erudite thinkers, from Uliva to Malpighi. But his career as a courtier and natural philosopher did not really begin until he dropped his subjects in Pisa in 1656 to be tutored in mathematics and natural philosophy for the next three years under Viviani.

This association between Magalotti and Viviani proved to be a fruitful one for both of them since, as Viviani revealed in the preface to his *De maximus et minimus* in 1659, the future Cimento secretary showed enough ability in geometry to be able to discuss Apollonius intelligently. Furthermore, Viviani also praised Magalotti for his knowledge of mathematics, philosophy, law, Latin, Tuscan, as well as for his candour and gentlemanly behaviour.[62] With these praises coming from one of the Court's most respected and talented natural philosophers, it would be fair to say that Magalotti had the right to prepare himself for a career as a courtier. Indeed, according to Stefano Fermi, Magalotti longed to be a member of the Tuscan Court and used his knowledge and charm on those who could facilitate his entry, such as Viviani. That is, he talked about his noble and gentlemanly background and he displayed his array of knowledge and his ability to discuss a variety of topics.[63] With such qualifications, and with the praises that came from Viviani, it would be quite fair to say, as does Fabroni, that Magalotti was 'born for the Court'.[64]

Late in 1659, Magalotti was admitted into the Tuscan Court and only a few months later, in May 1660, he replaced Segni as secretary of the Cimento.[65] During the next seven years he occasionally contributed to the group's experiments and, of course, spent much of his time compiling the *Saggi*. Following this publication in 1667, the Grand Duke sent Magalotti on a long journey across

[61] Fabroni (ed.), *Delle lettere familiari.*, i, xii.
[62] *Ibid.*
[63] Fermi, *Lorenzo Magalotti*, 27.
[64] Fabroni (ed.), *Delle lettere familiari,* i, xii.
[65] According to Fabroni, Viviani recommended Magalotti for the position. *Ibid.*, xiv.

Europe in order to present the Cimento's published work to other courts and institutions in Austria, Germany, Holland, Belgium, England, and France. When he completed these duties he was again required to accompany Prince Cosimo on a voyage to England, Spain, and Portugal, and during the next ten years he was also sent on various other diplomatic missions.[66] When he returned to Florence permanently in 1678, Magalotti continued to write his own philosophical manuscripts about religion and was eventually appointed as counsellor to Cosimo III in 1689.[67]

Therefore, Magalotti was just as much the classical courtier as Redi. They both had gentlemanly backgrounds, had spent their entire careers in service to the Tuscan Court, and were skilled in a variety of disciplines, including natural philosophy. In Redi's case we have also seen that despite the appearance from his courtly behaviour that he was committed to experimental and atheoretical fact-making, he was in actuality still concerned about the natural philosophical significance of his claims. Magalotti, too, participated in the natural philosophical speculations of his fellow academicians.

As we have seen earlier, Magalotti was responsible for compiling a text on behalf of the Cimento that some traditional historians have described as being representative of the origins of the scientific method. In accordance with this traditional view, Giuseppe Marchetti, a twentieth-century editor of Magalotti's letters, claims that during Magalotti's education in Pisa under Borelli and Viviani, the future Cimento secretary developed a 'scientific curiosity' from the 'experimental method' being used by these academicians.[68] Yet considering the arguments already put forward here about the supposed method of the Cimento and its members, this general and wide-sweeping summation of Magalotti's early training in natural philosophy does not seem entirely accurate. Indeed, instead of assuming that Magalotti learnt about an atheoretical experimental method that neither Borelli nor Viviani ever claimed to use, we should understand that Magalotti was being trained according to the skills of two of Galileo's most talented students. That is to say that Magalotti was well aware of the anti-Aristotelian natural philosophical commitments of the academicians.

When he was appointed secretary in 1660, Magalotti not only recorded what was happening in the Cimento's meetings, but occasionally he also participated in the group's experimental activities. In June 1660, for example, he suggested an experiment testing the rise in temperature of quicksilver when mixed with vitriol.[69] In 1663 he also assisted Viviani in an experiment testing the speed of light.[70] Furthermore, according to Fermi, it is possible that Magalotti might have even communicated to Leopoldo, his opinions about such topics as the movement of sound and light, the function of the barometer and the movement of the blood.[71] But the most interesting case study to involve Magalotti in the Cimento

[66] G. Marchetti, 'Introduzione', in L. Magalotti, *Lettere familiari*, Carnago, 1993, 8.
[67] Middleton, *The Experimenters*, 32.
[68] Marchetti, 'Introduzione', 7.
[69] BNCF, Ms. Gal. 262, f. 82v.
[70] *Ibid.*, f. 55r.
[71] Fermi, *Lorenzo Magalotti*, 103.

is his attempt to refract cold through a glass lens. Magalotti suggested this experiment in an undated note to Viviani, and it was carried out by the Cimento on 1 September 1660.[72]

At the top of Magalotti's note describing this experiment, the following words were written in large handwriting: 'E viva gl'atomi frigorifici'.[73] This loud declaration may indeed be demonstrative of Magalotti's personality, his natural ebullience as Middleton described him, but it also shows his commitment to atomistic beliefs. As we shall see in detail in Chapter Six, in such experiments pertaining to the properties and effects of heat and cold, typical of the academicians between 1657 and 1662, there was a clear conflict between scholastics claiming that heat and cold are no more than qualities of of various substances that interact with the qualities of the surrounding air, and the mechanists in the group insisting that heat and cold consist of material corpuscles. These contrasting natural philosophical beliefs formed the cornerstone of the academicians' workings, since the natural philosophical commitments of each of the academicians were at stake when debating this topic. So, Magalotti was participating in this debate and even contributing through the suggestion of an experiment. In fact, an analysis in Part Two of several passages scattered throughout the *Saggi*, will reveal that despite Magalotti's intention to present purely a narrative of the Cimento's experiments, he occasionally still hinted at the corpuscularian and mechanistic aims and interests of some of the academicians.

This then brings our analysis of the Cimento academicians to a conclusion. Experiments were indeed crucial to how the Cimento academicians carried out their work. As we saw from our analysis of the academicians, from Borelli to Magalotti, experiments were certainly not uncommon. But this is far from demonstrating that any of the group's members were actually using an experimental method, if by method we are to understand a programme for accumulating atheoretical and non-speculative matters of fact. Instead, we will continue to see that these experiments were constructed and interpreted according to much wider and far-reaching concerns in the intellectual circles of the seventeenth century, namely, natural philosophical skills and commitments.

[72] BNCF, Ms. Gal. 262, ff. 109r–v. See also Middleton, *The Experimenters*, 33.
[73] Florence, Bibl. Riccardiana, cod. 2487, f. 7r. As cited by Middleton, *The Experimenters*, 33.

PART TWO

THE ACCADEMIA DEL CIMENTO: 1657–1662

During their first five years of experimenting, the Cimento followed no formal structures. From 19 June 1657, a diary began to be taken of the daily experimental activities undertaken by the courtly philosophers and mathematicians under the Prince's commands, but there is no other evidence that a formal academy had been established in Florence. In fact, the existence of more than one diary and the possibility that both the Prince and the Grand Duke were running separate sessions, casts a confusing shadow over what exactly was happening with the Medici brothers, their courtiers and the possible plans for establishing a formal society.[1]

Nevertheless, in his analysis of the Cimento's foundation, Middleton attempts to show that the foundation of a well-organised academy had been on the minds of Ferdinando and his younger brother well before the first recorded meeting on 19 June. Middleton mentioned that in November 1656, Rinaldini compiled two reading lists requested by the Prince and intended for Leopoldo's 'anticipated program', as Middleton puts it.[2] Judging from Rinaldini's letter attached to the first list he sent to the Prince, these readings were indeed meant for study by the Court's experimental philosophers.[3] But there is no indication that the lists were contributions made specifically to Leopoldo's supposed plans for an academy to be set up seven months later, nor could we possibly deduce from this that Leopoldo had any type of research programme thought out, let alone had access through his advisers to some unique, transferable, and efficacious inductive scientific method. Instead, all we can confidently conclude is that by 1656, Leopoldo was continuing to refine his interest in natural philosophy and experiments by expanding his library.

What then might we instead find about the Cimento's formal structures and workings from letters and notes composed after 19 June 1657? To begin with, just

[1] The academicians' formal diary, transcribed by Targioni Tozzetti from the lost original manuscript, is held in the Biblioteca Nazionale Centrale in Florence amongst the Galilean papers. It will be referred to here as BNCF, Ms. Gal. 262. According to Middleton, the second diary manuscript labelled BNCF, Ms. Gal. 261, was actually a record of the experiments performed by Ferdinando's academy until January 1658. Middleton, *The Experimenters*, 42–45. See also Chapter One, n. 37.

[2] *Ibid.*, 47.

[3] BNCF, Ms. Gal. 275, f. 44r.

two days following the first recorded meeting, certain roles were established for some of the academicians. Besides Alessandro Segni's secretarial job, on 21 June, 'it was decided that Sig. Rinaldini, Borelli, and Uliva should meet at the Palace at the 21st hour every day to discuss the experiments to be made one after another on the following day, and to give the necessary orders'.[4] This is a strong indication that the Cimento was established as an academy with at least a minimum degree of formality, even though we may doubt whether Rinaldini, Borelli, and Uliva maintained these roles.[5] However, besides this rather slender piece of evidence in support of a formal structure, there are no clear signs that the Cimento had any regulations, guidelines, or schedule in place for their work. Meetings were often suspended because of the Court's travels around Tuscany and some members were often required to return to their teaching obligations at university – this was especially the case with Borelli and Rinaldini. Furthermore, there is no evidence suggesting that the academicians were concerned with a budget; it would seem that the princes took care of all the costs of obtaining the Cimento's extensive collection of delicate and expensive glassware. Finally, Leopoldo's society of mathematicians and philosophers did not even have a name until well after it was established.[6]

In addition to these ambiguities surrounding the existence of the Cimento and the formality of its activities between 1657 and 1662, I have not found any suggestion in either of the Cimento's diaries, that the academicians were following any rules associated with some type of inductive experimental programme, that dismissed the possibility of offering theoretical interpretations. There is no doubt that the academicians were dedicated to performing experiments, but we should not believe that this meant abandoning the natural philosophical beliefs that each of them had been accumulating throughout their careers. Indeed, there is no hint, not even in Leopoldo's correspondence, that he was enforcing the type of pure experimentalist activity that Magalotti, in the presentation of the Cimento's work, later claimed existed. In fact, Middleton reaches a similar conclusion when comparing the indifferent and anonymous style of the *Saggi*, with the informal structure of the Accademia during the first half of its existence:

> The fact that the publication that eventually resulted from the activities of the Academy is completely anonymous should not cause us to lose sight of the fact that the authors of a great many of the experiments, published or unpublished, are mentioned in the diaries .... Indeed, the requirement of strict anonymity seems to have grown slowly.[7]

So for the first five years, the academicians were certainly following the Medici's interests in experiments, but we have no evidence indicating that they were also

[4] As translated by. Middleton, *The Experimenters*, 53–54. 'Si determinò, che il Signore Rinaldini, Borelli ed Uliva dovessero ragunarsi ogni giorno a Palazzo alle 21 ora per discorrere e dare gli ordini necessari per le esperienze da farsi di mano in mano il giorno appresso'. BNCF, Ms. Gal. 262, f. 3v.

[5] As Middleton also notes, Rinaldini and Borelli had other obligations in Pisa, while the fact that Uliva earned very few mentions in the diary, does not help any speculation that he was a consistent contributor. Middleton, *The Experimenters*, 54.

[6] It seems that Leopoldo's group was never formally known as the Accademia del Cimento until its publication in 1667. However, there are some suggestions from letters written in late 1659 and early 1660, that the 'Accademia del Cimento' was used by some to describe Leopoldo's academy.

[7] Middleton, *The Experimenters*, 54.

following a programme which forbade individuals' expressions of natural philosophical opinion. As Paolo Galluzzi states: '[B]ehind the Accademia's serene façade, there unfolded a significant and lively confrontation based on principles.'[8] It will be the aim of the following two chapters to show that those 'principles' included the culture of natural philosophising examined in Chapter One, with particular reference to contrasting skills and commitments regarding mixed mathematics, its relation to mechanical natural philosophy, Aristotelianism, and anti-scholasticism. Furthermore, these are intellectual 'principles', or natural philosophies, that became endemically entangled in the academicians' experimental activities, creating a local field of competing natural philosophical interpretations between the group's mechanists and Aristotelians. This will be evident in our survey of the experiments performed by the academicians regarding the pressure of air and the existence of the vacuum, the freezing process, as well as the properties and effects of heat and cold. These experiments all took place between 1657 and 1662, before Leopoldo decided to cease the Cimento's meetings in order to focus on the publication of its experimental exploits.

[8] Galluzzi, 'L'Accademia del Cimento', 805.

CHAPTER FIVE

EXPERIMENTS CONCERNING AIR PRESSURE AND THE VOID AND A LOOK AT THE ACCADEMIA'S INTERNAL WORKINGS

In June and August 1657, the Accademia del Cimento began to perform numerous experiments involving Torricelli's barometer and questioning traditional beliefs about the pressure of air and the impossibility of the void. One could conduct an investigation of the academicians' work in this field relying solely on the experiments they reported in the *Saggi*, but this would hardly do justice to the natural philosophical issues involved in the construction and interpretation of the Torricellian barometer. We should instead consider the variety of positions surrounding studies in pneumatics during the seventeenth century. Therefore, this case study does not begin in 1657 when the academicians performed their first barometric experiments, but in 1644 when Evangelista Torricelli claimed to have constructed the first barometer, and sparked a flurry of activity among his colleagues in other parts of Europe.

Torricelli described how he made the instrument in a letter to his friend Michelangelo Ricci (1619–1682), on 11 June 1644.[1] He tells Ricci that he filled a tube, sealed at one end, with mercury and stopped it at the mouth with a finger (Figure 3). He turned the tube upside down in a bowl, also filled with mercury. When he released the finger, the mercury in the vessel descended slightly, leaving an empty space at the top. What followed was an attempt to demonstrate, first, that the space formed in the tube was vacuous and, second, that the mercury's descent was due to the weight of the surrounding air, and not because of the vacuum's force.

Torricelli added water to the mercury in the bowl and began to raise the tube slowly. When the mouth of the tube rose to the surface of the water, the mercury in the vessel poured out, and the water rushed in to fill the tube to its top, demonstrating that the space had indeed been empty and that there is a clear difference in the densities and movements of the two liquids.[2] This experience was enough

[1] BNCF, Ms. Gal. 150, f. 89r–90r. See also W.E.K. Middleton, *The History of the Barometer*, Baltimore, 1964, 23–24.

[2] E. Grant, *Much Ado About Nothing: Theories of Space and Vacuum from the Middle Ages to the Scientific Revolution*, Cambridge, 1991, 23. Mercury being a heavier and denser liquid than water, was also a more suitable substance for Torricelli's experiment, since it was not necessary to use long tubes that were normally required for water barometers.

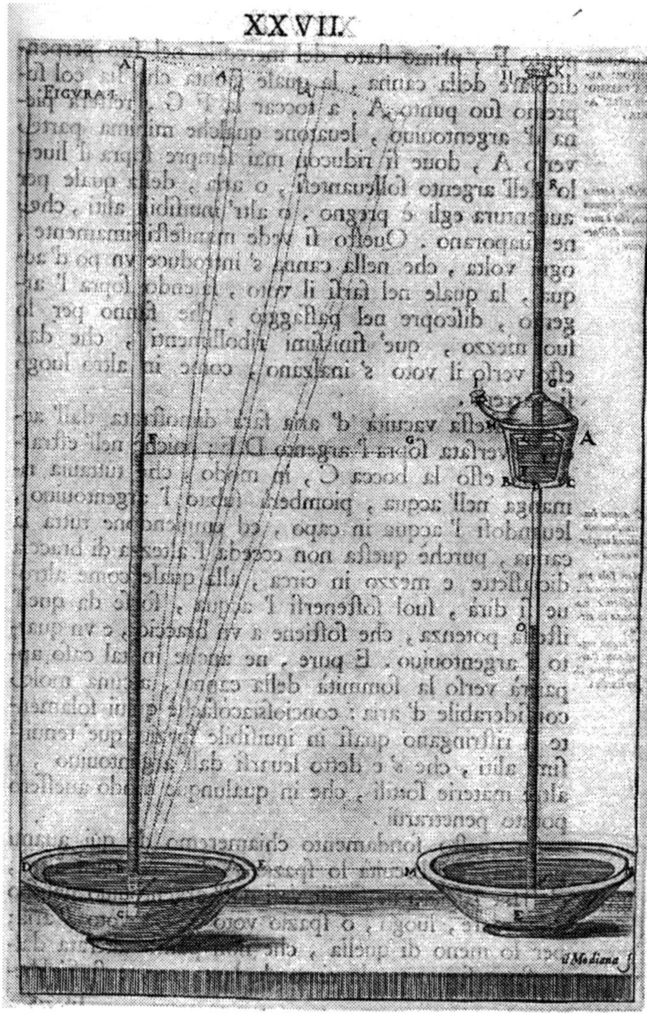

Figure 3. Torricelli's barometer and Roberval's barometer inside a barometer. (From L. Magalotti, *Saggi di naturali esperienze*, Florence, 1667, 27; courtesy of the IMSS Biblioteca Digitale.)

for Torricelli to believe that not only had he created a vacuum, but in the process had also produced an instrument that could measure the weight of its surrounding air.[3]

The implications of these claims were quite significant. Torricelli was directly opposing Aristotelian doctrine that nature abhorred the production of a vacuum.

[3] Middleton, *The History of the Barometer*, 25.

According to Aristotle, it is impossible for a vacuum to exist in nature. He, of course, believed in the existence of four terrestrial elements, earth, water, air, and fire, all moving naturally strictly in straight lines up or down, from or to the centre of the universe. Even the celestial realm was full of aethereal matter maintaining the planets in their perfect circular orbits. To admit the existence of vacuous spaces in either realm was to concede that these natural movements could be corrupted.[4] The argument from the perspective of Aristotelians, was that movement inside a vacuum could have no dimensions; meaning that matter could move in any direction, contrary to the Aristotelian belief in vertical straight-line motion. Furthermore, because of the lack of resistance to motion that would occur in a vacuum, bodies could move infinite distances, and at infinite speeds, contradicting the limits Aristotle placed on the size of the universe. Therefore, since the resuscitation of Aristotelian natural philosophy in the Middle Ages, it was widely believed by European scholastics that the creation of a void was impossible, except by God in certain interpretations. This position could not be compromised because of the danger of putting their entire natural philosophy under doubt.

By the sixteenth century, the restoration and translation of other ancient texts, including the work of atomistic vacuists such as Democritus, Epicurus, and Lucretius, led to challenges being put forward against Aristotelianism.[5] Figures such as Francesco Patrizi (1529–1597) and Bernardino Telesio (1509–1588) put forward their atomistic and vacuist beliefs against the Aristotelian plenist arguments. So by the beginning of the seventeenth century there was clearly some natural philosophical contention between Aristotelians and corpuscularians about whether it was possible for a vacuous space to exist in nature.[6]

But the weight and pressure of air were quite a different matter, as Middleton points out.[7] Since the very concept of the weight of air was simply contrary to Aristotelian beliefs, experiments attempting to capture the pressure caused by the weight of air would have been incomprehensible to scholastics.[8] This meant that the topic of air pressure was of much greater interest to seventeenth-century mechanists than to scholastics and those atomists belonging to what might suitably be termed the natural magic tradition.[9]

In Italy, Galileo and his colleagues became especially curious about the movements of water inside a suction pump. Galileo had always been prepared to accept the existence of the void even though it could not be proven, and by 1630 he insisted that it was in fact the 'force or resistance of the vacuum' that stopped

[4] These natural philosophical principles are explained clearly in Aristotle's *De Caelo*, written c.350 BCE.

[5] These ancient atomists believed that tiny vacuous spaces existed between the particles that made up all matter.

[6] C.B. Schmitt, 'Experimental Evidence for and against a Void: The Sixteenth-Century Arguments', *Isis* (1967), **58**, 363.

[7] Middleton, *The History of the Barometer*, 4–5.

[8] *Ibid.*, 5.

[9] For a sophisticated analysis and categorisation of the philosophy of early modern natural magicians, see J. Henry, 'Magic and Science in the Sixteenth and Seventeenth Centuries', in Olby et al. (eds.), 583–596.

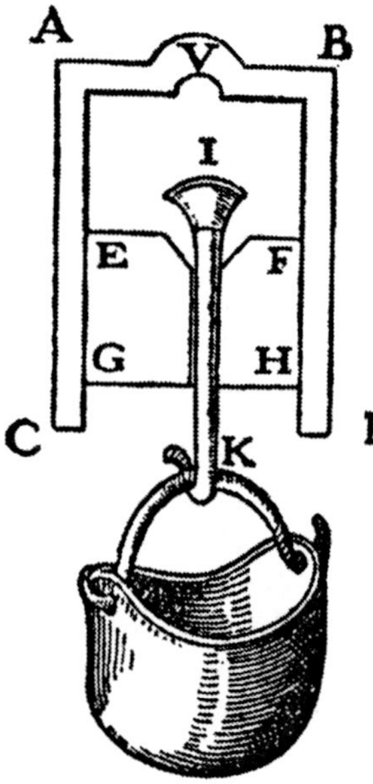

Figure 4. Galileo's experiment testing the 'force of the vacuum'. (From G. Galilei, *Discorsi e dimostrazioni matematiche intorno a due nuove scienze attenenti alla meccanica & i movimenti locali*, In Leida: Appresso gli Elsevirii, 1638, 15.)

the liquid in the water pump from rising beyond the height of approximately 10.5 m. Galileo expressed this view in Day One of *Two New Sciences* when he presented a hypothetical experiment with a piston to provide a calculation of the force that was resisting the piston from operating past a certain point (Figure 4). That force, he believed, could only be coming from the vacuum.[10] So Galileo still used a mathematical and mechanical demonstration to provide a quantifiable explanation of the vacuum, of its force, and why the water in a pump does not rise or fall beyond a certain point.[11]

Some of Galileo's colleagues and students, meanwhile, formed a slightly different point of view. For instance, Giovanni Baliani (1582–1666) agreed on the probable existence of the vacuum, although he claimed that it could only be created with difficulty. But he also believed that the pressure on the liquid came from the weight of the air rather than the force of the void.[12] 'We are under the

[10] Galileo talked about the 'resistenza del vacuo' in a letter to Baliani on 6 August 1630. Favaro (ed.), *Le Opere*, xiv, 127–130. He also expressed his support for the vacuum as early as 1612 in his Venetian publication: *De phenomenis in orbe Lunae*: 'se il vacuo non si può conoscer nè col senso nè coll'intelletto, come avete voi fatto a saper che non si dia?' *Idem.*, iii, 350.

[11] E.J. Dijksterhuis, *The Mechanisation of the World Picture*, (tr. C. Dikshourn), Oxford, 1969, 420–424.

[12] Middleton, *The History of the Barometer*, 9.

immensity [of air]', he explained to Galileo in a letter written in 1630, and 'the higher we go in the air the less heavy it is.'[13] In other words, it is only the pressure from this ocean of air that causes the liquid to rise in the pump, and the amount of pressure determined by altitude causes the liquid to rise only to a certain height.

In the meantime, there were some significant contributions being made to the debate in France and the United Provinces. According to Middleton, by 1618 Isaac Beeckman (1588–1637) had already proclaimed his belief in the elasticity, weight, and the compressibility of air, although he did not believe that the water pump could create a vacuum.[14] Soon afterwards, Gassendi not only believed in the atmospheric pressure exerted on bodies underneath the air, but also insisted that a large vacuous space could be created in instruments such as Torricelli's. Furthermore, despite the religious controversy that corpuscularian positions brought,[15] Gassendi openly employed the atomistic philosophies of the ancients to claim that tiny vacuous spaces existed between the particles that make up the different elements of nature.[16] Finally, René Descartes also expressed his alternative mechanistic opinion regarding the vacuum and the pressure of air. In 1631 he wrote a letter to an anonymous recipient declaring his acceptance of the notion that the pressure exerted by air sustained the level of water inside a pump. Yet he did not believe in the existence of the vacuum or any vacuous spaces between the particles of matter. Instead, he insisted that corpuscles of different sizes and densities occupied all spaces and may well be invisible to observers.[17] This is to suggest that a vacuum is not possible because the space in the water pump somehow always allowed tiny amounts of air to enter through the pores of the instrument or through the liquid. So Descartes believed that there was always some 'subtle matter' in the apparent empty spaces of these tubes, including Torricelli's barometer.

This was the intellectual environment in which Torricelli entered the debate on the pressure of air and the void. The proponents of each of the varying Aristotelian, Cartesian, Gassendian, and Galilean natural philosophies were asking themselves whether the vacuum could actually be created and whether it was the weight of the air that exerted pressure on the water inside the pump. Each of these groups interpreted the structure and movements of nature according to their own natural philosophical beliefs. During the late 1630s and early 1640s, Torricelli and others continued to construct and interpret empirical evidence to address their natural philosophical concerns. They were not only trying to support their commitments to a mechanical natural philosophy, mathematically articulated if possible, but were also attempting to destabilise the traditional, and still dominant, Aristotelian view on the topic.

[13] As translated by Middleton. *Ibid.* 'siamo nel fondo della sua [l'aria] immensità ... quanto l'aria è più alta, sia sempre più leggiera'. Favaro (ed.), *Le Opere*, xiv, 159.

[14] Middleton, *The History of the Barometer,* 6.

[15] See P. Redondi, *Galileo Heretic*, Princeton, 1987.

[16] Dijksterhuis, *The Mechanisation*, 426.

[17] M. Tamny, 'Atomism and the Mechanical Philosophy', in Olby et al. (eds.), 508.

1. TORRICELLI'S INTERPRETATION OF HIS BAROMETRIC INSTRUMENT

The natural philosophical tension surrounding Torricelli's work is evident in his report of the experiment in his letter to Ricci. Before launching into a description of the barometer, he clearly declares what he believed to be the theoretical aim of the experiment and the position that he adopted in the contemporary discussions on the pressure of air. He states that the purpose of the experiment was: 'not simply to produce a vacuum, but to make an instrument which might show the changes of the air, now heavier and coarser, now lighter and more subtle'.[18] He goes on to declare that this 'reasoning was confirmed by making the experiment'[19] with two barometers (Figure 5), the second, AE, with a larger vacuous area. If it were the vacuum that exerted pressure on the mercury, then the liquid would stop at different heights. Since the mercury, instead, reached the same height in both tubes, Torricelli claims 'that the force was not within'.[20] This went against Galileo's belief that the vacuum exerted a force on the liquid, and instead supported Baliani's notion regarding the exertion of pressure from the ocean of air above us.[21] So Torricelli's aim is clear from the start when he declares that he had certain theoretical expectations in support of the pressure of air before having constructed the instrument. Indeed, according to Carlo Dati, one of the Cimento academicians and the pseudonymous publisher of Torricelli's correspondence about the barometric experiment, Torricelli carried out his observations after he had speculated about the pressure of air. Dati states: 'Torricelli did not come across his experiment by chance, but was guided by a clear thought, and by the time he saw and experimented the effect, he had already speculated the cause.'[22]

When it came to expressing his beliefs regarding the possible creation of a vacuum, Torricelli claims that he was showing the existence of the void inside the barometer when the water in the basin rushed towards the space in the tube. In framing the significance of this experiment, he states:

> Many have said that [the vacuum] cannot happen; others that it happens, but with the repugnance of nature, and with difficulty. I really do not remember that anyone has said that it may occur with no difficulty and with no resistance from nature.[23]

In other words, Torricelli was attempting to strengthen his theory that the vacuum can be easily produced. In the process, he looked to refute both the Aristotelian

[18] 'non per far semplicemente il vacuo, ma per far uno strumento che mostrasse le mutazioni dell'aria, hora più grave e grossa, et hor più leggiera e sottile.' E. Torricelli, *Opere* (eds. G. Loria and G. Vasura), Faenza, 1919, iii, 186.

[19] 'Confermava il discorso l'esperienza fatta'. *Ibid.*

[20] 'la virtu non era dentro'. *Ibid.*

[21] According to Segre, this was Torricelli's principal aim: to design 'the barometer experiment to test previous theories rather than to generate new ones'. Segre, *In the Wake*, 87.

[22] Antiate, *Lettere a Filaleti*, 20. As cited by Segre, *In the Wake*, 87.

[23] 'Molti hanno detto che il vacuo non si dia, altri che si dia, ma con repugnanza della natura e con fatica; non so già che alcuno habbia detto che si dia senza fatica e senza resistenza della natura.' E. Torricelli, *Opere*, iii, 189.

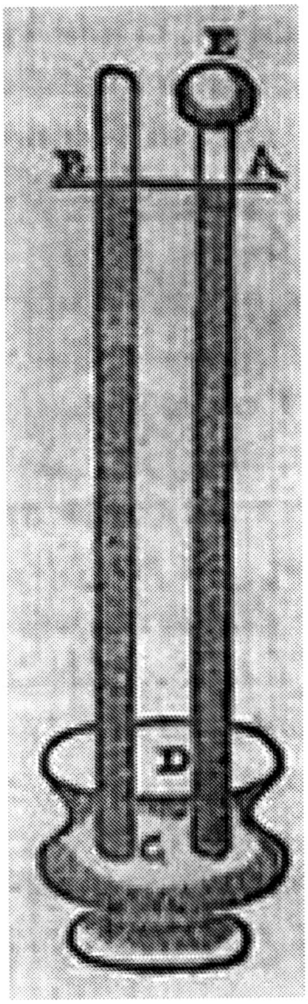

Figure 5. Torricelli's barometer testing the size of the vacuous space and the effect on the mercury. From E. Torricelli, *Opere* (eds. G. Loria and G. Vasura), Faenza, 1919, iii, 186.

claim that, since nature abhors the production of a vacuum, it simply 'cannot happen', and the opinion of Galileo and some of his followers, who regarded the vacuum as possible to produce, but 'with difficulty'. Now, despite Torricelli's opposition to Galileo on the supposed difficulty in creating a vacuum and the suggested force of the void on the limited height of the liquid, Torricelli was still insisting on an a priori experimental approach similar to Galileo's. That is, as we have just seen from his statement that his 'reasoning was confirmed by making the experiment', he was constructing an experiment that could provide a persuasive and authoritative presentation of his claim aimed against traditional Aristotelian views on the weight of air and the vacuum. Furthermore, through the construction of an instrument measuring the pressure of air, he was attempting to describe the mathematical and mechanical movements and measurements that he believed to be consistent in physics.

When it came to shaping his theoretical and natural philosophical expectations from his construction of the barometer, Torricelli was therefore calling upon mechanistic and anti-Aristotelian concerns of the period. He was, in fact, using and improving upon Galileo's natural philosophy, leaving scholastic opinion about the vacuum as his main target of criticism, and taking a mechanist view about the pressure of air as a central concern. Adding to this field of theoretical and natural philosophical contention entangled in the construction of the barometer, Middleton believes that Torricelli even acknowledged the Cartesian point of view.[24] Despite his belief that he had confirmed his theory by making this instrument, Torricelli still seemed to note, in a sceptical tone, that it was possible that the space in the tube only contained 'rarefied stuff', as was argued by Descartes and his followers.

This was the natural philosophical field of contention that continued to play through the construction, interpretation, and presentation of barometric experiments performed in France. During the 1640s and 1650s there was a great deal of activity, especially in Paris, as Descartes, Pecquet, Roberval, and Pascal, among many others, provided further debate regarding the capability of the Torricellian tube to measure the pressure of air and whether the space in the instrument was indeed vacuous.[25] So there existed a culture of Galilean, Cartesian, and Aristotelian discourses competing for widespread acceptance. Therefore, natural philosophy, rather than the strict application of an experimental method, was the main impetus behind the early modern pneumatic interests. Indeed, natural philosophical issues were entangled in the construction and use of Torricelli's instrument.

Some may argue that this was before the advent of the experimental, fact-making philosophy. For instance, Shapin and Schaffer claim that these theoretical debates over the barometer's function were 'a key example of scandal in natural philosophy' and how it was believed that factual knowledge was not achieved until such natural philosophical hypothesising had been dropped from experimental research.[26] Shapin and Schaffer claim that Robert Boyle (1627–1691) was the first to break away from offering any philosophies of knowledge or causal inquiries and instead came up with pure experimental matters of fact regarding air pressure.[27] Furthermore, according to accounts grounded in 'cultural' history, such experimental programmes were also adopted in seventeenth-century Tuscan institutions, exemplified by the Cimento, as the political and cultural advantages of supposedly neutral fact-making seemed to shift the knowledge-making landscape away from natural philosophising.[28] As has been stated earlier and as we must continue to see, particularly when we look at the Accademia's efforts to present their work in a theory-neutral manner in the *Saggi*, such 'cultural' studies are crucial to our understanding of the Cimento's foundations. However, apart from these social and political linkages we shall see that natural philosophising remained as a key concern for Galileo's followers, including the members of the Accademia del Cimento. So this case study is not an example of the

[24] Middleton, *The History of the Barometer*, 24.

[25] The competing natural philosophies of these men and how those philosophies were used in their barometric studies is discussed later in this chapter as we analyse the Cimento's retesting of the French pneumatic experiments.

[26] Shapin and Schaffer, 41–42.

[27] *Ibid.*, 49.

[28] Tribby, 'Dante's Restaurant', 320–321; Biagioli, 'Scientific revolution', 30; Findlen, 'Controlling the experiment', 39–41; Beretta, 136–137.

triumph of the 'new' seventeenth-century experimental philosophy. Instead, once we look into the natural philosophical aims of the Tuscan academicians and their French colleagues, we may recognise the type of cognitive interests, rather than an experimental method, that served as the catalyst behind the Cimento's work.

2. THE ACADEMICIANS' MECHANICAL UNDERSTANDING OF THE BAROMETER: WHAT THE *SAGGI* REVEALS

The first experiment discussed in the *Saggi* is the construction of Torricelli's barometer. The narration of the experiment is very similar to what we have already encountered in Torricelli's letter to Ricci, but our interests for the moment lie more in Magalotti's style of presentation. We have just seen Torricelli's report, where he admitted to constructing an instrument aimed at supporting his theoretical and natural philosophical concerns. In contrast to this a priori experimentation, Magalotti gave the impression to his readers that Torricelli used an inductive method. In keeping with the experimentalist rhetoric in the preface to the *Saggi*, Magalotti now wants to make it clear that Torricelli's and the Accademia's knowledge claims concerning the pressure of air derived purely from performing the experiment. In the opening sentence, Magalotti therefore suggests that Torricelli first constructed the instrument, and then reasoned upon the cause of the mercury's movement inside the tube:

> That famous experiment with the quicksilver that in 1643 presented itself before the great intellect of Torricelli is now known in every part of Europe, as is also the high and wonderful idea that he formed about it when he began to speculate upon the reason for it.[29]

Magalotti gives the reader the impression that factual knowledge is being attained purely through the use of an experimental programme. This rhetoric is seemingly supported by the air pressure and void experiments that follow the description of Torricelli's baromter in the *Saggi*, the majority of which were performed between late July 1657 until the closing months of 1658.[30] They included the repetition of some experiments performed by Torricelli himself,[31] Boyle,[32] Pascal,[33] and Roberval.[34] It is important to note how Magalotti gave the impression to his

[29] 'E' nota ormai per ogni parte d'Europa quella famosa esperienza dell'argentovivo, che l'anno 1643 si parò davanti al grande inteletto del Torricelli; e noto parimente è l'alto e maraviglioso pensiero che egli formò di essa, quand'ei ne prese a specular la ragione'. Magalotti, *Saggi*, 101. While Magalotti seemingly claimed that he performed the barometric experiment in 1643, according to Middleton, it was actually almost certainly carried out in 1644. Middleton, *The History of the Barometer*, 43.

[30] These experiments were recorded in the Accademia's unpublished diary, held in the Biblioteca Nazionale Centrale in Florence, in the folder labelled Ms. Gal. 262.

[31] Including placing animals inside the empty space of the barometer.

[32] Although Boyle was not mentioned in the *Saggi*'s first draft, he was regularly included in the subsequent versions written by Magalotti after Boyle's writings finally reached Italy.

[33] Including the climb of the Puy-de-Dôme, actually performed by Pascal's brother-in-law, to test the difference in the height the mercury reached at different altitudes and air pressures.

[34] Including several experiments with Torricelli's barometer, to test the pressure of air.

readers that they performed these experiments simply to fulfil their experimental aim of 'testing and retesting' the notions and experiments put forward by their colleagues in other parts of Europe.[35] Indeed, although there was considerable contention amongst some of the academicians about the interpretation of the experiments and whether they actually supported Torricelli's theory, Magalotti concludes in the academicians' report of their work, that Torricelli's claim concerning air pressure was actually proven to be true through the sheer weight of experimental evidence. In fact, at the conclusion of his narration of these experiments concerned with air pressure, Magalotti claims:

> Torricelli's concept of the pressure of the air on bodies beneath it now seemed well enough established by the series of experiments already described. Although it may be presumptuous and full of danger to make assertions about those things on which no lamp of Geometry shines to help our eyes, yet the presumption is never so excusable, nor the danger more certain to be avoided, than at the moment when, purely by way of many experiments all concordant, our intellect journeys to the attainment of its desire.[36]

This is another good example of how, in the presentation of the Cimento's work, Magalotti framed the construction of the Torricellian instrument within a rhetoric of experiment, dealing solely with theory-neutral artefacts, atheoretical matters of fact, and a loosely and vaguely articulated 'experimental method'. Although he was prepared to report the academicians' acceptance of Torricelli's claims regarding the pressure of air, this was supposedly based purely on the acquisition of facts through experimentation. Meanwhile, the notion of nature's abhorrence of the vacuum was a cornerstone of Aristotelian natural philosophy and to cast doubt on it publicly was a certain way of creating a great deal of controversy with scholastics and ecclesiastical authorities. Since the Medici patrons of the Cimento were unwilling to threaten the doctrines associated with the Catholic Church, as we shall see in Part Three, the acceptance of Torricelli's air pressure theory without explicit mention of corpuscles, was as far as Magalotti and his editors were willing to go. They would not dare to declare openly their corpuscularian and mechanistic beliefs regarding the causes of the mercury's movement or indeed the even more controversial anti-Aristotelian opinion that a vacuum was created inside the barometer. Magalotti therefore concludes the presentation of the Cimento's aims with the following words: 'It has been our intention only to discuss the space filled with mercury and to understand the true cause of the wonderful balancing of its weight, intending never to pick quarrels with those who oppose the vacuum.'[37]

[35] The rhetoric describing this approach was, of course, consistent with the Cimento's motto.

[36] 'Dalla serie delle narrate esperienze pareva oramai stabilito a bastanza il concetto del Torricelli, del premer dell'aria sopra le cose inferiori. Il che quantuque sia ardito e pieno di pericolo ad asserire di quelle cose ove a' nostr'occhi alcun lampo di Geometria non risplende, pure nè l'ardire è mai sì degno di scusa, nè 'l pericolo è più sicuro a chivarsi che allora che solamente per via di molte e tutte concordi esperienze cammina nostro intelletto al conseguimento del suo desiderio'. Magalotti, *Saggi*, 131.

[37] 'Conciossiacosachè sia stato solamente nostro intento discorrere sopra lo spazio pieno d'argento, ed intendere la vera cagione del maraviglioso libramento di quel peso, con animo di non imprender mai briga con gl'impugnatori del voto'. *Ibid.*, 105.

So, in these passages of the *Saggi*, Magalotti is fulfilling the aim he expresses in the Preface to tell about experiments with no hint of speculation. To show how far his account is from the field of natural philosophical contention in which the academicians were involved, I propose to pursue two arguments in what follows: First, despite the neutral style of the *Saggi*, there are strong hints in the text itself that natural philosophical aims and interests were present in the academicians' knowledge-making process. Second, the internal workings of the Cimento, as revealed in existing manuscripts and letters, show that the actual construction and interpretation of the experiments consisted of certain natural philosophical arguments competing for dominance in seventeenth-century Italy. Most of the academicians hoped that through the performance of these experiments they would be strengthening their anti-Aristotelian positions. Meanwhile, two members of the group voiced their objections to the mechanistic views of the pressure of air and the existence of the void. This indicates that far from representing a generation of new experimental scientists in the seventeenth century, the academicians were actually still very much making decisions, taking actions, and pursuing agendas regarding the structure and movements of nature according to their natural philosophical commitments.

3. FINDING EVIDENCE OF THE ACADEMICIANS' NATURAL PHILOSOPHICAL INTERESTS IN THE *SAGGI*

One passage in the *Saggi* that hints at the academicians' natural philosophical concerns, is the citation given above concluding on the certainty of air pressure 'by way of many experiments'.[38] The reader's attention may be immediately swept away with the experimentalist rhetoric dominating this paragraph, but one may also be curious about the obvious reference to the 'danger' surrounding those assertions that do not rely on the certainty of geometrical demonstrations – 'it may be presumptuous and full of danger to make assertions about these things on which no lamp of Geometry shines to help our eyes'. It is not by chance that Magalotti made this allusion to the 'lamp of Geometry' since nearly all the academicians regarded mathematical and geometrical demonstrations as the cornerstone of their natural philosophical pursuits. Indeed, the academicians were educated on the mechanistic example set by Galileo, Torricelli, and their colleagues. This, as we have already seen, was particularly the case with Borelli and Viviani, who based their careers on the successful restoration and application of ancient mathematical theories and their application to physical inquiries in opposition to Aristotelian natural philosophical beliefs. Furthermore, as we may recall from our analysis of Galileo's experimentalist image, as well as Torricelli's a priori construction of the barometer, the academicians were performing experiments to verify certain mathematical and geometrical beliefs. I am suggesting here that the above-mentioned reference to the 'lamp of Geometry' may be a clue in the *Saggi* that the academicians were following

[38] See p. 124, above.

the type of methodological approach we have seen in Galileo, Torricelli, Viviani, and Borelli. They were willing to perform multiple experiments on a topic, but ultimately, those experiments were constructed simply to verify their existing theories of a mechanical and physico-mathematical character.

The second clue we get to the academicians' cognitive interests is early in the *Saggi*'s section on pneumatics. In fact, immediately after Magalotti gives his opening phrase regarding 'that famous experiment with the quicksilver', and before he launches into a description of Torricelli's barometer, he provides a curious insight into the corpuscular reasoning behind the notion of the pressure of air.

> When we attempt to move solid bodies ... such as gravel, sand, and the like, or heaps of larger stones – they interfere with each other and pack together, thanks to the roughness and irregularities of their parts, in such a way that they hold and support each other so as to resist more strongly the force that is trying to remove them. Liquids, on the other hand – perhaps because of the slipperiness or the roundness of their very small corpuscles or from some other shape that may favour motion – though standing in equilibrium, yield in every direction and spread out as soon as they are pressed.[39]

For the first time in Magalotti's text, the reader is able to capture a glimpse of the academicians' corpuscularian natural philosophical concerns. It is evident from the above statement that they are interested in knowing the size, shape, mobility, and density of the corpuscles that they believe, because of their mechanistic natural philosophical backgrounds, are responsible for the movement of the mercury in the barometer. Furthermore, the suggestion that matter could move 'in every direction' was clearly opposed to the Aristotelian view that all terrestrial elements moved in vertical straight lines. It is strange to find such a corpuscularian statement openly expressed in the *Saggi*, and we can only assume that this paragraph somehow escaped the attention of the editors.[40] We shall continue to see this type of natural philosophical concern entangled in the academicians' construction and interpretation of barometric experiments.

[39] 'Poichè i corpi solidi, come verbigrazia la ghiaia sarebbe, la rena e simiglievoli, o pure le macíe de' sassi maggiori, nel far forza per muovergli anzi s'incastrano e stivansi insieme, congegnandosi per sí fatto modo mercè della scabrosità e irregolarità delle lor parti, e sì serrandosi in tutta la massa loro, ch'e' s'attengono l'un l'altro e puntellansi, onde più duramente resistono alla forza che tenta smuovergli. Ma al contrario i liquori, forse per lo liscio sfuggevole o per la rotondità de' lor minimi corpicelli o per latra figura ch'e' s'abbaiano inchinevole al moto, la qual mal posi e stia 'n bilico, via via che premuti sono, cedono per ogni verso e sparpagliansi'. Magalotti, *Saggi*, 101.

[40] Such hints about the natural philosophical concerns and contention inside the Cimento can be found on several occasions in the *Saggi*. These will be mentioned here and in the case study discussed in Chapter Six. However, the 'status' of these clues remains unclear; whether Magalotti intentionally wished to give his readers subtle reminders of the academicians' theoretical aims when experimenting. In this case, the first draft of the *Saggi* does not contain this introductory section to the academicians' baromoteric studies, and it is difficult to confirm who might have suggested its inclusion in the final published version. One may imagine that Rinaldini, an Aristotelian editor of the text, along with Marsili, one of the two Aristotelian voices in the Cimento opposing Borelli's and Viviani's mechanistic expressions, would have objected to this type of obvious reference to corpuscularianism.

4. 'EXPERIMENTS PERTAINING TO THE NATURAL PRESSURE OF THE AIR': ROBERVAL AND THE ARISTOTELIAN RESPONSE

On 2 August 1657, the author of the Cimento's diary reported how the academicians began their attempts to retest the French experiments regarding the pressure of air.[41] The first of these was Roberval's construction of a barometer inside a barometer, narrated in Jean Pecquet's 1651 publication, *Experimenta nova anatomica*.[42] A barometric tube, K was set up with its basin sitting inside the empty space of another barometer, D what we may call the 'outer' tube (Figure 3). Once the entire glass structure was filled with mercury, and the bottom of the outer tube was opened to create a normal barometric reading, the inner tube emptied itself completely of the liquid. Air was then allowed to enter into the empty space of the outer tube, causing the mercury to vacate that barometer completely while the liquid rose again to a regular height in the inner tube. So the mercury in the inner barometer set up inside the vacuous space only rose when air was allowed to enter that space and weigh down upon the liquid. Therefore, what Roberval and the academicians concluded was that the mercury rose to its normal height only when the weight of atmospheric air was permitted to press on the barometer.[43]

Pecquet claimed that after reading about this experiment, his audience 'should not continue to hold the opinion of the Ancients against the argument according to which the weight of the external air is balanced by the mercury inside'.[44] It can be clearly seen from Pecquet's words that there were some natural philosophical beliefs aimed against scholastic thought that were entangled in the construction of Roberval's work, and in proving the pressure of air. By taking Pequet's report of Roberval's experiment as a guide to the Cimento academicians' own work, we can be certain that they recognised the anti-Aristotelian value of their experiments. Indeed, while we receive no more detail in the *Saggi* regarding the theoretical significance of this experience, the Cimento academicians also played upon these natural philosophical concerns in the subsequent barometric experiments.

41 'Experiment referred to by Mr. Pecquet in the book containing his anatomical dissertations, in favour of the pressure of the air on bodies beneath it, and verified in our academy in the following way'. Abetti and Pagnini (eds.), 282. BNCF, Ms. Gal. 262, f. 22v. The Cimento's official diary was kept by Alessandro Segni until May 1660, when Lorenzo Magalotti replaced Segni as secretary. As has already been mentioned, separate diaries were also kept by Viviani and Rinaldini. See Chapter One, n. 37.

42 Although the academicians ascribed the experiment to Roberval, Pecquet suggested that it was Adrien Auzout's invention.

43 It must be noted how in this section and in some of the following experiments, it was assumed that the vacuum existed, or at least that the space consisted of extremely rarefied air. However, for the French thinkers, as well as our academicians, such assumptions, as we shall see later, were still laden with some very contentious natural philosophical arguments. In fact, the question of the vacuum became the main concern for these natural philosophers. According to Dear, the debate regarding the weight of the air was even of a 'secondary concern' to that of the vacuum. Dear, *Discipline and Experience*, 189.

44 J. Pecquet, *Experimenta nova anatomica*, Paris, 1651, 56. As cited by Middleton, *The History of the Barometer*, 49.

Immediately following this re-creation of Roberval's demonstration of air pressure, a determined attempt was made by the group's Aristotelians to counter the mechanist explanations of the barometer.[45] This refers to two experiments performed by the academicians on 4 and 6 August 1657. The first proposed placing a glass jar over the barometer, supposedly keeping the full weight of the atmosphere's air away from the instrument (Figure 6).

The second had the mercury in the vase sealed tightly covered so as to protect the liquid from the surrounding air. It was assumed that if the air-pressure theory were true, the mercury, under protection from the great weight of air by the glass jar would not rise to its usual height. Interestingly, these experiments found their way into the *Saggi* under the following title: 'Experiments adduced by some people against the pressure of air'.[46] Magalotti presented them by framing their theoretical significance:

> They [some people] persuaded themselves, that if it were true that it was the weight of all the region of air above that drove the quicksilver up the tube ..., then, if the stagnant quicksilver were protected by a glass wall from such great pressure, the imperceptible weight of what little air is included under the bell jar ought to remain unable to keep the mercury at the same height as that to which the weight of such a vast region of air had pushed it.[47]

Magalotti's narration of these activities clearly includes a prediction based on the theoretical conviction of those who constructed the described experiments. As it turned out, the liquid rose in the enclosed barometers to its usual measurement, suggesting that the Aristotelians may have been correct in their assertion that it is not the pressure created by the weight of the air that balances the height of the column of mercury.[48] But far from conceding the point, the mechanists in the Accademia, 'those who adhere to the doctrine of the pressure of the air', as Magalotti describes them, still claimed that 'the phenomena just recounted, far from contradicting their opinion, favoured it wonderfully' since this time the effect was believed to be caused by the compression of the air, rather than the weight of the air.[49]

> For, according to them, the immediate reason for the mercury being violently pushed to a height of an ell and a quarter and held there is not really the weight of

[45] Magalotti refers to them as 'some people' (see note 47, below) In a moment we shall see who these members of the Cimento were.

[46] 'Apportato da alcuni contro alla pressione dell'aria', Magalotti, *Saggi*, 108.

[47] 'Si persuadevano adunque, che se fosse vero che il peso di tutta la soprastante regione aerea pignesse l'argentovivo su per la canna, e col peso di esso s'equilibrasse, difendendosi quivi con l'argine del cristallo l'argentovivo stanante da così gran pressione, devrebbe l'insensibil peso della poc'aria rinchiusa sotto la campana rimanere inabile a mentener l'argento a quella medesima altezza alla quale il momento di così vasta regione d'aria l'avea sospinto'. *Ibid.*

[48] This is not to suggest that these Aristotelians were proposing an alternative explanation of the effects created by Torricelli's instrument. Rather, since the weight of air was not possible according to scholastics, they simply could not accept the efficacy of any experiment that attempted to demonstrate its existence. Middleton, *The History of the Barometer*, 5.

[49] 'Ma quelli che aderivano alla pressione dell'aria, respondevano a questa esperienze con dire, che i narrati avvenimenti, anzi di contraria, favorivano mirabilmente la loro opinione', Magalotti, *Saggi*, 110.

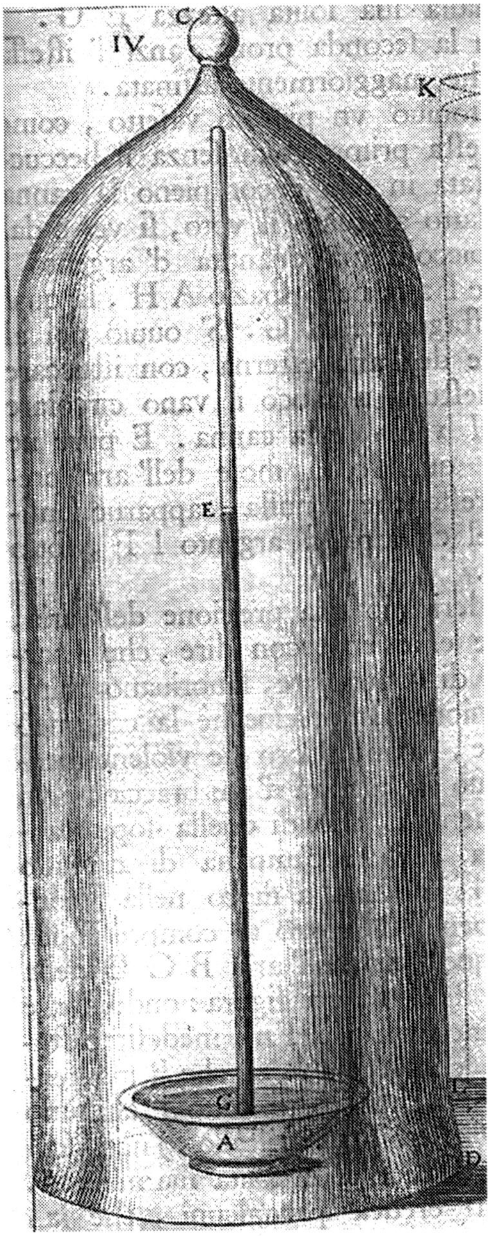

Figure 6. A jar placed by the academicians over the barometer was intended to test if the mercury would rise to its usual height in the tube even if the instrument were protected from the great weight of air. The second experiment testing this possibility placed a cover only over the vase. L. Magalotti, *Saggi di naturali esperienze*, Florence, 1667, 35. Courtesy of the IMSS Biblioteca Digitale.

> the superincumbent air, which is taken away by the bell jar in the first experiment and by the cement in the second, but on the contrary, the compression that had been produced by this weight .... It is therefore not to be wondered at that as the same state of compression is maintained, the elevation of the quicksilver does not fall below its usual measure.[50]

This is to say that the air that remained inside the vases enclosing the barometer dilated to create the same pressurising effect on the mercury. In other words, the compression caused by the height of the air above us is not lost when some air is trapped inside a vase. For the first time in the *Saggi*, the reader is introduced to the notion of the compressibility of the air. For readers of the Cimento's diary, however, it would seem that this notion had always been on the academicians' minds. For example, on 2 August 1657, when Alessandro Segni, the Cimento's secretary at the time, mentioned the academicians' re-creation of the French experiments in the diary, he also referred to the compression, rather than the simple weight of air.[51]

What is implied by the word 'compression' as opposed to 'weight'? Here we may take a quick glance back at the academicians' predecessors in the field of pneumatics. Isaac Beeckman was the first, according to Middleton, to suggest that air was like a sponge that is condensed near the ground by the sheer weight of all the air above it, but somewhat more dilated at higher altitude.[52] Descartes made a similar analogy using another compressible material, wool.[53] This was the same analogy Torricelli used in 1644 in his second letter to Ricci on the subject of the new instrument for measuring air pressure.[54] Furthermore, according to Middleton, it was this same notion of the compressibility and elasticity of the air that led Roberval to construct the above-mentioned 'void in a void' experiment, repeated by the academicians, and eventually conclude that it was the atmospheric pressure that balanced the weight of the mercury, rather than any force associated with the apparent void.[55] Indeed, Pecquet even used the wool analogy to describe this experiment. This was a very mechanistic notion because it strongly implied a corpuscularian structure of the universe and completely denied that any type of mystical attraction or repulsion caused the movement of the

[50] 'Imparciocchè la cagione immediata che pigne, secondo loro, e violentemente sostiene l'argentovivo all'altezza d'un braccio e un quarto, non è altrimenti il peso di quella sprastante aria che si leva con la campana di cristallo nella prima, e con la mestura a fuoco nella seconda esperienza; ma ben si l'effetto di compressione che fu prodotto da quel peso nell'aria ... onde non è maraviglia, che mantenendosi quella nel medesimo stato di compressione, non iscemi l'altezza dell'argenotvivo dalla solita sua misura'. *Ibid.*

[51] 'Si diede principio alle esperienze addotte dai Franzesi, ed altre aggiunte di nuovo nella questione della compressione dell'aria nei corpi inferiori'. BNCF, Ms. Gal. 262, f. 22v.

[52] Middleton, *The History of the Barometer*, 6.

[53] *Ibid.*, 7.

[54] *Ibid.*, 26. Although Torricelli and others used the same analogy of wool as Descartes, and although also Cartesian natural philosophers believed in the mechanistic notion of the compressibility of air, as we have already seen, they differed significantly in their views regarding the vacuum, to the other mechanists, such as Torricelli, Pascal, Roberval, and of course Borelli and Viviani.

[55] *Ibid.*, 49.

mercury, such as the Aristotelian suggestion that nature's abhorrence of the vacuum forces the mercury to rise in order to partly fill the tube. So references to the compression of air carried with them some weighty implications against Aristotelians. Moreover, the Cimento diary, as we have just seen, hints that this mechanistic concern was precisely what was on the academicians' minds as they embarked on their reconstruction of past barometric experiments.

The scholastics in the group were trying to prove an Aristotelian point, but the mechanists interpreted the experiment in favour of their own agenda. So, although there are no direct references in the *Saggi* to an anti-Aristotelian agenda, the implications surrounding the construction and interpretation of these experiments were that they were interested in promoting a mechanistic natural philosophy and dismissing scholastic opinions against the pressure of the air. But this natural philosophical contention certainly did not end there. There is still much to be learnt about the academicians' cognitive aims and interests from looking at their repetition of one other important French experiment, and the disputes that surrounded its interpretation.

5. 'EXPERIMENTS PERTAINING TO THE NATURAL PRESSURE OF THE AIR': RECREATING THE PUY-DE-DÔME EXPERIMENT

In 1648, Blaise Pascal orchestrated an experiment in which his brother-in-law, Florin Perier, climbed the Puy-de-Dôme with the barometer. As Perier ascended the mountain, he set up a barometer at different altitudes to measure whether the mercury in the tube would fall. Pascal and Perier obtained the following results: 'between the heights of the quicksilver in these two experiments, there was a difference of three inches and one-and-a-half lines'.[56] This was thus an experience not only verifying Torricelli's theory of the pressure of the air, but also obtaining a precise measurement of the height the mercury reaches at different altitudes. In Pascal's 1663 publication, *Traités de l'équilibre des liquers et de la pesanteur de la masse de l'air*, he even compiled tables indicating the changes in the height of the column of mercury as a result of air pressure.[57] So the barometer was finally being used to fulfil the instrumental role that Torricelli had intended for it, but only because the construction of Pascal's experiment was based on a mathematical and mechanical world view. We could draw similar conclusions from looking at the Cimento's attempt to recreate Pascal's demonstration of air pressure.

During the last weeks of September 1657, the Accademia del Cimento attempted to replicate Pascal's calculations. While the Medici Court was away from Florence at the nearby town of Artiminio, it is believed that Prince Leopoldo himself attempted observations similar to Pascal's by carrying the barometer up a hill.[58] Meanwhile,

56 'entre le hauters du vis-argent de ces deux experiences, il y eut trios pouces une ligne et demie de difference.' B. Pascal, 'Recit de la grande expérience de l'equilibre des liqueurs', in *Oeuvres completes* (ed. J. Mesnard), Paris, 1964, ii, 683.

57 This reflects Pascal's use of mixed mathematics in his knowledge claims. For a sophisticated analysis of the mathematical background to Pascal's representation of the experiment, see Dear, *Discipline and Experience*, 180–208.

58 The Court departed Florence on the 24th and left the Accademia officially in suspension until 3 October. All this is reported briefly in the diary manuscript. BNCF, Ms. Gal. 262, ff. 34v-35v.

back in Florence, Borelli also retested Pascal's experiment by taking the barometer to the top 'of one of the highest towers in the city'.[59] Both occasions were mentioned in the diary where the range in barometric readings was recorded as consistent. That is, the level of the mercury always varied in perfect proportion to the height that it was taken. As Magalotti reveals in the *Saggi*, they did witness significant variations in the readings, but these were considered to have 'occurred only because of the changes between hot and cold weather'.[60] So Magalotti concludes:

> Observations made in this way put it into the minds of some to make such an instrument serve as a very exact meter of the state of compression of the air, believing that the various heights of the cylinder of mercury ... ought to show without fail the changing pressure that it has on the stagnant surface ... thanks to the differing heights that it has in its region.[61]

Another clue is provided in this passage from the *Saggi* that a great deal was at stake in the construction of these experiments. The range in barometric readings on both occasions was believed to support Pascal's and Torricelli's claim that the instrument provided a 'very exact meter' for the pressure of air. So the academicians held the same mathematical concerns as Pascal. After all, Borelli twice recorded his performance of the experiment in Florence showing an appreciation for precise measurements and understanding the movement of the liquid according to the proportions of density and weight between air and mercury. On 26 September 1657, Borelli provided the following description of the experiment to Leopoldo:

> Having noted the degree that the mercury reached on the plain of the Ombrone, I would have walked up until the mercury had fallen exactly only one degree in the tube; and a firm mark had been placed here, like a stick fitted into the ground, or something else to be able later to measure comfortably the perpendicular height from this point to the level of the Ombrone.... Continuing to walk up, the spot will be similarly marked where the mercury falls exactly another degree. In this way, we shall realise that at unequal heights from the bottom level, the mercury falls in equal parts. ... I believe that next we would be able to presume the ultimate height of the atmosphere of air.[62]

[59] The diary reveals that this tower was that of the Palazzo Vecchio. BNCF, Ms. Gal. 262, f. 35r. 'una delle più alte torri di Firenze'. Magalotti, *Saggi*, 124. Borelli narrated this experiment in a letter to Leopoldo on 26 September 1657. Fabroni, *Lettere inedite*, ii, 62. According to Tozzetti, Borelli performed this experiment under the request of the Grand Duke. Tozzetti, *Notizie*, i, 206. In 1670, Borelli published his observations: 'Idipsum postea observavimus Florentia in altissima Turri Palatil in qua ascesnsis solummodo cubitis quinquaginta supra insimum Plateam, Palatii Atrium, depressus apparait Mercurius spatio unius gradus'. G. Borelli, *De motionibus naturalibus*, 238.

[60] 'che per la sola diversa temperie di caldo e di freddo accadevano'. Magalotti, *Saggi*, 124.

[61] 'Così fatta osservazione fece animo ad alcuni d'aversi a valere d'un tale strumento per misuratore esattissimo dello stato di compressione dell'aria, credendosi che le varie altezze del cilindro d'argento ... dovessero dimostrare senz'alcun fallo il deverso premere ch'ella fa sopra il livello stragnante ..., mercè delle diverse altezze che ell'à in sua regione'. *Ibid.*, 124.

[62] 'Notato nel piano dell'Ombrone il grado nel quale si solleva l'argento vivo, vorrei si camminasse all'insù sintanto che l'argento vivo calasse un sol grado precisamente del detto cannello; e quivi si ponesse un segno stabile, come un palo fitto in terra, o altra cosa per potere dopo comodamente misurare l'altezza perpendicolare da questo luogo al piano dell'Ombrone, la quale, posto che sia 100 braccia, seguitando a camminare all'insù, si contrassegni similmente il luogo, dove l'argento vivo cava un altro grado precisamente, e così appresso. Ci accorgeremo in questa maniera, che in altezze disguali dall'infino piano va calando l'argento vivo in parti eguali. Or se questa esperienza fosse fatta squisitamente, credo che assai prossimamente si potrebbe congetturare quanta è l'altezza suprema della sfera d'aria'. As cited by Fabroni, *Lettere inedite*, ii, 62.

Borelli's desire to provide precise measurements is quite evident here. He also recorded his performance of the experiment in Florence in his *De motionibus naturalibus* and claimed to have been quite careful in recording his measurements so that they were perfect accounts of the rise of the liquid in the barometer in proportion to altitude.[63] Clearly, the academicians were not only attempting to verify Torricelli's theory of the pressure of the air, but they were also constructing their experiments according to their mathematical and mechanical natural philosophical aims and interests. Not only is there a hint of this in the *Saggi*, but Borelli and the group's mechanists were also clearly interested in pursuing their work according to their natural philosophical backgrounds.

6. CONTROVERSY AND CONFLICT INSIDE THE ACCADEMIA DEL CIMENTO

Despite the reported certainty at having proven the pressure of the air, Magalotti was notably more cautious when dealing with the vacuum. Here, the author of the *Saggi* claims that the intentions of the academicians were 'never to pick quarrels with those who oppose the vacuum'.[64] So despite performing a number of experiments concerned with the existence of the void in Torricelli's barometer, Magalotti states:

> We do not presume to exclude from the space fire or light or aether or other very tenuous substances, either finely distributed with very small empty spaces between or filling the whole of the space that is called empty, as some would have it.[65]

Magalotti here seemed to be admitting the possibility that the space in the Torricellian tube may not be a void, that it could contain 'tenuous substances'. But such an ambiguous statement does not provide any real clue about the natural philosophical issues that were actually at stake in the debates over the vacuum and that dominated the academicians' meetings. We have already captured a glimpse of the Accademia's natural philosophical agenda regarding air pressure and now we shall continue to look further into their internal workings in order to appreciate how their cognitive interests remained entangled in their concern with the vacuum. The void had created a great deal of natural philosophical controversy for the French thinkers, and undoubtedly this concern was carried on by the academicians. In the meantime, it will be evident from our analysis of certain passages from the *Saggi*, such as the one above, that the Cimento accommodated objections to the void from Aristotelians in the presentation of their experiments, while attempting to conceal the mechanistic agenda of the Accademia's leading contributors.

By 1648, after Perier climbed the Puy-de-Dôme with the barometer, it was widely agreed that Torricelli's barometer could successfully measure the weight

[63] Borelli, *De motionibus naturalibus*, 238.
[64] See p. 124, above.
[65] 'Non si presume già d'escluderne o'l fuoco o la luce o l'etere o altre sottilissime sustanze le quali, o in parte con finissimo spargimento di minimi spazzi vacui, o in tutto quello spazio che si chiama voto impiendo, altri vi vogliono'. Magalotti, *Saggi*, 105.

of air. Nevertheless, whether the space in the tube was vacuous remained a topic of great natural philosophical controversy. As an example, we should recall that the opposing camps of Aristotelians and Cartesian mechanists both provided different plenist explanations of the barometer. Aristotelians continued to believe that the air was rarefied in the tube, while Descartes claimed that the space contained 'subtle matter'. Meanwhile, other mechanists and corpuscularians, including our academicians, as well as the French mathematicians Pascal and Roberval, called upon the corpuscularian philosophies of Epicurus, Democritus and Lucretius, together with their skills in mathematics and geometry, to maintain that the space was indeed vacuous.

According to Middleton, French thinkers such as Pascal, Petit, and Perier speculated whether any matter entered the space through the glass or the mercury, as Cartesian mechanists suggested. But they believed that if this were the case, then plenty of matter should be continually entering the tube, thus continuing to push the mercury down. So Pascal concluded in his *Expériences nouvelles touchant le vuide*, that 'none of the substances that can be perceived by the senses, or of which we have knowledge, fill this apparently empty space ... that it is truly empty, and destitute of all matter'.[66] Meanwhile, Cartesian and Aristotelian philosophers still purported to show that one way or another, subtle matter or air remained in the tube. As an example, Roberval created the barometer with a bladder inside the Torricellian space. The bladder inflated as the apparent void was made. For Roberval, this supported the notion of the pressure of air, as it did for the mechanist academicians who repeated this experiment in August 1657 and had it published in the *Saggi*.[67] Yet, Cartesian mechanists, although in support of the pressure of air, still suggested that the bladder inflated because of the 'subtle matter' that filled the Torricellian space.[68]

While Descartes' explanations regarding the impossibility of creating a vacuum may have been viable for many French thinkers, it is unlikely that he had the same authority in Italy.[69] As we have already seen, in the *Saggi* Magalotti vaguely acknowledged the possibility of the presence of some type of 'subtle matter' in the space of the Torricellian tube, so as not to quarrel with scholastic and Cartesian plenists. But this was as close as the Cimento came to acknowledging Descartes'

[66] B. Pascal, *Expériences nouvelles touchant le vuide*, Paris, 1647, 73. As cited by Middleton, *The History of the Baromoter*, 44.

[67] The experiment was performed by the academicians on 9 August 1657. BNCF, Ms. Gal. 262, ff. 24r-v. and also published in the *Saggi*. Magalotti, *Saggi*, 106.

[68] Middleton, *The History of the Baromoter*, 49.

[69] Despite the struggles that Cartesian natural philosophers faced in gaining acceptance during the mid to late seventeenth century, Descartes' mechanical philosophy was not forgotten. For his supporters, the vortex theory continued to provide reason for doubting the vacuity of the Torricellian space. Just like the Aristotelian view, it could not be denied convincingly by all the barometric experiments performed at that time. An example of how seriously Cartesian mechanism continued to be considered in Paris, and how such natural philosophical issues did not die down with the advent of Boylean experimental philosophy, was the work by Jacques Rohault who introduced Cartesian principles to those who attended his weekly meetings and then provided experimental evidence in support of those principles. Desmond Clarke provided an excellent account of how Rohault and others in late seventeenth-century Paris continued to disseminate Cartesian natural philosophy after Descartes' death. Clarke, *Occult Powers and Hypotheses*, 18.

theories on space and matter. In all their discussions about Torricelli's barometer, they never seemed willing to accept Descartes' 'subtle matter'. In fact, in July 1660, Magalotti wrote to Ricci, criticising the Cartesian stance against the void.[70]

In the meantime, inside the Accademia, the debate about the vacuum was centred solely on the Aristotelian and post-Galilean corpuscularian/mechanist positions and experiments were performed and interpreted according to these competing natural philosophical beliefs. Viviani and Borelli, both corpuscularians and firm believers in the vacuist theory, argued in the Accademia's manuscripts that the results of their barometric experiments complied with their natural philosophical beliefs. Borelli in particular, strenuously contended that from the movement of the liquid in the Torricellian barometer, and from the mathematical, corpuscularian and mechanical principles taught to them by the ancient authors, the space had to be vacuous. In the meantime, Rinaldini and Marsili remained unconvinced by mechanical explanations regarding the cause of the mercury's movement in the tube and the description of the space in the barometer as vacuous. They were determined to defend the Aristotelian view which maintained the plenist argument that nature simply abhorred the production of a vacuum and that the space remained full of either air or some sort of tenuous substance – no experiment, they argued, could prove otherwise.

Rinaldini and Marsili put forward arguments that give good reason to believe that Magalotti was referring to these two members of the Accademia when he mentioned 'those people' to have come up with some experiments in opposition to the notion of the pressure of the air. Our argument therefore now turns to the primary sources regarding the Accademia del Cimento's cognitive interests and conflicts. These are the sources that are not mentioned in the recent literature concerned with the cultural history of early modern Tuscan natural philosophy, but it is important to note that most of the arguments put forward here regarding the academicians' disputes, were discussed by Paolo Galluzzi as early as 1981. Galluzzi cited from the academicians' letters and manuscripts to show how conflicting 'principles' were heavily involved in their barometric experiments.[71] Furthermore, it will be argued that the Accademia's members contended about

[70] Fabroni, *Lettere inedite*, i, 88. Borelli was also quite critical of Cartesian natural philosophy throughout most of his career, despite adopting a physiology actually quite similar to Descartes'. See Marquet (tr.), *On the Movement of Animals*, 426–427; Galluzzi, 'G.A. Borelli', 346. The problem for Descartes, according to Desmond Clarke, was that he could not provide convincing mathematical calculations for his 'subtle matter'. In other words, Descartes' metaphysical foundations for natural philosophising may not have seemed adequate for those who demanded more mathematical and geometrical demonstrations. This may sound like a strange accusation to make against Descartes considering his belief that mathematics was the only reliable source of knowledge making. However, Clarke claims that when it came to explaining the existence of the vacuum, Descartes and his followers virtually painted themselves into a corner by making strict distinctions when defining the properties of space and matter. That is, Cartesian theories about matter and extension demanded that some 'subtle matter' had to be completely filling the space. It therefore followed that that matter had to be lighter than the atmospheric air supporting the mercury. This meant that Descartes and his followers had to hypothesise about the 'heaviness' of the matter aether, but they had no way of defining its density, one of the so-called primary qualities for characterising nature's properties according to Descartes. Clarke, *Occult Powers and Hypotheses*, 76.

[71] Galluzzi, 'L'Accademia del Cimento', 803–811.

the interpretation of the experiments according to the contrasting natural philosophical positions of that time.

The difference in natural philosophical approach inside the Cimento came to the surface as the experiments on air pressure and the void were being carried out. Viviani and Borelli found themselves confronting the peripatetic arguments of the only two Aristotelians in the group, Marsili, and Rinaldini. The first clear sign of Borelli's frustration with scholastic opposition in Florence comes from his letters to Paolo del Buono and Viviani during the last three months of 1657. As an indication of how significant natural philosophical issues were to the dynamics of the Cimento's internal workings, it took Borelli no more than one year in Tuscany, and only some three months collaborating with fellow academicians, to show his impatience with the group's Aristotelian sympathisers. Since late July of that year the academicians had been occupying much of their time with barometric experiments and by September they had retested Pascal's demonstration of air pressure on two occasions, a culminating point in their support for Torricelli's theory. They even rejected the attempts in August to disprove experimentally the pressure of air. Yet this was still not enough evidence for Marsili and Rinaldini to abandon their Aristotelian interests. In a letter to Paolo del Buono, written on 10 October 1657, Borelli showed his frustration with what he believed to be 'disorder' inside the Accademia caused by the arguments coming from the group's peripatetics:

> Regarding our Accademia, which you call lycée, I wish that the laws that you imagine were in place; but the unfortunate thing is that all that is found is disorder; and this is because of the ambitions of one of the academicians, a rotten and mouldy peripatetic, who wants to appear in the gowns of a free and sincere philosopher. ... I have a very great desire for these few days of October to pass quickly so that I may return to Pisa, and there occupy my time advancing the studies of my liking.[72]

There is little doubt from this letter that Borelli's frustration was aimed at the persistent Aristotelian opposition coming from one of the members of the Cimento. The participation of scholastic natural philosophers, according to Borelli, was not allowing for the Accademia's progress, and was even a waste of his own time. He confided his frustrations to fellow moderns in the group, including del Buono, and later Viviani.

On 28 December Borelli wrote to Viviani stating that the peripatetics in the group were 'denying the compression of the air on the mercury, something which should by now be admitted by any stubborn mind'.[73] Clearly, the opposition from Marsili and Rinaldini was quite determined and these natural philosophical concerns were continually playing through the minds of the academicians as they constructed and interpreted their air pressure and void experiments. Perhaps the

[72] 'Intorno alla nostra Accademia, che Ella chiama Liceo, vorrei che in essa avessero luogo le Leggi da VS immaginate; ma il male è che solamente vi si trovano i disordini; e questo dipende dalla troppa ambizione di alcuno degli Accademici, il quale essendo Peripatetcio marcio e muffo, vuol comparire con una toga tolta in prestito di Filosofo libero e sincero ... sto con grandissimo desiderio che passino presto questi pochi giorni d'Ottobre, per andermene a Pisa, e quivi occupare il tempo che mi avanzerà, in studi di mio gusto'. Tozzetti, *Notizie*, i, 440; Fabroni, *Lettere inedite*, i, 94.

[73] BNCF, Ms. Gal. 283, f. 37r; Galluzzi, 'L'Accademia del Cimento', 807.

strongest piece of evidence demonstrating this contentious natural philosophical culture entangled in the Cimento's activities, is Borelli's statement: 'one cannot expect any profit whatsoever, nor can we ever walk together in agreement along the path of philosophical speculation, when we are so opposed in our very principles'.[74] This is the manuscript evidence to which Galluzzi also refers when discussing the academicians' competing 'principles' (or natural philosophical commitments), as the basis for: 'confrontation inside the Accademia between Aristotelians and innovators'.[75] Therefore, in a blow against historiographies discussing the rise of an atheoretical experimental philosophy inside the Accademia del Cimento, Borelli was saying not only that natural philosophical theorising was a crucial part of the academicians' work under Medici patronage, but also that they contended over the significance of their experiments according to those competing natural philosophical commitments.

7. MARSILI'S DEFENCE OF THE PLENUM

From June 1660 to late 1662, the academicians made several observations of natural phenomena inside the apparent void, such as the movement of heat and smoke, whether they could detect the sound of a ringing bell inside the void or the transmission of light through the apparent empty space. They even tested the reaction of many different animals placed inside the barometer. On every occasion in the *Saggi*, Magalotti gave no indication of the natural philosophical issues at stake in these observations. But it is worth noting that these were all experiments performed or suggested by the academicians' predecessors. In particular, since Aristotelians claimed that a vacuum would not be able to transmit certain qualities that we normally see in nature, seventeenth-century studies in pneumatics had attempted to prove whether this was indeed true with the space in the barometer. As an example, when Gasparo Berti (d. 1642) attempted the construction of a water barometer he faced criticism from Aristotelians against the void. Two Jesuits who witnessed Berti's work questioned whether the space created in the long tube was vacuous. As Emmanuel Maignan's narration of this episode reveals, it was believed that qualities such as light and sound that contain properties transmitted though air, should not be felt through a void.[76] This was why Berti attached a bell inside the empty space of his instrument. The bell was later also used by the academicians for the same natural philosophical purpose of refuting scholastic criticisms against the vacuum.[77] Similarly, Torricelli himself

[74] 'non si può sperare frutto nessuno, ne possiamo mai camminare d'accordo nel corso delle specu-lazioni filosofiche quando siamo tanto contrari ne' princìpi stessi'. BNCF, Ms. Gal. 283, f. 27v; Galluzzi, 'L'Accademia del Cimento', 807.

[75] Galluzzi, 'L'Accademia del Cimento', 807.

[76] Middleton, *The History of the Baromoter*, 13–14.

[77] Unfortunately for both Berti and the academicians, this experiment did not work because the bell created vibrations within the glassware that transmitted the sound, regardless of what was, or was not, in the space.

attempted to witness the reaction of animals inside the space of the tube and prove that he had created a vacuum.[78]

All of these experiments, therefore, had natural philosophical agendas embedded in their original construction and their subsequent re-creations by the Accademia del Cimento. But rather than focus on these experiments published in the *Saggi* under a façade of the Cimento's atheoretical rhetoric, it will be far more revealing to investigate the construction and interpretation of an experiment suggested by Alessandro Marsili which was excluded from the final publication. Although this experiment was originally intended for publication amongst those performed 'inside the vacuum', it was actually constructed by Marsili *against* the notion of the vacuum inside the barometer.

In the academicians' diary, the entry for 13 August 1660, while the group was performing many of the above mentioned experiments inside the apparent vacuum, mentions that 'Alessandro Marsili proposed an instrument to investigate whether the void left by the mercury was refilled by the evaporation from the mercury itself.' (Figure 7)[79] This instrument involved a rather complicated construction, including a bladder, D, fastened at the end of a tube. FMC, and two spouts emerging from that tube, IL, adjoined to a bulb placed over the bladder, AE. Marsili was hoping to see the bladder inflate after the void was constructed. As the diary mentions: 'If from the sustained column of mercury, a mercurial substance should evaporate to refill the empty space ... the bladder would have to inflate.'[80] This experiment was included in an early draft of the *Saggi*, providing a narration similar to that in the diary, but in Magalotti's draft, he elaborated on the 'mercurial substance'. That is, Magalotti mentioned that they could have expected 'insensible breaths of very tenuous exhalations'.[81] Importantly, as was stated in the diary and in Magalotti's draft, they had to ensure that no air could possibly enter the instrument, otherwise they could easily deduce that the inflation of the bladder was caused by the dilatation of the air entering the tube. Marsili's aim then was clearly to demonstrate that the mercury emitted a 'tenuous' substance and therefore the space repeatedly created by the academicians with Torricelli's barometer, could never be vacuous.

Both Viviani and Borelli were critical of the clumsy construction of this instrument. Viviani mentioned that Marsili's tenuous substance would also rise through the spouts on the tube and thus that the experiment would not work without somehow cutting these off after the vacuum was created. Borelli questioned whether the mercurial emissions would pass through the pores of the thin bladder. The difficult construction and execution of Marsili's experiment was obviously a concern for the academicians. This could begin to explain its eventual

[78] This did not work for Torricelli either since animals died as they passed through the mercury. The Cimento, meanwhile, had more success here since they practiced how to seal the barometer with a bladder at the end where they introduced the animals.

[79] 'Propose il Signore Alessandro Marsili un instrumento per assicurare se i vuoti lasciati dall'argento vivo fossero ripieni dalle evaporazioni dell'istesso argento'. BNCF, Ms. Gal. 262, f. 104v.

[80] 'Se dunque dal cilindro d'argento sostenuto svaporeranno acquosità mercuriali a riempire lo spazio vuoto ..., dovrà gonfiare la vescicetta'. BNCF, Ms. Gal. 262, f. 105r.

[81] As translated by Middleton, *The Experimenter*, 264. 'insensibili filatamenti d'esalazioni più tenui'. Abetti and Pagnini (eds.), *Le Opere*, 305.

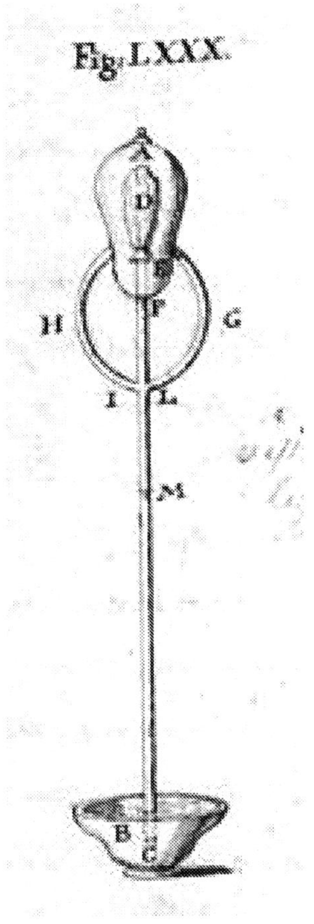

Figure 7. Marsili's experiment testing the vacuity of the space in the Torricellian tube. BNCF, Ms. Gal. 262, f. 104v. Courtesy of the Ministero per i Beni e le Attività Culturali/Biblioteca Nazionale Centrale di Firenze. Protected by Copyright.

exclusion from the *Saggi*.[82] But we are yet to consider the natural philosophical concerns implicated in this experiment.

[82] This negotiation surrounding the efficacy of Rinaldini's experiment reflects some of the issues discussed by prominent sociologists of scientific knowledge since the 1970s. In particular, Harry Collins has argued in his analyses of contemporary scientific experiments regarding the detection of gravitational radiation, that the success of an experiment is measured according to the social processes involved in its construction and interpretation. That is, what comes to be known as a 'well-done experiment' depends upon what the rest of the relevant scientific community come to establish as a consensus about the criteria for competency in this experimental practice. Therefore, the efficacy of an experiment must be negotiated by scientists according to established structures of knowledge-making. H.M. Collins, 'The Seven Sexes: a study in the sociology of a phenomenon or the replication of experiments in physics', *Sociology* (1975), **9**, 205–224; H.M. Collins, 'Son of Seven Sexes: the social destruction of a physical phenomenon', *Social Studies of Science* (1981), **11**, 33–62. This is also the type of social constructivist position used by Trevor Pinch to argue that instruments and experimental apparatus are also entangled in social processes of knowledge-making. See T. Pinch and W.E. Bijker, 'The Social Construction of Facts and Artefacts: or how the Sociology of Science and the Technology might benefit each other', *Social Studies of Science* (1984), **14**, 399–441; T. Pinch, 'Towards an Analysis of Scientific Observation: the externality and evidential significance of observational reports in physics', *Social Studies of Science* (1985), **15**, 3–36. We may contend that the points put forward by these sociologists are certainly applicable to how Borelli and Viviani negotiated the efficacy of the experiment performed by their natural philosophical opponent, Rinaldini.

According to Middleton, the importance of this suggestion regarding the possible exhalations from the mercury, is revealed in a letter written by Magalotti to Leopoldo on 24 April 1659. Middleton argues that the academicians may have been asked to consider if the mercury could evaporate. Magalotti's response was as follows: 'it is the most universal subterfuge of all those who deny the vacuum, to have recourse to these exhalations of the mercury, violently extracted from its bulk in some way, to fill the space left open by its fall'.[83] In addition to Middleton's excellent scholarly work on this issue, we may recall how those who objected to the existence of the vacuum were Aristotelians and Cartesian mechanists. In this case, Marsili, as we have seen, was a supporter of Aristotelianism and he wished to dispute the mechanist notion of the vacuum by having 'recourse to these exhalations of the mercury'. So this experiment was not only a failure because of its poor construction, but was also unsuitable to the corpuscularian and mechanist academicians because of the Aristotelian principles it attempted to support. In other words, the suggestion that tenuous substance invaded the vacuum was of concern to the group's mechanists and continued to fuel the natural philosophical contention amongst the academicians.

Pneumatics was only one field that the Cimento explored. The next chapter will look at their work concerning the natural and artificial freezing process, as well as the properties and effects of heat and cold. These will continue to tell us much about this group's processes of knowledge-making, including how they played upon the natural philosophical concerns of the period, and how those concerns also played upon the academicians, in the construction and interpretation of their experiments. In the process, we shall be putting an end to any suggestions that throughout its existence, the Accademia del Cimento was constructing atheoretical matters of fact.

[83] As translated by Middleton, *The Experimenters*, 267. 'essendo universalmente subterfugio di tutti coloro che negano il vacuo di ricorrere a questa esaltazioni del mercurio, estratto in un modo violentemente dalla massa di esso per riempire lo spazio lasciato nella sua caduta'. BNCF, Ms. Gal. 275, f. 147r.

CHAPTER SIX

THE ARTIFICIAL FREEZING PROCESS OF LIQUIDS, AND THE PROPERTIES AND EFFECTS OF HEAT AND COLD

On 22 June 1657, just three days after the first formal meeting of the Accademia del Cimento, the academicians were already showing an interest in the effects of heat and cold on water placed in varying conditions and mixed with different substances.[1] During the following two months they suspended any such investigations, preferring instead to concentrate on testing and measuring the weight of air.[2] In September, with great enthusiasm, they returned to their investigations regarding the effects of heat and cold and particularly the freezing process of liquids. The friction that was beginning to show between the group's mechanists and Aristotelians during their experiments in the field of pneumatics was to escalate once they seriously began to dedicate themselves to this second topic. Indeed, the freezing process of liquids, as well as the properties and effects of heat and cold, were the academicians' most rigorously explored and debated topics; their experiments in this field came to dominate the *Saggi*'s pages.[3] Furthermore, as Middleton points out, the published experiments represent only a portion of all the work they carried out on the topic, including the hours that each academician spent negotiating the interpretation and significance of each experiment.[4]

The first clue regarding the Cimento's cognitive interests here is in the *Saggi*'s opening two paragraphs on the subject. Here, Magalotti refers to 'the opinion of Galileo' and finding results 'in conformity with the words of Galileo'.[5] There is

[1] They performed four experiments on that day, as recorded in the Cimento diary (BNCF, Ms. Gal. 262, ff. 4v–5v). The first was to test whether water was cooler or warmer after it had been shaken. The second examined if the water was warmer after it had been mixed with ashes. And finally, by the end of the day, they tested the rates at which water mixed with vinegar and wine, became cooler.

[2] This does not include two heat and cold experiments that had little value. On 20 July they tested the sounds that hot and cold water make when splashed on the ground (BNCF, Ms. Gal. 262, f. 16r). On the 27th of that month they experimented on whether water approached room temperature when placed in a container and left in a room for three days (BNCF, Ms. Gal. 262, ff. 16r,19r).

[3] In fact, about one-third of the text is dedicated to their experiments on freezing and the effects of heat and cold.

[4] Middleton, *The Experimenters*, 270–271.

[5] 'opinione del Galileo'; 'in conformità del detto del Galileo'. Magalotti, *Saggi*, 164.

only one Galilean publication that Magalotti could have been referring to as so crucial to the academicians' work on the freezing process, his 1612 explanation of floating bodies, *Bodies that stay atop water or move in it*.[6] Galileo opposed the fundamental Aristotelian claim that water reduces in volume and increases in weight when frozen. He suggested instead that water rarefies upon freezing and therefore becomes lighter, explaining why ice floats. This suggestion that water expands, rather than contracts, when frozen, clearly opposed scholastic doctrine of the interaction between the four fundamental elements of nature and their qualities, in this case water and cold.

Galileo was not mentioned in any of the diary entries for September 1657, when they carried out most of their debates on freezing. However, it will become clear in our analysis of the academicians' experiments that the group's mechanists were interested in building upon some of Galileo's anti-Aristotelian work on hydrostatics. Furthermore, the academicians relied heavily on the atomistic work of Pierre Gassendi, who is also mentioned on occasion throughout the *Saggi* and the academicians' diaries and manuscripts. Gassendi's atomism was to play a vital role in the natural philosophical concerns behind the Cimento's work on freezing and the effects of heat and cold. In the meantime, the scholastic arguments continued to be defended by Rinaldini.

This second case study contains many of the same issues regarding natural philosophical skills, commitments, and agendas that we have already explored through the individual careers of the academicians, and their work on pneumatics. By analysing the construction of their experiments concerned with freezing liquids and the effects of heat and cold, we shall continue to see that there existed a great deal of friction between the Cimento's contrasting personalities and their commitments to competing natural philosophical beliefs. So the picture that will continue to emerge here of the Cimento's work between 1657 and 1662, is not one that has been depicted by traditional or 'cultural' historians. Instead of a formal programme or routine for gaining factual knowledge through the application of an inductivist, atheoretical experimental method, we are seeing an academy constructing and debating knowledge claims according to the competing natural philosophical aims and interests of its members.

Despite the similar lessons that both case studies provide, the second, presented in this chapter, will develop one more argument of particular importance to our study of the Cimento's foundations, purpose, and workings. That is, that the group's patron and protector, Prince Leopoldo, was heavily involved in the Cimento's investigations into natural and artificial freezing, as well as thermal dilation. Leopoldo even suggested experiments to lend further support to the mechanist/corpuscularian beliefs that were upheld by several of the academicians against the Aristotelians in the group. This supports the notion that Leopoldo was not enforcing a no-theorising policy on the Accademia from 1657 to 1662, before they embarked on the publication of their experiments. In fact, for the first

[6] G. Galilei, 'Discorso al Serenissimo Don Cosimo II, Gran Duca di Toscana, intorno alle cose che stanno in su l'acqua o che in quella si muovono', in Favaro (ed.), *Le Opere*, iv, 63–141.

time in this analysis of the Cimento's activities, we shall see that Leopoldo was in fact a participant in the natural philosophical contention amongst the academicians. While studying the freezing process of liquids, evidently one of Leopoldo's preferred topics, the Prince assisted in formulating the mechanical and corpuscularian theories that were intended to displace the traditional beliefs in Aristotelian natural philosophy.

So this second case study, much like the first, does not begin with the academicians themselves, but rather with their predecessors in Tuscany and across Europe. However, thinkers such as Galileo and Gassendi were themselves acting and making decisions based upon previous sixteenth-century discussions between Neoplatonists, including those of a natural magic bent, and Aristotelians, regarding the freezing process of water and the ever-pertinent question of the vacuum. Therefore, in order to grasp the entire natural philosophical controversy that this topic encapsulated before the academicians' involvement, this case study will begin with a look at the sixteenth-century discussions regarding the freezing process and the vacuum.

1. SIXTEENTH-CENTURY ATOMISTS: FREEZING AND THE VACUUM

At the beginning of the seventeenth century, it was still widely believed in natural philosophically literate circles that liquids, when frozen, condense. Scholastics believed that in this process, water, one of the four fundamental elements of nature, could be transformed by one of its qualities, cold, into ice, a supposedly denser and heavier substance. According to this theory, condensation is a characteristic of cold, as seen in the creation of ice through freezing water. To support their claim, scholastics argued that ice has the appearance of being denser and heavier than water, and that any volume of water could be seen to take up less space when frozen.[7] A sixteenth-century Jesuit scholar, Franciscus Toletus (1533–1596), designed a hypothetical experiment to support these Aristotelian beliefs. Toletus proposed that a container be filled with water, hermetically sealed and placed overnight in a very cold environment. Once the water is frozen, he believed that the newly formed ice, having condensed from its former liquid state, would occupy a smaller volume.[8]

During the sixteenth century, these Aristotelian claims came under closer scrutiny from atomists and followers of natural magic, especially Bernardino Telesio (1509–1588) and Francesco Patrizi (1529–1597).[9] However, rather than

[7] S. Drake, *Cause, Experiment and Science: a Galilean dialogue Incorporating a new English translation of Galileo's 'Bodies That Stay Atop Water or Move in It.'* Chicago, 1981, xvi–xvii.

[8] Schmitt, 357.

[9] Although both were Neoplatonists and held similar views about the creation of the vacuum, Telesio and Patrizi actually disagreed on various crucial methodological points. In particular, since Patrizi argued for the priority of mathematics over the physical sciences, he was also opposed to Telesio's idea that he could rely solely on sense experience to observe matter and forces at work in nature. N.C. Van Deusen, *Telesio: The First of the Moderns*, New York, 1932, 10–13.

question whether the water became a denser substance after freezing, Telesio and Patrizi pursued their own natural philosophical agenda. Although they agreed with scholastics that freezing water condenses, they insisted that this occurred because the vacuous spaces between the atoms of water would be pushed out from the liquid due to the cold. This would mean that a large empty space, void of any matter, could be created inside the container.[10]

Furthermore, in response to the scholastic arguments made against the creation of the void, Telesio and Patrizi proposed that vaporous atoms were also frozen and condensed with the water. Moreover, they insisted that in any contrived experience such as Toletus', a strong and thick container should be used to ensure that the supposed force of nature's abhorrence of the vacuum would not break the container.[11] This would mean that without the influx of air through a crack in the container, the vacuum would be easily produced.

At this point, contentious arguments by vacuists such as Telesio and Patrizi failed to make any sort of impression on the dominance of scholasticism in Europe. Although there was considerable debate during the late sixteenth century about the existence of the vacuum, there was still no discussion that could challenge the Aristotelian notion that the quality of cold in water condenses the liquid into a denser and heavier substance. In fact, although Telesio and Patrizi did not refer to heat and cold as essential qualities, as claimed by scholastics, the effects that they perceived them to have on nature were still compatible with a qualitative natural philosophy. That is, heat and cold act as virtues that bodies possess and react to. So, Telesio, as well as Patrizi, were actually eclectic qualitative atomists, and although anti-Aristotelian, particularly with regard to the vacuum, according to Van Deusen and Shumaker, they still agreed that water condenses when frozen because of qualities inherent in nature.[12]

So during the late sixteenth century, there was a clear conflict of natural philosophical interpretation regarding the freezing process of water. Telesio and Patrizi agreed with scholastics about the notion of condensation, but disagreed about the possibility of creating a void. Underlying these similarities and differences was their atomistic natural philosophy. Additionally, Telesio and Patrizi obviously encountered opposition from Aristotelians who did not wish to resign a crucial aspect of their natural philosophical principles, that is, nature's abhorrence of the

[10] Telesio believed that cold was one of the only two active forces in nature, the other being heat. As forces and not corporeal bodies, heat and cold are impressed on matter and while heat causes matter to expand, cold makes it contract. Meanwhile, although perhaps not supportive of Telesio's reliance on heat and cold as forces, Patrizi also argued for the condensation of freezing water on the basis of his beliefs in the movements and properties of atoms. J. Henry, 'Patrizi's Concept of Space and its Later Influence'. *Annals of Science* (1979), **36**, 563.

[11] Schmitt, 358; Van Duesen, 46.

[12] Van Deusen, 35–37; W. Shumaker, *Natural Magic and Modern Science. Four Treatises: 1590–1657*, New York, 1989, 1998. We will find later in this chapter that similar descriptions may be applied to the early seventeenth-century corpuscularians such as Galileo and Gassendi. Although they were of a more mechanistic bent, their claims with regard to the effects of heat and cold, still implied the existence of qualities in nature. In this sense, we shall see, they were eclectic atomists, since they drew from the qualitative beliefs of scholastics and Neoplatonists in this field, while still attempting to construct a mechanistic explanation for the effects of heat and cold.

vacuum, to the atomistic beliefs that were highly contentious, and often even seen as heretical.[13] Therefore, as Schmitt quite rightly points out, the observation of the same natural phenomena by scholastics and atomists, led to contrasting interpretations. This reflected the conceptual framework in which those observations were carried out and interpreted.[14]

Now, while the possibility of creating a vacuum was the central topic of concern for some sixteenth-century natural philosophers, seventeenth-century inquiries into the freezing process had quite a different goal. This is evident in the Cimento's *Saggi*. In introducing the experiments concerned with freezing, Magalotti refers to the great amount of work produced on the topic by the Cimento's predecessors.[15] Yet, rather than mention the vacuum, Magalotti makes it clear that his fellow academicians were instead interested in 'the power that nature makes use of in her process of freezing: whether in doing so she contracts or expands water and other liquids; whether she transmutes them slowly, taking time, or really with instantaneous speed'.[16]

Therefore, for the Accademia del Cimento, the aim of the experiments with freezing liquids was certainly not to discuss merely the vacuity of a container filled with freezing water. Since the Cimento's mechanists believed that the water expands when frozen, the vacuist arguments were no longer applicable to this topic. Instead they were to draw from Gassendi's atomistic and mechanical natural philosophy, and particularly from Galileo's anti-Aristotelian work on floating bodies, in order to question the qualitative ontology that Aristotelians relied upon so heavily for their natural philosophising. So our immediate task, before embarking on an analysis of the intellectual concerns and contentions entangled in the academicians' work, is to examine exactly what contributions Gassendi and Galileo made to the debate regarding the freezing process.

2. GASSENDI, GALILEO, ATOMS, AND FREEZING

In his physics, Gassendi once again revived the ancient atomistic works of Epicurus, Democritus, and Lucretius. He also took over the atomistic works of his predecessors, Patrizi in particular, as well as other magical Neoplatonists and atomists such as Tommaso Campanella, to devise his own corpuscularian theories concerning the structure and movements of nature.[17] Gassendi insisted that

[13] The reason why atomism was so abhorrent to scholastics was that, in antiquity, Epicurus believed in an atomistic ontology that minimised the role of religion in natural philosophy. By the seventeenth century, this reputation still surrounded Epicurean atomism and scholastics accused atomists of wishing to abolish God's interaction with nature. Dijksterhuis, 424; Redondi, 9–27.

[14] Schmitt, 363.

[15] 'sono andati in ogni tempo variamente speculando gl'ingegni'. Magalotti, *Saggi*, 162.

[16] 'il magistero di cui si val la Natura nel suo agghiacciare, s'ella ciò faccia strignendo o rarifacando l'acque e i liquori, se lentamente e con tempo o vero con istantanea velocità gli trasmuti'. *Ibid.* Curiously, Magalotti added to this passage: 'Or else cold may be nothing but a total absence and expulsion of heat'. This hints at the view held by Galileo and Borelli, that cold corpuscles do not exist. This issue will be examined later in this chapter.

[17] Henry, 'Patrizi's Concept of Space', 567.

nature consists of only matter and motion. More specifically, he accepted first the Epicurean and anti-Aristotelian notion that all material substances are made of indivisible particles, separated by tiny vacuous spaces. Second, he proposed that the particles move through those spaces and collide with one another. This provides a wholly materialistic concept of atoms, which can be subjected to mathematical and mechanical scrutiny.[18]

Gassendi also relied on his atomism with regard to the freezing process. In this case he argued that frigorific atoms, that is atoms that produce cold, put pressure on the particles of water that help to push out the vacuous spaces between them and condense the water into ice.[19] As Barry Brundell points out, this was a theory dedicated to replacing an Aristotelian natural philosophy based on elemental qualities, with an Epicurean belief in matter and motion, illustrating the freezing process according to a corpuscularian and mechanistic approach.[20]

The situation regarding the effects of heat and cold on water was quite different for Galileo who believed that water rarefied, rather than condensed, during the freezing process. But at the same time, Galileo's philosophy regarding atomism held some similarities to Gassendi's. In 1625, and again in 1636, Gassendi wrote to Galileo seeking to establish a dialogue regarding what Gassendi believed was a link between Epicurean atomism and Copernicanism. According to Brundell, Gassendi wished to make it clear to Galileo, that they were both after the same thing: the overthrow of Arisotelian natural philosophy. Furthermore, Gassendi believed that an atomistic theory of matter and motion could support the suggestion that the Earth is just another planet orbiting around the Sun.[21] The two never actually met, but Gassendi's atomism and his studies of

[18] Osler, *Divine Will*, 191–195. As an example of this atomistic philosophy, we may recall the explanation provided by Gassendi for the speed and movement of sound mentioned in Chapter Two. He relied on the notion of atoms of sound travelling through, and colliding with, atoms of air until they reached our senses. We may also recall that this was the philosophy adopted by the Cimento academicians when they decided to study acoustics briefly in 1656. The involvement of Borelli and Viviani in those sound experiments in 1656 and their corpuscularian framing, may provide us with a valuable clue as to how they were to approach their work regarding the freezing process of liquids and Aristotelian qualities such as heat and cold only months later.

[19] Dijksterhuis, 428.

[20] Brundell, 51–59. Gassendi's central agenda when compiling his major treatises regarding atomism was to present Epicurean philosophy of nature in an acceptable light for scholastics and religious authorities. Much as Thomas Aquinas intended to merge Aristotelianism with Christianity during the Middle Ages, so too did Gassendi attempt to make Epicurean atomism acceptable to Catholic dogma. According to Brundell, Gassendi's means for achieving this aim were to equate Epicurus' claims about matter and motion with Aristotle's. Gassendi argued that, in some respects, both ancient thinkers presented similar theories about nature's structure and movements, with Epicurus making some notable improvements to Aristotle's work. In particular, Gassendi noted that identifying qualities in nature was equally important to Epicurean atomism as it was for Aristotelians. The difference between the two ancient natural philosophers, according to Gassendi, was that instead of nature inherently possessing qualities such as heat and cold, these qualities are produced through the position and motion of atoms. Gassendi's work was unmistakably mechanistic and anti-scholastic, but it was still eclectic in that he attempted to adopt and reinterpret parts of Aristotelianism, including his elemental qualities. This does not mean that he was in the same school of qualitative atomists as the sixteenth-century natural magicians, but simply that there was a certain ambiguity concerning the role of qualities in his ontology.

[21] Brundell, 52–53.

matter and motion were still quite compatible with Galileo's philosophy of nature, at least in the view of Gassendi and his followers. This was despite the fact that Galileo never professed to subscribe to the views of the ancient atomists, quite unlike Gassendi. In addition, unlike Gassendi and the sixteenth-century atomists, Galileo did not believe in the condensation of freezing water. Nevertheless, he still used a corpuscularian philosophy in his studies about floating bodies in order to refute Aristotelian principles.

In 1611, Galileo held a discussion with colleagues about the condensation and rarefaction of water.[22] It was here, according to Drake, that Galileo resolved to write a treatise opposing the Aristotelian notion that ice, being condensed water, is therefore a heavier and denser substance than water. Galileo made several observations of the formation of ice that year, which gave rise to a treatise entitled *Bodies that Stay Atop Water or Move in It* (1612). In it he made the following anti-Aristotelian claim:

> I should have thought ice to be rarefied water, rather than condensed, since condensation gives rise to shrinkage in volume and increase of heaviness, but rarefaction to greater lightness and increase in bulk; now, when ice is formed, it is lighter than water, since it floats thereon.[23]

Galileo then referred to Archimedes' claim that, when a solid is condensed and made smaller, its specific weight increases in ratio to its decrease in volume. Although with this reference Galileo was already contesting Aristotelian natural philosophy, no longer recognising condensation as a characteristic of cold, this was by no means the full extent of his argument. Galileo's discussion with his colleagues regarding rarefaction and condensation, incorporated more wide-reaching issues regarding the nature of floating bodies.

Scholastics supposed that, regardless of its weight, a heavy body such as ice cannot be made to penetrate the surface of a body of water simply because of its shape. Since ice is regularly of a broad and flat shape, they argued, it can neither divide the water nor overcome its resistance, and for these reasons cannot sink.[24] In response to this view, Galileo used the argument cited above to insist that instead ice floats on water because it is a more rarefied and lighter substance. Galileo therefore concluded that 'all bodies heavier than water ... would indifferently go to the bottom, while those lighter would indifferently float regardless of shape'.[25]

[22] Drake, *Cause, Experiment and Science*, xvi; G. Bonera, *Galileo Oggi*, Pavia, 1995, 1989.

[23] As translated by Drake, *Cause, Experiment and Science,* 22; 'avrei creduto più tosto il ghiaccio esser acqau rarefatta, che condensata; poi che la condensazione partorisce diminuzion di mole e augumento di gravità, e la rarefazione maggior leggerezza e augumento di mole, e l'acqua nel ghiacciarsi cresce di mole, e 'l ghiaccio già fatto è più leggier dell'acqua, standovi a galla'. Galilei, 'Discorso intorno alle cose che stanno in su l'acqua', in Favaro (ed.), *Le Opere*, iv, 65.

[24] Aristotle also contended that ice floats because it contains air and is therefore subject to positive levity. However, considering that ice is supposed to be a condensed substance, this argument does not carry much credibility even within Aristotle's own reasoning.

[25] As translated by Drake, *Cause, Experiment and Science*, 23; 'tutti i corpi più gravi di essa, di qualunque figura si fussero, indifferentemente a galla'. Galilei, 'Discorso intorno alle cose che stanno in su l'acqua', in Favaro (ed.), *Le Opere*, iv, 66.

So Galileo was not only using Archimedes' work on hydrostatics to describe the ratios of weight and volume that allow some bodies to stay atop water, but he was also elaborating on Archimedes' work in order to question the principles of Aristotelian physics by demonstrating the notion that the density of floating bodies is less than the density of water.[26] He denied that water condenses as a result of the interaction between it, a fundamental element of nature, and one of its qualities, cold, providing further examples from Archimedian hydrostatics to oppose the Aristotelian notion that water resists being divided by flat objects such as blocks of ice. In developing his own account of the freezing process, Galileo included corpuscularian claims which further opposed Aristotelian natural philosophy, arguing that the effects of heat and cold involved the existence of geometrically shaped atoms that penetrate the vacuous spaces of other solids and liquids. Gassendi's later theories about hot and cold atoms resembled this approach.[27]

In the 1612 publication on floating bodies, Galileo had failed to mention the existence of corpuscles, but in his later works, he endorsed the notion of mathematical indivisibles and the existence of particles of varying types. In the *Assayer* (1623) he provided a more thorough account of these physico-mathematical, mechanical, and anti-Aristotelian beliefs.[28] Here Galileo attacked the Aristotelian notion that qualities are inherent in bodies. Instead, just like Gassendi, he proposed that the interaction between nature's quantifiable parts, meaning its smallest atoms, to produce motion and qualitative sensations, could be demonstrated mathematically.[29]

Finally, in Day One of *Two New Sciences*, Galileo not only suggested that the existence of vacuous spaces determined the resistance of the void, but also speculated upon the infinite number of atoms that are present in all solid bodies and explain why bodies condense or rarefy. After providing a lengthy geometrical demonstration of indivisibles, involving the infinite sides of a circle, Galileo concluded that 'the expansion of infinitely many indivisibles with the interposition of indivisible voids ..., can be said to explain the condensation and rarefaction of bodies'.[30] In this way Galileo provided a geometrical demonstration for the expansion and condensation of all bodies that implied the existence of corpuscles.[31]

[26] Bonera, 90.

[27] It is worth noting that unlike Gassendi, Galileo did not believe in the existence of cold atoms, preferring instead to claim that cold was merely caused by the absence of hot atoms. This is important since, as we shall see later, it was precisely the position adopted by Borelli in 1657. Grilli and Sebastiani, 312–313.

[28] *Ibid.*, 309–313.

[29] Again, despite the impression that Galileo was constructing a natural philosophy based purely on a mechanistic corpuscularianism, we may still be critical of his treatment of qualities, since he insists that the creation of heat, caused by the rapidly moving corpuscles, can even convert particles of other substances into heat. Grilli and Sebastiani, 311–312; Dijksterhuis, 424. That is to say that Galileo faced much the same problem as Gassendi when trying to formulate a mechanical atomism; they both refuted the existence of Aristotelian elemental qualities, but they still argued that atoms produce the sensation, for humans, of qualities in nature.

[30] Drake (ed.and tr.), *Two New Sciences*, 57; Favaro (ed.), *Le Opere*, viii, 96. See also Dijksterhuis, 422.

[31] According to Segre, Galileo's indivisibles also inspired his students, including Buonaventura Cavalieri and Torricelli, to divise similar concepts in mathematics. Segre, *In the Wake*, 71.

Therefore, for Galileo, experiments to do with the freezing process, whether hypothetical or actual, contain much more than just the sixteenth-century arguments concerned with the void. Galileo was also attempting to undermine the validity of Aristotelian elemental qualities. Furthermore, together with the rising interest in Epicurus, Democritus, and Lucretius that accompanied the humanist movement and aided the natural philosophical interests of atomists such as Gassendi, Galileo wanted to replace Aristotelianism with a type of mathematical, mechanical, and corpuscularian natural philosophy. Galileo's skills and commitments, including his anti-Aristotelian agenda, were adopted by many of the Cimento's members, including the patron Leopoldo, during their experiments on the freezing process of liquids and the effects of heat and cold. Indeed, the academicians' correspondence, together with the draft manuscripts of the *Saggi*, show that the Cimento performed their experiments on freezing specifically to further their own natural philosophical opinions based on either scholastic or mechanistic principles.

3. ARTIFICIAL FREEZING

On 1 September 1657, the academicians recorded their first attempt to freeze water in an enclosed jar.[32] The aim of this experiment, as noted in the Cimento diary entry for that date, was 'to recognise whether it is true if ice is rarefied water'.[33] More specifically, as revealed through Magalotti's description of this experiment in the *Saggi*, the academicians were attempting to verify the claim, 'in conformity with the words of Galileo, that ice, whether formed into huge slabs or broken into the smallest pieces of whatever size and shape you will, always floats on water'.[34] With these words, Magalotti showed that, from the beginning, the academicians wanted to test Galileo's contentious claim that water expands and becomes lighter when frozen, contrary to the scholastic belief that it contracts and increases in weight. The entire first set of experiments that Magalotti reported in the *Saggi* were devoted to testing Galileo's contention.

Yet the passage of the text introducing these experiments still reveals much more about the theoretical aims of some of the academicians. In fact, these two short paragraphs in the *Saggi* provide us with a valuable clue regarding the natural philosophical concerns involved in the construction and interpretation of these experiments. Magalotti continued: '... [C]onsidering the whole volume that is being frozen, it acquires lightness, either by the interposition of minute empty spaces or by a very finely divided admixture of particles of air or other similar material.'[35]

[32] The use of salt and other substances to increase the freezing power of the ice led the academicians to label all these experiments as examples of 'artificial freezing'. Meanwhile, their 'natural freezing' experiments involved exposing liquids to naturally cold conditions, which often required many long nights of observational work.

[33] 'Per riconoscere se fosse vero che il ghiaccio sia un acqua rarefata' BNCF, Ms. Gal. 262, f. 29v.

[34] 'in conformità del detto del Galileo, che l'acqua tanto formata in ampie falde di ghiaccio, quato rotta in minimi pezzi di qualsivoglia grandezza e figura sta a galla sopra all'atr'acqua'. Magalotti, *Saggi*, 164.

[35] ...[A]ttesa tutta la mole che s'agghiaccia, se le arroge leggerezza, o sia per interponimento di minimi spazi vacui o per un minuto permischiamento di particelle d'aria o d'altra simil materia. *Ibid.*

Magalotti was therefore providing the first clue regarding the corpuscularian interests pursued by some of the academicians in performing these experiments.

However, Magalotti was not yet fully endorsing a corpuscularian natural philosophy. As usual, he was being careful with how much he revealed about the natural philosophical agenda pursued by the majority of the academicians or whether they supported any corpuscularian and mechanist interpretations of their experiments. This passage, nevertheless, provides a valuable clue regarding the Cimento's interests when examining artificial freezing.[36] Here in the first paragraph of the presentation of the Cimento's freezing experiments in the *Saggi*, the text that is supposedly the exemplar of their strict experimental and theory-free method of research, we find a cautious reference to the academicians' interests in revisiting the anti-Aristotelian implications of Galileo's work on floating bodies. It will not be the last such passage that we come across in the *Saggi* on this topic. So just as we have seen with the case study on pneumatics, once again Magalotti hinted in the *Saggi* about the natural philosophical issues that were at stake in the construction of the Cimento's experiments and their interpretation. Within two short paragraphs, he allowed the reader to acquire a taste of the intellectual concerns the academicians actually maintained during the late months of their inaugural year.

So on 1 September 1657, with their theoretical and natural philosophical concerns firmly in mind, the academicians poured water into a vessel made of a thin sheet of silver. The container was filled with water, closed with two screw-on lids, and left in ice overnight (Figure 8). The following morning the academician found that the water had pushed out and burst through the inside lid, leaving a crust of ice sitting on top.[37] This experiment echoes that proposed by Toletus, in which sixteenth-century scholastics expected that the container would indeed be cracked open, owing to nature's abhorrence of the vacuum. However, although their observation agreed with this Aristotelian notion, the academicians had a different natural philosophical agenda and thus a different interpretation. The conclusion that was apparently reached by the majority of the academicians, and that was reported in the *Saggi*, was that water, when frozen, expands rather than contracts. That is, although having nowhere to go, the water became rarefied and overcame the resistance of the tightly sealed container.[38]

[36] Once again, this reference to the corpuscularian interests within the Cimento was not in the first draft of the *Saggi*, and it is difficult to determine when this passage was included in the text, who might have suggested its inclusion, and why (see Chapter Five, note 40). In other words, there is no evidence whether Magalotti deliberately intended to hint at some of the details of the academicians' natural philosophy – were such hints considered necessary for conveying the Cimento's aims when experimenting, or was Magalotti genuinly attempting to provide his readers with clues about the academicians' natural philosophical concerns? In any case, Magalotti was careful not to describe theoretical positions too explicitly and stuck to the rhetorical aim he set out in the *Saggi*'s preface.

[37] BNCF, Ms. Gal. 262, ff. 29r–30v.

[38] 'mentre essendo violentata dalla virtù del freddo a ristrignersi in minore spazio, essa per paura di lasciar voto il luogo, di cui andava a mano a mano ritirandosi, era sempre venuta serrandosi addosso il coperchio, finchè non potendo quello distendersi maggiormente era venuto a schiantarsi. Non à luogo dico un simil discorso; poichè in tal caso averemmo auto a trovare il coperchio affossato in dentro, dove lo trovammo sforzato in fuori, e di piano ch'egli era, vedemmo esser divenuto colmo notabilmente e colma osservammo la superficie del ghiaccio ritrovato nel vaso. Di più gli orli dell'apertura erano arrovesciati in fuora; onde si raccoglie che grandissimo dovess'esser l'impeto con cui fu fatta'. Magalotti, *Saggi*, 166.

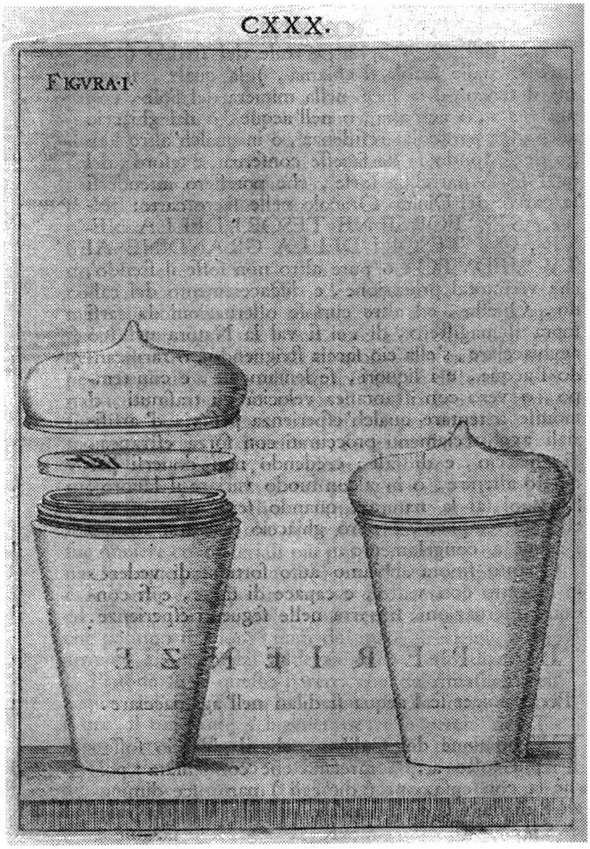

Figure 8. Experiment by the Accademia del Cimento testing the expansion of freezing water in a tightly sealed container. L. Magalotti, *Saggi di naturali esperienze*, Florence, 1667, 130. Courtesy of the IMSS Biblioteca Digitale.

Yet at this point, they still needed to verify Galileo's opinion that ice floats because it is lighter and more rarefied than water. As Magalotti claimed in his first draft of the *Saggi*, they were seeking 'to arm [Galileo's claims] with experience'.[39] So the following day, they froze water inside a thicker silver egg-shaped ball, closed with a screw in the middle (Figure 9). The ice appeared not to have forced its way out on this occasion, although upon repeating the experiment one month later, and after closer examination, they found that the water did indeed leak out through the screw as it froze. The academicians also noted that despite

[39] 'Questo dettato di quel grand'uomo quantunque per sè solo ed ignudo ci si rendesse autorevole, abbiam tentato nulla dimeno d'armare con esperienza'. Abetti and Pagnini (eds.), *Le Opere*, 308. This expressed interest in performing experiments in order to verify Galileo's claims about floating bodies reflects the theory-laden experimentalism that the academicians practised.

Figure 9. Cimento experiment demonstrating the rarefaction of freezing water. L. Magalotti, *Saggi di naturali esperienze*, Florence, 1667, 132. Courtesy of the IMSS Biblioteca Digitale.

the curious formation of a hole in the centre of the newly formed block of ice, it still floated just above the surface of the water. A passage in the *Saggi*'s draft, omitted from the final publication, indicates the general opinion of the Cimento that Galileo was indeed justified in his beliefs: 'From this effect it would seem that Galileo's opinion remains firm, since, while the force of rarefaction seemed beaten by the resistance of the vase, the ice still took on its effect.'[40] The 'effect' Magalotti referred to here is, therefore, not only the rarefaction of water when freezing, but also, as the final confirmation of Galileo's 1612 publication, that the ice floats because of its lighter and rarefied condition.

[40] *Ibid.*, 309.

So 'in conformity' with Galileo's opinion that ice floats, together with his anti-Aristotelian agenda, the academicians sought to disprove the scholastic belief that condensation is a characteristic of the interaction between the element, water, and its quality of cold. At this point the Cimento's mechanists became determined to strengthen the mathematical and mechanical framing of their work by attempting to measure the force of the expansion of freezing water. Indeed, four of the experiments that followed, performed during early October, were concerned with this task.

4. THE FORCE OF EXPANSION OF FREEZING WATER

The academicians observed changes effected in bronze, glass, brass, and even gold containers, when filled with water and frozen. On each occasion, regardless of the density of the container, the water always either burst through any possible opening, or distorted the container's shape. These experiments demonstrated the impressive physical force of expansion. However, while reading through Magalotti's draft account of these experiments for the *Saggi*, Borelli insisted that a more precise mathematical and mechanical description should be provided in the presentation so as to leave no doubt in the mind of the reader about the precise force exerted by the freezing water.[41]

In his editorial notes for the *Saggi*, Borelli proposed that 'in the first place, the exact measurements of the thickness of the lead, silver, gold, and bronze containers that were broken by the force of the water in the act of freezing, should be expressed with great precision. Second, Borelli continued, 'I would believe that the measurement of the growth of the frozen water should be added.'[42] Intending to provide precise mechanical accounts of the expansion of freezing water, by August 1662, when the academicians were retesting the experiments intended for publication, Borelli and Viviani concentrated on this aim, displaying their commitment to a mathematical and mechanical natural philosophy. In particular, this shows that, in their attempt to deal with the dynamics of the freezing process of water, the Accademia's mechanists were drawing from their knowledge of statics, a mixed mathematical field. What this means, is that they were not in a position to determine the actual force of freezing water in terms of a dynamics they did not possess. Rather, they could accumulate data about the parameters of the freezing process and the thickness of the containers required to resist the water's expansion, in hopes of giving some statically based qualitative representations of the forces involved.

[41] During the second half of the year 1662, Borelli, Viviani, and Rinaldini edited Magalotti's first draft of the *Saggi*. All three editors made several notes for the Cimento's secretary, now published in Abetti and Pagnini 280–348. This part of the *Saggi*'s editing process will be discussed further in Part Three.

[42] 'nel primo luogo, si dovessero esprimere con molta esattezza le misure precise della grossezza dei vasi di piombo, argento, oro e bronzo che furono rotti dalla forza dell'acqua nell'atto dell'agghiacciare. Nel secondo luogo, crederei che si dovesse aggiugnere la misura dell'accrescimento nell'acqua agghiacciata'. *Ibid.*, 333.

Sometime after July, 1662, the academicians performed an experiment designed to measure 'the force of expansion of water that is confined while freezing'.[43] They had a metal ball made thick enough to withstand the force of the water's expansion. It was filled with water and frozen and, as described in the *Saggi*, they then used a lathe to shave away thin layers of the ball until they detected a crack in the metal. The academicians believed that this observation provided a measurement for 'the greatest thickness overcome by the expansion of water that is confined while freezing'.[44] This demonstrates the language of force used by the Cimento members. They were not providing a theory about the dynamics of the expansion of freezing water, but were rather simply providing a measurement of this physical phenomenon based on their skills in mixed mathematics and statics. In fact, such was their commitment to a statical mechanics in this case, that after performing this experiment, they 'began to wish to reduce this force to that of a dead weight' that would provide them with another type of static measurement of the water's force of expansion during the freezing process.[45] This was, in fact, an experiment proposed by Borelli in his editorial notes for the first draft of the *Saggi*.[46]

Borelli suggested that a ring be constructed using metal of the same hardness as in the experiment described above and cast of the same thickness as the ball used to measure the greatest thickness overcome by freezing water. This ring 'of a conical form', as described in the *Saggi*, would be held up by two wooden supports (Figure 10). Finally, a solid iron cone would be made to fit perfectly through the hole of the ring, with some of the iron extending above the ring. Borelli then proposed that ever-greater weights should be placed progressively on top of the iron cone, uniformly putting the metal ring under greater strain until it would eventually begin to break. It was supposed that the force exerted by the minimum dead weight which cracked the ring would be the same as the force of the expanding water.[47]

[43] 'Esperienza per misurare quanta sia la forza della rarefazione dell'acqua serrata nell'agghiacciarsi'. Magalotti, *Saggi*, 172. This experiment seems never to have been recorded in the Cimento's diary, or in the first draft of the *Saggi*. It only appeared in the final published version of the text, which could lead us to believe that the experiment was performed after July 1662, following Magalotti's completion of the first draft of the *Saggi*.

[44] 'Si che ci parve di poter dire esser quella la massima grossezza superata dalla rarefazione dell'acqua serrata nell'agghiacciarsi'. *Ibid.*

[45] *Ibid.*, 173.

[46] BNCF, Ms. Gal. 267, f. 21v; Abetti and Pagnini (eds.), *Le Opere*, 334.

[47] Once again we may be able to utilise the work by some twentieth-century sociologists of scientific knowledge to illustrate the social negotiation of the efficacy of this experiment. See Chapter Five, note 82. In this case the academicians were not only using a statical mechanics to measure a dynamical force, which is the force of expanding water, but the success of their experiments also depended upon what they all agreed and negotiated as being reasonable grounds for assuming that the results were correct. See Collins, *Changing Order*, 5–28. This means that the academicians argued that the dead weight equalled the force of expansion of freezing water, on the basis of what they generally believed to be correct parameters for measuring natural phenomena. This is how they negotiated whether a static weight was the same as a dynamic force, because they were the accepted modes of thought for many seventeenth-century mechanists.

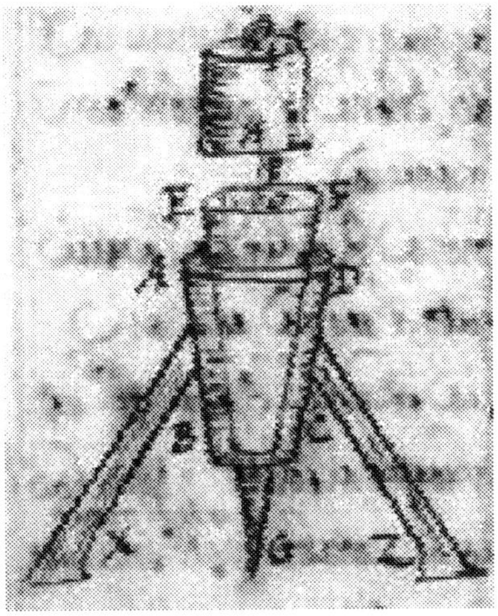

Figure 10. Experiment suggested by Borelli to measure the water's force of expansion during the freezing process. BNCF, Ms. Gal. 267, f. 21v. Courtesy of the Ministero per i Beni e le Attività Culturali / Biblioteca Nazionale Centrale di Firenze. Protected by Copyright.

This was the third of a series of experiments Borelli proposed in his editorial notes to Magalotti which attempted to provide a precise measurement of the force of expansion of freezing water. It is a perfect example of the mathematical and mechanical skills and commitments that Borelli and his fellow academicians were using to construct knowledge claims. More specifically, this comparison between the force exerted by freezing water on a thick metal ball, with the force exerted by a dead weight on a ring of the same thickness, is an example of the statical mechanics the academicians were using in this case. In other words, this experiment still consisted of the mixed mathematical skills required in studies of statics, and showed no relation to the more dynamically oriented mechanics shortly to emerge in the work of Newton and Huygens, who respectively focused on momentum and, in effect, 'energy', aspects of dynamical systems.

The problem with this experiment as Magalotti noted in the final version of the *Saggi*, was that the thickness of the ring would be so great and the weights required so large, that they had to perform the experiment on a smaller scale. Their intention was to calculate the force on the original ring in proportion to the reduced dimension of the experiment, although this would only obtain an approximate measurement. Magalotti recorded that they experienced further difficulties obtaining the desired thickness of the metal ball, owing to 'inequalities in the resistance of the metal'.[48] Incidentally, although Borelli proposed this experiment in his editorial notes in 1662, a remark in the manuscript's margin, in what appears to be Viviani's handwriting, mentions that Viviani had performed

[48] 'inducenti nel metallo varie disuguaglianze di resistenza'. Magalotti, *Saggi*, 173.

the experiment first.[49] However, since it is written here that this experiment 'was never discussed again', it would seem that Viviani too had little success with it.[50]

In any case, regardless of whether or not it succeeded, the experiment demonstrated both Borelli's and Viviani's mechanistic skills. They put these skills to work again on 7 August 1662, again in all probability while revising what was to be published, when Viviani suggested some 'Experiments to measure the greatest expansion that water receives in freezing', eventually included in the *Saggi*.[51] This was simply comparing the height that the water reached in a glass tube before and after freezing.[52] The conclusion was that the proportion of the measured increase upon freezing was 8–9.

In summary, when studying freezing in September and October of 1657, and again in 1662, Viviani and Borelli, the Cimento's brightest and most outspoken mechanists, exploited their mathematical and mechanical interests through the construction and interpretation of these experiments. If Leopoldo was trying to maintain an institution free from any controversies in natural philosophy, he certainly was not showing it here. In fact, he too was deeply interested in describing the freezing process according to mathematical and mechanical principles.

5. LEOPOLDO'S EXPERIMENT MEASURING THE FREEZING PROCESS OF WATER

The Galilean manuscript folder labelled Gal. 263, containing drafts of the Cimento's experiments, many excluded from the *Saggi*, includes a document, published also by Abetti and Pagnini, with the following title: 'Events observed in the freezing of water'.[53] It contains the description of observations made by Leopoldo regarding the process of freezing water (Figure 11). The following translation consists of all but the last sentence of that manuscript, which we shall return to later when analysing the academicians' corpuscularian concerns regarding the effects of heat and cold.

The glass vessel similar to AB, capable of holding an ounce of water, with a very thin neck AC, divided into 200 degrees and open at A, was filled with water up to 42

[49] According to Targioni Tozzetti, Borelli acknowledged (in a manuscript that seems to have gone missing since Targioni Tozzetti's reading of it), that this was on 7 November 1660. Targioni Tozzetti, *Notizie*, i, 426. If this were true, it would have been after the Cimento suspended its meetings for that year on 25 October.

[50] The whole note reads: 'Già questa esperienza fu proposta da quel coglion del Viviani, e messer Filippo prese l'ordine di far questi et altri anelli e maschi a cono, ma non se ne discorse mai più'. BNCF, Ms. Gal. 267, f. 21v; Abetti and Pagnini (eds.), *Le Opere*, 334. In my opinion, what casts some considerable confusion over these words, is that if Viviani did indeed write them, then for some incredible reason, he referred to himself as an 'ass [coglion]'.

[51] 'Esperienze per misurare la massima dilatazione che riceve l'acqua nell'agghiacciare'. Magalotti, *Saggi*, 174.

[52] BNCF, Ms. Gal. 262, f. 136r.

[53] 'Accidenti osservati nell'agghiacciamento dell'acqua'. BNCF, Ms. Gal. 263, f. 50r; Abetti and Pagnini (eds.), *Le Opere*, 389.

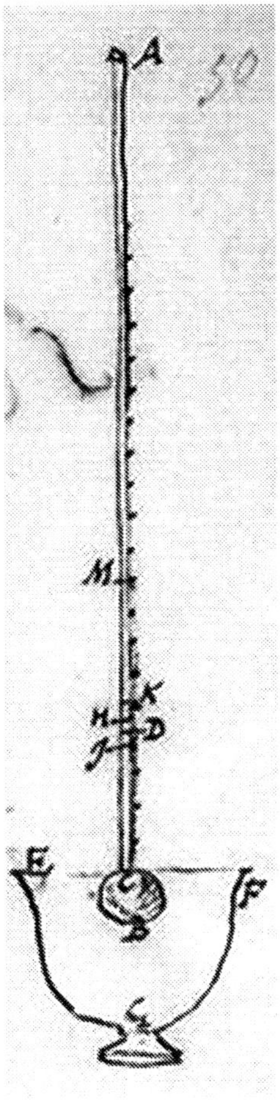

Figure 11. Leopoldo's experiment describing the freezing process. BNCF, Ms. Gal. 263, f. 50r. Courtesy of the Ministero per i Beni el le Attività Culturali / Biblioteca Nazionale Centrale di Firenze. Protected by Copyright.

degrees at D, and the said decanter was immersed up to the base of the neck C in a basin, EGF, full of pounded ice mixed with common salt, saltpetre, and spirits (without which the water would not freeze from the ice). The Most Serene Prince Leopoldo observed five notable effects.

Firstly the water quickly rose 3 degrees from its initial level D to 45 degrees at the mark H.

2°. Without stopping at H, in a movement opposite to, and slower from, the first, it began to fall successively slower, plunging beyond the limit D as far as 40 degrees at I.

3°. After an apparent stillness at I it began to rise again regularly with an accelerated motion up to 47 degrees at K.

4°. Then, almost instantaneously it jumped the space of 40 degrees from K to M, and at the same time, just as swiftly, the water contained in the decanter CB, was seen to cloud up and freeze.

5°. Then when the said decanter was taken out of the vessel EFG, the second that the water was reduced to its natural fluid state it became smaller and fell to 3 degrees below the limit D, and in the passing of time the water returned to 42 degrees at D, at which it began.

And replicating the said experiment the first jump DH was seen to be greater and greater as the water became warmer, so that in a natural state the jump DH was ... degrees and with the water moderately heated it was ... degrees, and notably heated was ... degrees; what's more when two instruments were used of the same capacity but with glasses of different thickness, the one with the thickest glass made the jump higher but more quickly.

The second effect was opposite to the first since the drop HI was observed to be smaller as the water became more and more heated.

The third effect of the rise IK was the same; whether hot, warm or cold water was used.

The fourth effect of the rise KM in another vessel capable of holding a greater volume of water was slower, and the freezing was not as quick as the first time.[54]

Leopoldo had created what amounted to be a very sensitive thermometer.[55] More importantly, he provided his academicians with the opportunity to document the freezing process of water with precision. Indeed, in October 1657 they repeated Leopoldo's experiment on various occasions using a variety of liquids and compiled tables based on the Prince's observations.[56] The *Saggi* contains 20 such tables and the diary several more, each one showing varying results, but clearly documenting the same freezing process. The academicians always recorded a 'jump upon immersion', a 'fall', a 'point of rest', a 'rise', and a 'jump upon freezing' (Figure 12).[57]

As Middleton observes, traditionally historians of science have not placed a great deal of value on these tables.[58] After all, the accuracy of these measurements is questionable and their practical benefits are ambiguous to say the least. Middleton even claims that the compilation of such tables was 'completely futile simply because it was completely empirical'.[59] However, this position fails to consider the cultural and intellectual issues involved in such an experiment. This was one of Leopoldo's preferred topics of research, and he would have most certainly been calling upon the resources at his disposal, including his courtiers, to assist him in his countless investigations into the natural and artificial freezing process.

[54] BNCF, Ms. Gal. 263, ff. 50r–51r; Abetti and Pagnini (eds.), *Le Opere,* 389–390.

[55] This is, in fact, how Magalotti describes it in the *Saggi*. 'potendosi in tal caso considerar tutto il vaso com'un termometro gelosissimo per la gran capacità della palla e per l'estrema sottigliezza del collo'. Magalotti, *Saggi*, 179. In the first draft of the *Saggi*, Magalotti had credited this experiment to Leopoldo, but later crossed out the Prince's name. See Abetti and Pagnini (eds.), *Le Opere,* 313.

[56] The precise observations recorded in the *Saggi* in relation to Leopoldo's experiment, do not seem to have been mentioned in the diary. However, in his editorial remarks to the *Saggi*'s draft, Borelli stated that the above observations 'occurred in October 1657 when His Serene Highness became aware, before anyone, of this wonderful effect'. Abetti and Pagnini (eds.), *Le Opere,* 335. It is also clear from the diary that by 15 October, the academicians did start looking at the phases of the freezing process. The diary in Ms. Gal. 261, shows that similar experiments to Leopoldo's were performed on 12 October.

[57] Magalotti, *Saggi*, 182–193.

[58] Middleton, *The Experimenters,*186, n.204.

[59] *Ibid.*, 272.

PRIMO AGGHIACCIAMENTO

Dell' acqua di fonte.

		Gradi del vaso.	Differenze.	Gradi del term.	Differ.	Vibraz.	Differ.
Agghiacciamento dell'acqua di fonte. Primo.	Stato naturale	142		139		—	
	Salto dell' immersi	143	1	133	6	23	[illegible]
	Abbassamento	120	23	69	64	255	[illegible]
	Quiete	120	—	49	20	330	[illegible]
	Sollevamento	130	10	33	16	462	[illegible]
	Salto dell' agghiac.	166	36	33	—	—	—

Figure 12. Table compiled by the Cimento documenting the freezing process according to Leopoldo's experiment and containing the five phases of freezing as identified by the academicians. L. Magalotti, *Saggi di naturali esperienze*, Florence, 1667, 156. Courtesy of the IMSS Biblioteca Digitale.

At the very least, this begins to tell us something about the courtly setting of the Cimento and the power Leopoldo had over the decisions and actions of the group. Additionally, the Prince was clearly committed to providing a mathematical and mechanical description of the physical phenomena of freezing water. There is no known document where he openly declared a mechanical allegiance when it comes to freezing, but why would he perform such an experiment, compile such tables, and drag his academicians into such lengthy investigations, if not to satisfy his mathematical and mechanistic natural philosophical commitments? In fact, as was argued in Part One and in the lead up to the description of the academicians' experiments on freezing, these were commitments common to most natural philosophers of the seventeenth century across Europe.

Indeed, the mathematical and natural philosophical value of Leopoldo's observation of the freezing process was accentuated when Borelli revised it for publication. Borelli's notes suggesting some changes to Magalotti's draft emphasised the need to bring more precision to the academicians' observations of the freezing process. For this reason he proposed using a pendulum in this experiment, in order to time each phase of the water's fall and rise, and a thermometer to show the relation between the temperature of the water, and the changes in its movements.[60] Furthermore, Borelli again suggested that a precise measurement be provided of the thickness of the glass of the tubes and vessels used in the experiments.[61] This is another indication of how Borelli and most of the academicians measured, recorded, and conceptualised the force of the expansion of freezing water according to their skills in mixed mathematics and statics.

[60] 'Una tale Accademia non par che soddisfaccia al suo debito se in questa storia dell'agghiacciamento non v'aggiugne la curiosa notizia de'tempi orari ridotti a minuzie per via dei pendoli, nei quali tempi si fanno tutte le sopradette operazioni dello scemare, crescere ecc. Di più richiede la curiosità saper anche quali gradi di freddezza producono simiglianti effetti, i quali squisitamente possono misurarsi con termometri gelosissimi d'aria o d'acqua arzente'. Abetti and Pagnini (eds.), *Le Opere*, 334.

[61] *Ibid.*, 334–335.

Accordingly, on 31 July 1662, the same day the academicians gathered at Magalotti's house to retest the experiments intended for publication, they observed 'through the oscillations of the pendulum, the times in which the following changes of condensation and rarefaction that natural water makes upon freezing, take place'.[62] During almost the whole month following, these observations continued to be repeated with the precision that Borelli had called for and that was eventually included in the final publication of the *Saggi*.

These observations with the annotated tables and insistence on precise measurements are reminiscent of Pascal's contemporary work with the barometer, discussed in Chapter Five. We may recall that Pascal orchestrated the climb of the Puy-de-Dôme with Torricelli's barometer and compiled a series of tables showing precise measurements of the height the mercury reaches at different altitudes.[63] Similarly, with the help of his Cimento academicians, Leopoldo compiled a series of tables that provided standard measurements of the freezing process of water. This demonstration of mixed mathematical skills and their application to physics provides us with a firm understanding of how the Cimento academicians, just like their colleagues in other parts of Europe, constructed knowledge claims based on a physico-mathematical and mechanical natural philosophy. Furthermore, the very basis of these observations, the assertion that freezing water expands rather than contracts, contradicted scholastic Aristotelian beliefs in elements and their qualities.

The Cimento's construction and interpretation of experiments to do with the effects of cold on water were therefore deeply rooted in the academicians' commitment to the mechanistic natural philosophy shared by their princely patron. But what we have seen of this case study thus far is only the beginning of such natural philosophical concern and contention. Following Leopoldo's experiment and the Cimento's quantification of the artificial freezing process, the group's mechanists began to question the effects of heat and cold according to the even more sensitive issue of corpuscularianism. Their venture into this domain raised more wide-reaching natural philosophical questions regarding matter theory and led to controversy between the Cimento's mechanists and Aristotelians.

6. 'EXPERIMENTS ON A NEWLY OBSERVED EFFECT OF HEAT AND COLD, RELATING TO CHANGES IN THE INTERNAL CAPACITY OF METAL AND GLASS VESSELS'[64]

The document containing Leopoldo's observations of the freezing process contains the first signs of the academicians' debates regarding the properties and effects of heat. Having distinguished five phases of freezing, Leopoldo had

[62] 'Si volle vedere il progresso dell'agghiacciamento altre volte veduto nella solita palla, e di più, osservare per via delle vibrazioni del pendulo i tempi, nei quali accadono le seguenti alterazioni di condensarsi e rarefarsi, che fa l'acqua naturale nel suo gelare'. BNCF, Ms. Gal. 262, f. 132v.

[63] See Chapter Five, notes 56 and 57.

[64] 'Esperienze intorno a un effetto del caldo e del freddo nuovamente osservato circa il variare l'-interna capacità de'vasi di metallo e di vetro'. Magalotti, *Saggi*, 203.

observed that the various rates at which the water moves depend on the warmth of the environment surrounding the vessel. One effect of heat on liquids was that the water's movement was less violent when the atmosphere was mildly heated. At the end of the manuscript documenting this experiment, the author noted that one of the Cimento's members tried to provide a causal explanation of this effect. 'Of these unexpected phenomena, one of our academicians attempted to render its causes.'[65]

Reading through the collection of the Cimento's works and manuscripts, we find that this is not the only reference to 'one academician' who was prepared to make some rather controversial theorising. In fact, in the pages of the *Saggi* introducing their experiments on the effects of heat and cold, we are again told that one academician had attempted to persuade his colleagues and his patron to adopt his corpuscularian beliefs. According to Magalotti, this academician made his claims after they had observed the effect of heat on a tube filled with water. Instead of seeing a 'jump upon immersion', such as when the water is subjected to cold, they saw the water drop suddenly, only to return slowly to its normal graduation and then continue to rise. Magalotti recorded that while some academicians justified the changes in the height of the water in the tube by referring to the Aristotelains notion of antiperistasis, others did not agree.[66] More specifically, Magalotti stated that 'it occurred to one of us to attribute it to a cause that various later experiments appeared to favour admirably'.[67] That causal explanation is reported in the following passage from the *Saggi*.

> Dealing first with the fall that follows the immersion of the vessel in hot water ... this occurs by the thrusting corpuscles of fire that evaporate from the water into the external pores of the glass, forcing it apart as if they were so many wedges. In this way the internal volume of the vessel is necessarily expanded, even before the corpuscles are transmitted through the hidden passages of the glass to the liquid inside. Cold too, contracting the same pores, makes the vessel too small for the volume of water that is in it, before this volume, still lacking the new cold, becomes smaller.[68]

So, according to the academicians, the result of the intrusion of corpuscles into the glass is that the volume of the container is momentarily increased until those

[65] 'Di questi inaspettati accidenti alcuno de' nostri accademici tentò di renderne le caggioni'. Abetti and Pagnini (eds.), *Le Opere*, 390.

[66] It was traditionally thought that a body possessing a quality such as heat could at times intensify when surrounded by a body bearing the opposite quality, cold. According to scholastics, on such an occasion the water's sudden initial movement, the 'jump upon immersion', was therefore a result of it being surrounded by the quality of either heat or cold. This effect was referred to as antiperistasis.

[67] 'Questo effetto veduto fece cader nell'animo a qualcuno d'applicargli una tal cagione che poi deverse esperienze parve che mirabilmente favorissero'. Magalotti, *Saggi*, 204.

[68] 'ma bensì (trattandosi in primo luogo dell'abbassamento che segue nell'immergere i vasi nell'acqua calda) vogliono più tosto che ciò avvenga per lo ficcamento de'volanti corpicelli del fuoco che dall'acqua svapora nell'esterne porosità del vetro; i quali a guisa di tante biette sforzandolo, ne vien necessariamente dilatata l'interna capacità del vaso, anche prima che per l'occulte vie dello stesso betro si trasmettano nel liquor contenutovi: che il freddo poi ristrignendos gli stessi pori faccia divenir misero il vaso alla mele dell'acqua che v'è dentro, prima che la mole dell'acqua ancor digiuna del nuovo freddo non si diminuisce'. *Ibid.*

corpuscles pass through the glass and reach the liquid. There is a clear distinction here between the atomistic and mechanistic philosophy put forward by the academicians in this passage, and the scholastic position regarding 'antiperistasis'.

The author of this reasoning, and, in all probability, the subject of the anonymous references to 'one of our academicians' and 'one of us', is in fact, identified in the Cimento's diary entry for 3 December 1657. Not surprisingly, this entry states that it was Borelli who had noted the effect of heat on vessels containing liquid and who had claimed that the dilation of the glass container was due to 'the intrusion of the heat atoms in the minute parts [particelle] of the glass'.[69] In proposing a causal explanation for the effects of freezing and heating on all solids and liquids, Borelli rejected Aristotle's qualitative natural philosophy and drew on the Epicurean atomism that we have seen in Gassendi's work, and in some parts of Galileo's work.[70]

This natural philosophising was reflected in the first experiment narrated in the *Saggi*, where 'several small hollow enamel balls, hermetically sealed, were enclosed in a glass bulb full of water' (Figure 13).[71] These balls would normally rise or fall in the water depending on whether the water was heated or cooled. Yet when the bulb was submitted to either hot or cold conditions, the balls actually failed to move for some time, despite the initial rise or fall of the height of the water. It was only after that initial sudden jump or drop that they moved. So, as the academicians concluded in the *Saggi*: 'This phenomenon really appears to suggest more strongly that water, and other liquids as well, do not move of themselves in these first motions, but merely in obedience to the changes of the vessels.'[72] In other words, it was believed that this experiment verified the notion that the atoms of heat and cold moved through the pores in the vessel to either amplify

[69] 'per l'intrusione degli atomi ignei nelle particelle del vetro'. BNCF, Ms. Gal. 262, f. 47r.

[70] However, it is also clear from this citation that the Cimento still employed a qualitative atomism. That is, while the 'corpuscles of fire' act mechanically when they penetrate the glass owing to their shape and movement, they still seem to be bearing the quality of heat. This was not stated explicitly by Magalotti, but there appears to be an ambiguity in the theoretical framing of the academicians' work on the effects of heat and cold, as presented in this passage, that suggests a belief in qualitative atomism, of the type that was noted in Galileo's and Gassendi's work. In particular, in the latter part of the passage, Magalotti mentioned how the corpuscles of cold penetrating the glass take effect on the volume of the container before the 'new cold' reaches the liquid. This clearly assumes that cold and heat are properties of those invading corpuscles. In other words, just like Galileo and Gassendi, the academicians attempted to construct a wholly mechanical and corpuscularian philosophy of nature. Yet in their efforts to present this philosophy in their work on heat and cold, they still referred to the existence of qualities and their role in the freezing and heating processes. This is not to suggest that the academicians were still purposefully hanging to the remnants of Aristotelian ontology, but rather that in their ambiguous utterances concerned with the atomistic properties of heat and cold, they simply fell short of being what we might consider pure mechanists dealing only with mathematical explanations of natural phenomena. It should be noted that, with regard to the main argument of this chapter, this did not stop them from pursuing physico-mathematical, mechanical, and anti-Aristotelian principles.

[71] 'Si chiusero in una palla di vetro piena d'acqua parecchie palline di smalto vote e sigillate alla fiamma'. Magalotti, *Saggi*, 204.

[72] 'Riprova in vero di qualche apparenza per insinuar maggiormente che l'acqua, e così gli altri liquori, in quei primi movimenti non si muovono per loro stessi, ma obbediscono meramente all'alterazioni de'vasi'. *Ibid.*, 206.

Figure 13. First experiment testing the effects of heat and cold. L. Magalotti, *Saggi di naturali esperienze*, Florence, 1667, 180. Courtesy of the IMSS Biblioteca Digitale.

or restrict the space for the water. This is why, so the Cimento corpuscularians believed, the water made a sudden movement before actually being affected itself by the atoms. This also supposedly explained the delay in movement of the enamel balls.

This experiment was included in Magalotti's first draft of the *Saggi* and in his editorial notes for this draft, Borelli suggested that they could reveal more in the text about how this experiment, and those that followed on the effects of heat and cold, were conceived by him and the Prince. Borelli argued: 'I would believe that it would not bring dishonour to the Accademia to narrate how this matter was pursued.'[73] He then revealed that in October 1657, the Prince had realised, 'before anyone', the effect of heating on vessels, such as is described in the experiment above.[74] Furthermore, he stressed that since the above experiment was only to demonstrate an effect, he became determined to find the causes behind the changing conditions of liquids and solids submitted to heat or cold. Borelli was particularly inspired, so he stated, by an experiment that the Prince had described to him in a letter, concerned with the dilation of a heated metal ring.[75]

73 'Crederei che il raccontar questo fatto come seguì non arrechi disonore all'Accademia'. Abetti and Pagnini (eds.), *Le Opere*, 335.

74 It must be made clear, however, that there appears to be no evidence that substantially verifies that the experiment with the enamel balls was indeed the Prince's suggestion.

75 'io ebbi la fortuna di trovarne la vera cagione, la quale trovando ostacolo si fece la sperienza dell'anello infuocato, della quale S.A. mi fece onore di scrivermi a Pisa, da dove io proposi e predissi l'effetto degli anelli di legno con altre circostanze'. Abetti and Pagnini (eds.), *Le Opere*, 335.

This praise for the Prince's role in these experiments and this show of determination to find the causes for the vessel's dilation, are also reflected in two letters that Borelli wrote on 13 and 14 November 1657. The first correspondence was to Viviani in which Borelli once again referred to a letter he had received from Leopoldo describing an experiment where a metal ring was seen to dilate when heated.[76] A similar experiment testing the dilation of a bronze ring, probably suggested by Borelli, was performed by the academicians on 3 December, the same day that the diary's author recorded Borelli's claim that fiery atoms cause the expansion of a glass container. It is likely that Borelli also proposed that the academicians perform this experiment on thermal dilation that Leopoldo had suggested to him. The bronze ring was made to fit perfectly on a plug, but after being subjected to fire, it was instead seen to fit quite loosely. In contrast, the fit tightened dramatically when the ring was frozen (Figure 14).[77]

Figure 14. Second experiment testing the effects of heat and cold. L. Magalotti, *Saggi di naturali esperienze*, Florence, 1667, 180. Courtesy of the IMSS Biblioteca Digitale.

[76] BNCF, Ms. Gal. 283, f. 14r; Galluzzi, 'L'Accademia del Cimento', 811.
[77] BNCF, Ms. Gal. 262, f. 47r.

Borelli was overjoyed with this experiment, not only mentioning it to Viviani, but also praising the Prince for it in a letter dated 14 November. This is arguably the most important of Borelli's letters we have seen so far, since it was the first to reveal explicitly the corpuscularian natural philosophy that he had evidently felt inspired to pursue following the freezing experiments performed in October, 1657, and the role of experiments in his investigations.

> Both experiments made by Your Most Serene Highness to clearly convince that fiery bodies dilate the glass vessel, and the deprivation of them tightens it, appeared to me to be so beautiful, delicate and measured to the necessity, that it seemed like a shame to me not to laud them and to praise them highly, as they deserve. I do not know how to find enough commendations to celebrate the generous Patron, Promoter, and Author of an Academy so useful and necessary for the acquisition of the true Philosophy .... I found myself believing, by virtue of these very clear experiments ..., that I should not find any difficulty in persuading others of the truth of such a conclusion,[78] and also to have such efficacious proof that heat is a substance, and that on the contrary, cold is the deficiency of it [heat] (since always seeing, from any significant degree of cold, the restriction of the volume of the vessel, and never its dilation ..., it appeared to me to be very reasonable to conclude that Gassendi's cold atoms were nothing more than the deprivation of heat).[79]

Clearly Borelli was constructing and interpreting experiments performed within the Accademia del Cimento according to mechanistic and corpuscularian concerns. But unlike the conclusion that was reached in the *Saggi* where atoms of both heat and cold were considered, Borelli argued that cold atoms do not actually exist. This issue will be addressed later when we examine how Borelli devised this notion, and how he explained it explicitly in another letter to Leopoldo almost one year later. For the moment we will follow how the academicians' heat and cold experiments, performed mostly during November and December of 1657, were used to argue for contrasting and competing natural philosophical positions.

[78] This statement amounts to Borelli telling the Prince that he needed experiments as a tool of authority and persuasion in favour of his natural philosophical beliefs. This was a crucial revelation by Borelli since, as we shall soon see, at times he even preferred to rely more on his experiments than mathematical demonstrations to convince his rivals. This point will also be mentioned later in this chapter.

[79] 'Ambedue le sperienze, fatte da V.A.Sma per evidentemente convincere, che i corpi focosi dilatano il vaso del vetro, e la privazione di essi lo ristringe, mi hanno pavuto tanto belle, gentili, et accommodante al bisogno, che mi parrebbe peccato a non le lodare e massimamente commendare, come esse meritano, non sapendo trovare encomi sufficienti per celebrare il generosissimo Mecenate, Promotore, et Autore di una Accademia tanto utile e necessaria per l'acquisto della vera Filosofia. ... io mi credea in virtù di queste evidentissime sperienze ... non dover trovare difficoltà veruna a persuadere la verità di tal conclusione, e anche avere un'assai efficace prova che il Calore sia assolutamente corpo, e che per il contrario il freddo sia mancamento di esso (poichè vedendosi sempre mai da qualunque efficace grado di freddo ristringersi la mole del Vaso, e non mai dilatarsi ... mi parea poter assai ragionevolmente concludere, che gli Atomi frigorifici del Gassendo fossero non altro che privazione di calore)'. BNCF, Ms. Gal. 275, f. 84r; Targioni Tozzetti, *Notizie*, 404–405.

7. HEAT AND COLD: QUALITY VERSUS SUBSTANCE

Not all the academicians were convinced of the validity of the corpuscularian beliefs that Borelli had become so dedicated to proving. In fact, as was recorded in the diary, the ring experiment was performed on 3 December in response to objections from someone in the group. Those objections, once again, came from the only two members of the Cimento who seemed determined to defend scholastic thought, Marsili and Rinaldini. In a letter to Leopoldo written from Pisa on 11 November 1657, Rinaldini made part of his objections clear. He expressed his doubts regarding the mechanistic interpretation of the experiments testing the dilation of the heated ring. He claimed that rather than the heat causing the dilation of the ring itself, it made the surrounding air lighter, thus allowing for the ring to be more loosely applied to the plug than when the air is dense. Therefore, Rinaldini concluded, they could not be sure that the effect on the ring was due to the intrusion of fiery corpuscles, as Borelli insisted. Indeed, they could not even say for certain that the inner diameter of the ring became greater after being subjected to fire.[80]

But this was not Rinaldini's first contribution to this topic. Two months prior to his letter to Leopoldo, on 7 September 1657, well before Borelli or Leopoldo had carried out any experiments to do with the effect of heat on solids or liquids, Rinaldini had suggested an experiment which, he believed, supported the Aristotelian notion that heat is nothing more than a quality. As is revealed in the Cimento diary, the aim of this experiment was 'to identify whether the expansion of heat and cold is spherically uniform'.[81] A metal ball was frozen in ice for almost two days, and when it was taken out, small thermometers were supported above, below, and to the sides of the ball. It was found that the thermometer below the frozen solid was the most affected by the cold, while those on the side showed little change, and the thermometer above the ball even less. The opposite effects were visible when a ball was used that had been exposed to fire instead of ice. In this case, the thermometer to register the greatest change in degrees was the one held up above the ball.[82]

With this experiment Rinaldini tried to verify that the quality of heat was transmitted in some manner through the air, but without any explicit theoretical framing, the Cimento's mechanists seemingly saw little value in his experiment, and initially paid little attention to it. So, several weeks later, on 11 November, Rinaldini composed a letter to Leopoldo from Pisa attempting to explain the

[80] BNCF, Ms. Gal. 275, ff. 83r–83v.

[81] 'Per riconoscere, se l'espansione del caldo, e del freddo fosse sfericamente uniforme'. BNCF, Ms. Gal. 262, f. 31v.

[82] 'Per riconoscere, se l'espansione del caldo, e del freddo fosse sfericamente uniforme doppo molte maniere di accetarsene con porre instrumentini di 10 gradi attorno due palle di metallo, cioè sotto sopra e dai lati, l'una stata 46 ore sepolta nel diaccio, l'altra arroventata nel fuoco si trovò dei detti strumentini del caldo più alerato quello di sopra, meno quei dei lati, e meno di questi quello di sotto, e per lo contrario, di quei del freddo si mutò più degli altri quel di sotto, meno quei dei lati, e meno di questi quello di sopra'. BNCF, Ms. Gal. 262, ff. 31v–32r.

significance of the experiment. He insisted that it had not been adequately discussed when it was performed in September, and he begged the Prince that the experiment be repeated with the following explanation in mind:

> when the ball is cooled the little instrument below is affected the most, that is, it contracts more than the upper one, not because an effusion of corpuscles takes place, which, being heavy, descend, so that the cold works downwards more than upwards, as heat on the other hand, because its lighter corpuscles go upwards etc, but because the cold of the ball cools the air around it, makes it heavier, and makes this heavier part descend. ... On the other hand if the ball is heated it warms the air and makes it lighter so that it rises ....[83]

In the same letter to Leopoldo, Rinaldini also stated that he had some doubts regarding the experiments testing the dilation of the heated ring. Rinaldini was clearly expressing his opinion that a corpuscularian understanding of the effects of heat and cold is out of the question. For him, as for all scholastics, heat and cold are qualities that can be transmitted from substance to substance; they are not qualities of atoms (which do not exist) nor effects caused by the motion or arrangement of such atoms.[84] This natural philosophical contention is also revealed in a manuscript, published by Abetti and Pagnini, describing Rinaldini's experiment and the controversy over interpreting its signifcance. This document begins by clearly stating the natural philosophical issues at stake in the construction and interpretation of Rinaldini's experiment.[85]

> It is commonly said among the Peripaticians that a natural agent diffuses its action equally in every direction, this proceeding with a uniformly difform motion, as they assert.[86] It therefore occurred to us to find out about this by experiment, as far as this

[83] As translated by Middleton. *Ibid.*, 302. 'Essendo la palla AB, gli strumentini CD, EF, mentre la palla sia aggiacciata, lo strumentino di sotto opera più, cioè più si costringe che quello di sopra; non perchè si faccia effusione di corpuscoli, li quali per esser grave vadino all'in giù, onde più operi il freddo all'ingiù che all'insu, come all'incontro il caldo, perchè I suoi corpuscoli son più leggieri, vadino all'insù ec, ma perchè il freddo della palla raffreddando l'aria ambiente, e rendendola più grave, fa che queste parti più gravi descendino. Onde le più fredde attorniando lo strumentino EF, non è meraviglia che lo rendino più raffredato, si che l'acqua dentro di esso più si costringa. All'incontro essendo riscaldata, e riscaldando l'aria la rendono più leggiera; si che questa andando all'insù, ed attorniando lo strumento CD maggiomente lo riscaldono, de quello possino fare allo strumento EF'. BNCF, Ms. Gal. 275, f. 82v.

[84] Middleton claims that when Rinaldini suggested this experiment, and a variation of it was performed on 11 September 1657, 'he saw reason for postulating corpuscles, either of heat or of cold, but was satisfied with simple transfer of heat (whatever heat may be) from the ball to the air'. That is to suggest that Rinaldini was somehow interested only in the pursuit of a scientific insight, regardless of the contrasting corpuscularian and scholastic beliefs that competing members of the Cimento upheld. Middleton further excludes Rinaldini from the natural philosophical debate by suggesting that it was only Marsili who offered resistance to Borelli's claims. However, this was clearly not the case. Middleton, 'Carlo Rinaldini and the discovery of the convection in air', *Physis* (1968), **10**, 304.

[85] 'Esperienza nel modo che si dirà per venire in cognizione se il calore si diffonda sfericamente'. Abetti and Pagnini (eds.), *Le Opere*, 390.

[86] Note the continued appeal to the language of the latitude of forms deriving originally from the fourteenth century, and as Marshall Clagett points out, also used by Galileo. Clagett, *The Science of Mechanics*, 219, 253.

> might be possible. The principal object, then, was to see in some manner whether heat is a quality, or whether it is nothing but corpuscles, of any shape whatever.[87]

Suddenly, there was plenty at stake in the repetition of Rinaldini's experiment inside the Cimento. The mechanists in the group were hijacking Rinaldini's search for the qualities of heat and cold in nature, in order to promote their own natural philosophical agenda. This established an interesting and tense situation inside the Accademia that was reflected in the remainder of this narrative of Rinaldini's experiment. The document describes how the academicians carried out the experiment with thermometers, yielding the observation that the top thermometer is affected by heat and the bottom by cold. Finally, a conclusion was reached that was not surprisingly pro-corpuscularian and against the notion that heat is a quality.

> From this it appears that we might gather that heat does not diffuse equally in every direction, but more upward than downward. From this it similarly appears that we might conclude what was chiefly being sought, that heat is not a quality, because if it were, it would appear that it should diffuse equally in every direction, exactly as is asserted by the Peripatetics. But if it consists of corpuscles, as is claimed by Democritus, it does not seem difficult to understand that they would move mainly upward.[88]

The reasoning behind this conclusion is not entirely clear in this passage. More specifically, the author does not explain why the heat corpuscles 'would move mainly upward'. However, we can assume from the letter cited earlier from Rinaldini to Leopoldo written on 11 November, that Rinaldini explained the corpuscularian belief that corpuscles carrying heat were presumed to be light while cold corpuscles were supposedly heavy. Now, the ambiguity surrounding this claim lies with the fact that this position did not seem to harm Aristotelians at all. That is, the mechanists' own explanation of what occurred with the thermometers actually did not deny the scholastic notion of positive levity. As Middleton points out, they had not considered 'that the postulated upward motion of their particles of heat was a variant of the Peripatetic dogma of the positive lightness of fire'.[89] It was perhaps because of this ambiguity in the mechanist explanation of the experiment that 'one of the academicians', as the document continues, formulated a reply defending Aristotelian elemental qualities.

[87] As translated by Middleton, 'Carlo Rinaldini', 303. 'E' cosa assai vulgare appresso i Peripatetici che l'agente naturale diffonda la sua azione ugualmente d'ogni intorno procedendo quella con una uniforme difformità come egli asseriscono. Perciò cadde in pensiero d'accertarsene per quanto fusse possibile con l'esperienza. Il fine poi principale fu per veder in qualche modo se il calore sia qualità, o pure altro non sia che corpiccioli d'una tale qual si sia figura'. Abetti and Pagnini (eds.), *Le Opere*, 390–391. Although Magalotti claimed that this was the traditional theory regarding the diffusion of heat as it is 'commonly said among the Peripaticians', we should note that it is not the theory that Rinaldini, in defence of Aristotelianism, provided in his 11 November letter to Leopoldo.

[88] As translated by Middleton, 'Carlo Rinaldini', 303. 'Di qui parve che si potesse raccorre ch'il calore non si diffonda ugualmente per ogni parte, ma più all'insù, che allingiù. Donde poi similmente parve che si potesse conchiudere quello che principalmente si cercava, cioè che il calore non fusse qualità, perciò che, parrebbe che dovesse diffondersi ugualmente per ogni parte com'appunto da'Peripatetici viene asserito. Ma essendo corpiccioli come da Democrito si pretende non par difficile intendere ch'egli abbino il lor movimento maggiormente all'insù'. Abetti and Pagnini (eds.), *Le Opere*, 390–391.

[89] Middleton, 'Carlo Rinaldini', 305.

> It appeared to one of the academicians that he could reply to this experiment by saying, in favour of the Peripatetics, that the quality of heat, warming the ambient air, makes the hotter parts ascend, and the cooler remain below; so that the warmer parts surrounding the upper thermometer make it move more than the lower one, which is surrounded by the parts of the air that are less warm.[90]

The academicians were clearly constructing and interpreting an experiment according to the competing and contrasting natural philosophical positions of the leading academicians. Borelli and Viviani were trying to dismiss any Aristotelian beliefs in elemental qualities to the point where they did not seem to mind making a concession to the scholastics that fire corpuscles contain the property of heat, as long as their central mechanistic and anti-Aristotelian argument could be maintained. In the meantime, Rinaldini was evidently quite vocal and effective in his defence of Aristotelians against Borelli's and Viviani's corpuscularian and mechanistic proposals.[91] It is therefore not surprising to find, as we shall now see in more of the academicians' correspondence, that Rinaldini, as a defender of Aristotelian principles, continued to object strenuously to the corpuscularian interpretations of the effects of heat and cold on all solids and liquids. Furthermore, he attempted to negotiate the significance of the Cimento's experiments on the topic, according to his own scholastic views.

8. RINALDINI STANDS HIS GROUND

In his 13 November 1657 letter to Viviani, Borelli did not take lightly Rinaldini's outspoken scholastic objections to corpuscularianism. After mentioning that he had received Leopoldo's letter regarding the dilation of a metal ring when heated, Borelli informed Viviani that someone still had the 'malice' to oppose the belief that the internal diameter of the ring does not expand.[92] We can only assume that Borelli was referring to Rinaldini, who had just written his letter to Leopoldo providing an explanation of the effects of heat and cold based on the transmission of qualities through the air. In his letter to Leopoldo on 14 November, Borelli provided some further commentary, showing awareness that Rinaldini opposed the atomistic point of view.

> [T]here are those who are satisfied with those fertile, fair, and virtuous Peripatetic words, referring to the quality of heat and cold because *caloris est rarefacere, et frigoris condensare*. They still say that they can defend their entire appearance, and moreover, their reasoning, without resorting to atoms of fire. Thanks to this noble

[90] As translated by Middleton, 'Carlo Rinaldini', 303. 'Parve ad uno degl'Accademici potersi rispondere all'esperienza fatta, dicendo in favore de'Peripatetici, che la qualità del calore riscaldando l'aria ambiente fa che le parti più clade ascendino, e le men calde rimanghino di sotto; di modo che quelle più calde attorniando il termometro superiore lo facci operare più dell'inferiore, circondato dalle parti men calde dell'aria'. Abetti and Pagnini (eds.), *Le Opere*, 390–391.

[91] With such controversy evident even in the presentation of this experiment, it is not surprising that this narrative was not admitted into the *Saggi*. As we shall see in Part Three, this was precisely the type of contention and ambiguity that Leopoldo was eager to avoid displaying in the Cimento's only publication.

[92] BNCF, Ms. Gal. 283, f. 14r. See also Galluzzi, 'L'Accademia del Cimento', 811.

> Opponent, you will need to do other experiments, which are recorded on the attached sheet. So, others in no way wish to admit, that because of the intrusion of atoms, or calorific wedges, the internal concave surface of the glass can dilate, even though the volume of the said glass increases notably.[93]

Evidently Borelli proposed further experiments that, he believed, would put to rest the Peripatetic opposition from Rinaldini. In other words, Borelli proposed that the effect of heat on solids be made more obvious to the senses and thus harder for critics of the corpuscularian theory to deny. He was calling upon the authority that his experiments carried, in order to persuade his opponents of the validity of his natural philosophical beliefs. This is the persuasive and authoritative role of experiment that was discussed in Chapter One and that we have seen mentioned by Borelli himself when he praised the Prince for his experiments in the letter dated 14 November 1657.[94]

We cannot be sure what experiments Borelli described in this attachment, but it would not be unreasonable to assume that they included those experiments performed inside the Cimento in early December. We have already seen that one of those experiments was a replication of Leopoldo's suggestion that the inner diameter of a bronze ring expands to fit more loosely on a plug. On 6 December, two more experiments were performed which assisted the corpuscularian academicians tremendously in grounding their anti-Aristotelian claims.[95] Both were eventually published in the *Saggi*.

The first involved the use of a hollow glass ring with two funnels (Figure 15). As hot water was poured into one funnel, the academicians assumed that the air could escape quite easily through the other, and the ring could expand because of the heat that fills its inside. To prove that the inner diameter of the ring expanded, they placed a cross over the ring, made from two enamel rods, and adjusted so that their extremities were only barely able to lean on the glass ring. As the ring heated and expanded, the cross came into less and less contact with the glass, until it fell through the inner diameter.[96]

The second experiment demonstrated that the inner diameter of a ring expands, not only because of the intrusion of corpuscles of fire, but also because of moisture. A ring was made out of boxwood (Figure 16), and a conical frustum (the lower part of a cone with the top cut off) out of steel with several circles inscribed on its surface parallel to its base.

[93] 'Vano è stato il mio credere, poichè vi ha chi si appaga di quei fertili, sufficienti e virtuosissimi vocaboli Peripatetici, cioè di qualità calda, e fredda perchè *caloris est rarefacere, et frigoris condensare*, e però dicono potersi salvare senza ricorrere ad atomi di fuoco, tutta l'apparenza et il resto della ragione, che'io ne adduco. In grazia di questo nobile Oppositore, bisognerà fare alcun altre sperienze, le quali nel collegato ricordo vengono registrate. Altri poi non vogliono in niun modo ammettere, che per l'intrusione delle biette, o cunei calorifici possa mai dilatarsi la superficie concava interna del vetro ancorchè la mole di detto vetro venga notabilmente accresciuta'. BNCF, Ms. Gal. 275, f. 84v.

[94] See n. 78, above.

[95] Neither of these experiments were mentioned in the official Cimento diary, except for the inclusion of their diagrams along the margins. But they were recorded in the second Court diary. BNCF, Ms. Gal. 261 ff. 60v–61v.

[96] Magalotti, *Saggi*, 208.

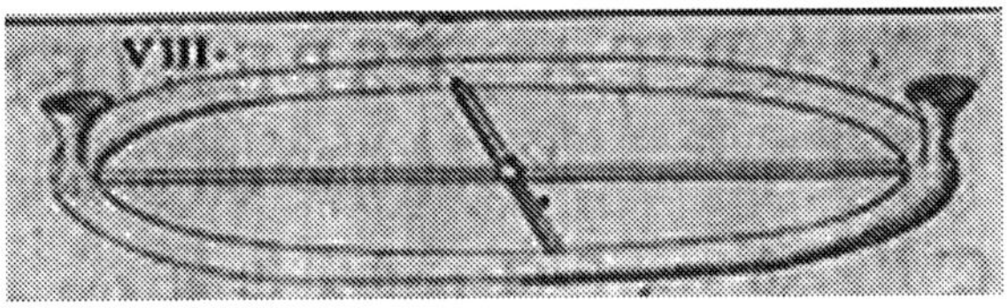

Figure 15. Third experiment testing the effects of heat and cold. L. Magalotti, *Saggi di naturali esperienze*, Florence, 1667, 180. Courtesy of the IMSS Biblioteca Digitale.

Figure 16. Fourth experiment testing the effects of heat and cold. From L. Magalotti, *Saggi di naturali esperienze*, Florence, 1667, 180. Courtesy of the IMSS Biblioteca Digitale.

The ring was fitted over the frustum and the academicians made a careful note of which circle marked the bottom of the ring, which was then left to soak in water over three days. When the entire wooden ring was penetrated by moisture, it was once again placed over the frustum. The academicians noted that the inner concave surface had apparently expanded, 'for the base of the ring went down a considerable distance below the former circle'.[97]

It would appear that both these experiments served to demonstrate, according to the majority of the academicians, that heat, like moisture, is a substance. This explanation still contained vague references to heat as a property of fire corpuscles, but the academicians' main natural philosophical agenda was far more pertinent: to provide a persuasive and authoritative experimental demonstration of the notion that heat is not an elemental quality, but rather that it is a substance that causes the expansion of solid bodies. Therefore, in opposition to the notion of the positive levity of fire, as well as the transmission of the qualities of heat and cold, Borelli was attempting to persuade others of the validity of his mechanistic and corpuscularian views.

Nevertheless, Rinaldini remained unconvinced. He had already made his anti-corpuscularian sentiments quite clear in his letter to the Prince of 11 November. On 17 and 26 November 1657, Viviani wrote to Rinaldini declaring his support for Borelli's demonstrations of thermal dilation and offering new observations and a mathematical demonstration that would persuade the Cimento's most outspoken scholastic thinker to accept the corpuscularians' claims.[98] Replying to both letters, on 19 November and 3 December, Rinaldini refused to accept the efficacy of Viviani's proposed observations and demonstrations. In typical Aristotelian fashion, he claimed that the abstractness of mathematics and geometry could not possibly provide a valid basis for studies in physics.[99] Of course, Rinaldini could have, and possibly may have, applied this argument to most of the academicians' work analysed in this chapter, regarding freezing and the effects of heat and cold. But this appears to be the only occasion on which he expressed this typical Aristotelian position, perhaps even demonstrating the pressure he was under defending Aristotelianism and the lengths he went to in order to discredit his natural philosophical opponents.

This dispute within the Cimento shows the extent to which experiments can be constructed to favour one or the other position within natural philosophy. While Rinaldini and Marsili defended the Aristotelian position regarding heat and cold

[97] 'la superficie concava era dilatata, calando la base dell'anello per notabile spazio sotto il cerchio di prima'. *Ibid.*, 207. This experiment is also described in a manuscript published by Abetti and Pagnini (eds.), *Le Opere*, 417–418.

[98] Viviani proposed that glass and wooden rings of varying thickness and diameter be heated. At different stages of the heating process, the internal diameter of the rings could be measured and an exact measurement of the heating process would be obtained. Abetti and Pagnini (eds.), *Le Opere*, 415–416. This is, of course, reminiscent of the academicians' application of their skills in mixed mathematics and statics when measuring the freezing process and reflects their commitment to a physico-mathematical and mechanical natural philosophy. See also Galluzzi, 'L'Accademia del Cimento', 812.

[99] *Ibid.*, 812–813.

as qualities, the notion of positive levity, and even antiperistasis, Viviani and Borelli had the opposite aim. As mechanists and corpuscularians, they were determined to interpret and negotiate the significance of their experiments in accordance with the type of Gassendian atomism and Galilean terrestrial mechanics that were described earlier in this chapter.

9. BORELLI'S CONCLUSIONS: THE DEPRIVATION OF HEAT

By the end of these few weeks of correspondence, experimenting and natural philosophical contention between the academicians, Borelli expressed his exasperation at the refusal of Rinaldini to accept the dilation of solids due to corpuscles of fire. On 28 November 1657, he wrote the following to Viviani:

> I cannot understand how there can be found such stubborn minds that they do not allow themselves to be persuaded .... In short, I cannot find any other excuse for him other than maybe from the beginning he committed himself to contradicting my reasoning and now out of politics, by which he tends to philosophise and to make use of philosophy, he continues to hold my reasoning in contempt.[100]

Not only was Borelli annoyed at Rinaldini's refusal to accept his point of view on the topic, but also several months later, he seemed determined to strengthen his corpuscularian beliefs. In particular, he wished to argue that only atoms of heat exist, and that cold is caused only by the absence of heat atoms. Borelli had already mentioned this belief in his 14 November letter to Leopoldo cited earlier, but in September of 1658, he provided a detailed explanation of this concept. This came after Leopoldo once again weighed into the controversy, probably also during September of that year, by suggesting his own experiment that was also laden with natural philosophical aims and interests.[101]

In the *Saggi*, Magalotti provided a brief report of an experiment testing 'whether the cooling of a body results from the entry of some kind of special atoms of cold'.[102] An empty glass flask with a very long and thin neck was sealed with a flame and placed in ice. Then, breaking the neck of the flask underwater, the academicians observed that water was sucked in to fill the space in the instrument. In other words, during the cooling of the flask, no substances, such as atoms of cold, had made their way through the glass and into the instrument. Instead, the air

[100] 'io non so capire come si possan trovare cervelli così contumaci che non si lascino persuadere ... In somma, non so trovar altra scusa per lui se non che forse egli sul principio s'impegnò a contraddire tal mia ragione et hora per politica, colla quale ei suol filosofare e servirsi della filosofia, continua a disprezzare così fatte ragioni'. BNCF, Ms. Gal. 283, ff. 22r–23v. See also Galluzzi, 'L'Accademia del Cimento', 812–813.

[101] It is not known exactly when Leopoldo performed this experiment. While it was described in the diary, no date was given; it was only entered in between the entries for 7 September 1658 and 20 May 1660. Judging by when Borelli wrote what appears to be a response to this experiment, we may be willing to believe that it was performed in 1658.

[102] '... se il raffreddarsi d'un corp derivi da insinuazione d'alcuna spezie d'atomi particolari del freddo'. Magalotti, *Saggi*, 252.

inside the flask became thinner, and when the instrument was broken open, water rushed in to fill the space occupied only by rarefied air. When the flask was heated instead of cooled, and the glass was again broken underwater, the opposite seemed to occur. Bubbles were seen to emerge in the water, suggesting to the academicians that something was evacuating the flask that had not been there when the instrument was cooled. That is the extent of the description of this experiment in the *Saggi*. Magalotti once again hinted at the corpuscularian concerns of the academicians, yet as always, he avoided openly declaring the academicians' natural philosophical position or whether they drew any explicit theoretical conclusions from the experiment.

Nevertheless, much more detail is given in another description of this experiment, possibly a draft manuscript for the *Saggi*, entitled: 'Experiments made in the following manner to make sure if cold consists of some corpuscles, or if is it a mere deprivation of heat, that is to say, of calorific corpuscles.'[103] To begin with, the final line of this manuscript reveals that 'the author of this experiment was the Most Serene Prince Leopoldo'.[104] So, just as we have seen previously that the Cimento's patron was clearly a mechanist and was applying experimental techniques and mixed mathematical skills in order to pursue his own interests in freezing and natural philosophy, he was also eagerly assisting to construct experiments suggestive of the existence of corpuscles in the effects of heat and cold. Therefore, once again we are assured that from 1657 when the Cimento began, until 1662 when the group devoted itself to carrying out a publication, Leopoldo was not enforcing any type of atheoretical and inductivist policy that restrained the academicians from pursuing and debating their natural philosophical aims and interests. In fact, the Prince himself was engaged in this natural philosophical contention. The opening paragraph of this manuscript also reveals a great deal about what issues were at stake for Leopoldo and those Court members interested in corpuscularian physics.

> Those who finally took themselves to be convinced by experiments and by reason, that cold is no more a quality than heat, found the same difficulty that provided so much trouble [travaglio] among those who believed that heat is a quality. That is to say, if cold is purely the deprivation of heat, or should they say calorific corpuscles, a body is made cold only because of the deprivation of such corpuscles. This was the aim of the following experiment.[105]

[103] 'Esperienze nella forma che segue per assicurarsi se il freddo consista in alcuni corpiccioli, o pure sia una mera negazione del caldo, cioè a dire de' corpuscoli calorifici'. BNCF, Ms. Gal. 263, ff. 59r–60r. Also published by Abetti and Pagnini (eds.), *Le Opere*, 392.

[104] 'L'autore di questa esperienza fu il S.mo Principe Leopoldo'. BNCF, Ms. Gal. 263, f. 60r; Abetti and Pagnini (eds.), *Le Opere*, 392.

[105] 'Quelli che finalmente credettero esser convinti dall'esperienza e dalla ragione, ch'il freddo non fusse qualità come nè meno il caldo, incontrarono quella difficoltà medesima che diede molto travaglio a quelli presso de' quali il calore era qualità; cioè a dire se il freddo sia una pura negazione di calore, o voglian dire di corpiccioli calorifici, sì che un corpo esser freddo, altro non sia che esser privo di corpiccioli somiglianti. Questo fu il fine della seguente esperienza'. BNCF, Ms. Gal. 263, f. 59r; Abetti and Pagnini (eds.), *Le Opere*, 392.

The experiment was then described as in the *Saggi*, but clearly it was constructed with the aim of resolving a natural philosophical question grounded firmly in particular corpuscularian principles (first propounded by Galileo in *The Assayer*), according to which only corpuscles of heat exist, cold being due to the mere absence of heat corpuscles.[106] However, as Paolo Galluzzi argues, the notion that only heat corpuscles exist, created a problem for the corpuscularian academicians who had earlier implied that cold corpuscles were responsible for the rarefaction of water during the freezing process.[107] Unless this problem was resolved, serious questions could be asked about the validity of the Cimento's corpuscularian beliefs. In effect, Leopoldo's experiment was a challenge to his academicians to provide hypotheses that could comply with the effect of heat and cold on solids. Borelli responded to the Prince's experiment in a letter addressed to him from Rome on 21 September 1658.[108] Here Borelli mentioned a sheet attached to the letter which describes his three hypotheses, 'each of which can comply with the marvellous appearance of the dilation of the water in the act of freezing'.[109] What follows in this document (published by Abetti and Pagnini) is a detailed description of Borelli's atomistic speculations, focused particularly on the assumption 'that cold is the deprivation of heat'.[110]

To begin with, Borelli noted several suppositions and three propositions. He argued that the minimum components of water, the tiny corpuscles, take the form of perfect Platonic geometrical solids. They are much smaller than atoms of air (and often are found to be inside the air atoms, as can be seen on many occasions when air and water seem to be mixed together), and as Galileo believed, are capable of coming together delicately yet strongly, as if by magnetic attraction. However, Borelli argued, the interposition of fiery atoms among the water atoms, does not allow for the water atoms to attract in a magnetic fashion. During freezing, he suggested, those fiery atoms are ousted and in their absence many water atoms unite and the smooth liquid appearance of water gives way to a more solid substance, ice. Also during this process, many water atoms that are normally sitting inside the large atoms of air escape to unite with each other, leaving pure air atoms still mixed in with the water. Borelli then suggested that the larger atoms of air occupying the spaces formerly taken up by fiery corpuscles, give the frozen water a greater volume. According to Borelli, it was clear that cold atoms were not responsible for the increase in the volume of the freezing water, because most other frozen substances, especially solids, were seen to diminish in volume, rather than increase.

106 This link between the corpuscularian beliefs of Galileo and the Cimento is also noted by Grilli and Sebastiani, 326.

107 Galluzzi, 'L'Accademia del Cimento', 816.

108 Since Borelli was in Rome during this time collaborating on the translation of Apollonius' lost books, the Cimento's meetings were suspended.

109 'Tre ipotesi con ciascuna delle quali si può sodisfare alla maravigliosa apparenza della dilatazione dell'acqua nell'atto dell'addiacciarsi'. Abetti and Pagnini (eds.), *Le Opere*, 412.

110 'supposto che il freddo sia privazione del calore'. *Ibid.*

Furthermore, frozen water becomes lighter, not heavier, showing that it had not absorbed any other bodies, such as atoms of cold.[111]

Having declared these suppositions, Borelli finally provided his three hypotheses. It is not necessary for us to examine these in detail. It will suffice to note that all three theories were based on speculation concerning the geometrical dimensions of water atoms, and the way that, during the freezing process, the shape and sizes of these atoms effect the interposition of atoms of air. For example, in the first hypothesis, after referring to the regular shape of water atoms as octahedrons, Borelli stated that if '20 or more octahedron solids are made of wood, smooth and all of the same size, a small collateral bump will be observed to greatly increase the interposed spaces between the said octahedrons'.[112]

The material discussed here shows that Borelli was determined to apply geometrical principles to his studies in physics, including his atomistic structure of nature. Not only did he insist, as we have seen throughout this case study, that Aristotelians such as Rinaldini were mistaken in their qualitative beliefs, but he also pursued a corpuscularian natural philosophy based on the geometrical, mathematical, and mechanical skills and commitments that he had developed throughout his career. As we saw in Part One, these were the intellectual concerns that he and Viviani had developed as members of the mechanistic tradition in natural philosophy in seventeenth-century Italy, and particularly as students of Galileo.

10. CONCLUSION

To conclude this analysis of the atomistic natural philosophical tradition pursued in the Tuscan Court, we may note the discussions that were carried out at that time inside the Accademia della Crusca, Tuscany's grandest and most prestigious academy devoted to literature. In 1666, Orazio Ricasoli Rucellai, the Crusca's leading member, published a text in which, drawing on his knowledge of ancient atomistic writings, together with the Cimento's work on atomism, he speculated about the existence of cold atoms.[113] With this in mind, and considering also that Viviani, Magalotti, and Carlo Dati were all members of the Crusca, Galluzzi suggests that the literary academy almost constituted a second academy for the Cimento's members, where they were free from 'the humiliating experimental discipline imposed by Leopoldo on the Cimento'.[114]

[111] Despite the mechanistic appeal to Platonic solids and to the shape and movement of corpuscles, Borelli still incorporated the qualitative aspect of corpuscularianism that, as was argued earlier, was also in Gassendi's and Galileo's work. Furthermore, he jeopardised his mechanistic principles by relying on the description of magnetic attraction (much like his description of the movement of planets) to explain the condensation of water during freezing. In any case, these issues should not distract us from the main agenda behind the academicians', and especially Borelli's, work – to establish a type of physico-mathematical and mechanical natural philosophy that could replace Aristotelianism.

[112] 'Si può sensatamente sperimentare l'effetto che produce la varia dispozione di detti corpi ottaedrici con fabbricarne 20 o più dei detti ottaedri di legno, lisci e tutti della medesima grandezza, si vedrà ad una minima scossa collaterale accrescersi grandemente gli spazi interposti tra i detti ottaedri'. Abetti and Pagnini (eds.), *Le Opere*, 414.

[113] This text was titled *Contro il freddo positivo*. See Galluzzi, 'L'Accademia del Cimento', 817–818.

[114] *Ibid.*, 818.

Given that the *Saggi* was written with minimal references to the type of theoretical speculations that the academicians made in their letters and manuscripts, it would seem that Galluzzi is justified in stating that Leopoldo was holding the Cimento to an experimental programme free from any natural philosophical theorising. However, there is no evidence anywhere to suggest that while the academicians were performing experiments in any discipline from 1657 to 1662, they were not permitted to discuss natural philosophy. Furthermore, if the Prince himself suggested experiments that clearly supported a mechanistic and corpuscularian natural philosophy opposed to Aristotelianism, how could we possibly believe that he was also enforcing some type of atheoretical programme? The fact is that we should not believe so. From 1657 to 1662, the academicians were clearly dedicated to resolving natural philosophical questions of central importance to the debate between Aristotelians and mechanists.

I suggest that we be more cautious when assessing the value of Rucellai's publication. As we have seen throughout this thesis and in particular in our analysis of Borelli's and Viviani's careers, when it came to individuals publishing their own work in Tuscany, there was no policy in place to obviate controversy arising from theories expressed in publications. However, things were different for the Cimento, not because of a pure scientific desire to be the first to put in practice a programme of atheoretical experimentation, but instead because the experimental rhetoric in the *Saggi* served as a very particular cultural and political tool for the Prince and his royal family. As Borelli himself put it in his very telling 14 November letter to the Prince, he needed experiments as a tool of authority and persuasion in favour of his natural philosophical beliefs.[115] This is what makes the publication process of the Cimento's work from 1662 to 1667 so interesting, and so crucial in arriving at a complete understanding of the group's foundations, purpose, and workings. This publication process will be discussed in Part Three.

[115] See note 503, above.

PART THREE

THE ACCADEMIA DEL CIMENTO: 1662–1667

We saw in Part One of this book that throughout their individual careers, the members of the Accademia del Cimento maintained certain natural philosophical skills, commitments, and agendas. As if the lives and works of the academicians were not enough evidence of the natural philosophical concerns behind the Cimento's experiments, we have also found in Part Two that even the group's princely patron, Leopoldo de' Medici, actually participated in the natural philosophical speculations that were being debated inside his academy.

However, as important as natural philosophy is to our understanding of the Cimento's activities, the Medici Grand Duke and Prince Leopoldo did not establish the Cimento merely to pursue their intellectual interests. It was argued in Chapter One that the status of mathematicians in Europe's Courts and intellectual communities had been increasing during the seventeenth century as their skills contributed to practical military and engineering innovations and to the credibility of an alternative natural philosophy to Aristotelianism. As 'cultural' historians Mario Biagioli, Jay Tribby, and Paula Findlen have helped to establish, this created a more prominent socio-cultural status for natural philosophers that was to be exploited by royal and princely courts, such as the Tuscan Medici Court, in order to raise their own status and reputation. There is no better example of this than the exposure the Medici family received across Europe in return for their patronage of Galileo. The widespread use of the telescope, the naming of the Medicean Stars surrounding Jupiter, and the continued work by Galileo on celestial and terrestrial phenomena, were all 'status-carrying' gifts for the Medici which helped them to lift their reputation across Europe as protectors of credible knowledge-making. This push for status and reputation was the reason behind Leopoldo's and Ferdinando's support for Viviani's hagiography of Galileo, their interest in the supposed practical applications of knowledge by such Galilean followers as Torricelli, Viviani, Borelli, Redi, and others, the experimenting conducted informally under Ferdinando's supervision, and finally the foundation of the Accademia del Cimento.

The history of the Cimento, therefore, also involved other wider social and political issues aside from the ways in which natural philosophical beliefs shaped the construction and interpretation of experiments. The next two chapters will reveal that Leopoldo and his academicians strove to put together a rhetoric of unbiased and uncontroversial fact-making in the presentation of their experiments in the *Saggi*, in order to achieve the authority and persuasiveness in their work that would gain them the reputation as reliable knowledge-makers that they were seeking. Additionally, it will be shown that an experimental rhetoric, free from any anti-Aristotelian natural philosophical speculations also averted the type of condemnation from the Catholic Church that Galileo had received, thus preserving the academicians' reputations as uncontroversial thinkers. It is therefore the aim of the final part of this book to see how and why the academicians presented their work veiled with the rhetoric of experimentalism and the supposed atheoretical and factual reporting of knowledge claims.

CHAPTER SEVEN

THE CIMENTO'S PUBLICATION PROCESS AND PRESENTATIONAL TECHNIQUES: FORMULATING A POLICY OF SELF-CENSORSHIP

The Accademia del Cimento never maintained a consistent schedule for its meetings. At times, some of the academicians could not go to Florence regularly because of their professional obligations in other parts of Tuscany. Borelli, Rinaldini, Marsili, and Uliva, for example, were often occupied with their positions at Pisa, while Viviani was also required to travel around Tuscany in order to fulfil his obligations as chief engineer to the Medici Court. Additionally, political duties and poor health seemed to restrict the time and effort that the Prince could dedicate to the academy. This meant that during their first five years as a formal institution, the academicians met only during several months of the year, especially in 1657, 1658, 1660, and 1662. Despite this inconsistency, during the months when they did hold meetings, they were prolific in their investigations, performing hundreds of experiments and regularly engaging in debates regarding the natural philosophical significance of their observations. However, during the last of those five years the Cimento went through a dramatic change in direction. Instead of just experimenting, the academicians became determined to ensure that their work should be published. As a result, most of their efforts were directed towards this end.

This change in scope is clearly evident in the Cimento's manuscript and unpublished material. To begin with, on 31 July 1662, it was reported in the official Cimento diary that the academicians met at Magalotti's house 'with the aim of repeating some experiments that seemed the most necessary for the completion of the work that is to be printed'. Furthermore, the point is made that 'when these become easy by practice, they all have to be made again in the presence of His Most Serene Highness'.[1] From that day until 9 September, the academicians

[1] 'Si ragunò l'Accademia in casa del Signore Lorenzo Magalotti a fine di replicare alcune esperienze che parevano più necessaria per dare compimento all'opera, che debba stamparsi, le quali tutte, quando ne venga agevolata la practica hanno a rifarsi alla presenza dell' A. V. Serenissima.' BNCF, Ms. Gal. 262, ff. 132r–132v.

repeated and refined many of the experiments that they had already performed, especially those regarding the effects of heat and cold and the production of the vacuum. Following this period in 1662 when they revisited and repeated past experiments and arguments, the academicians ceased to operate as a group. For example, during the entire year of 1663, the diary reports only one experiment carried out in July in which Magalotti, not usually a contributor to the performance of experiments, assisted Viviani in an attempt to measure the speed of light. More experiments concerned with light were never carried out, and even this sole experiment did not include all of the Cimento's members.

This minimal amount of activity continued until 1667 when finally three of the academicians, Borelli, Rinaldini, and Uliva, sought Leopoldo's permission to depart Tuscany.[2] Besides losing the biggest contributor to the group and the most talented member of the Court in Borelli, towards the end of 1667, Leopoldo was made a cardinal and was eventually required to fulfil his duties in Rome.[3] So as the Prince revealed in a letter to Christian Huygens on 10 February 1668, with his new duties and without three of his regular academicians, he had little hope of keeping the Cimento alive.[4] From January to March 1667, Leopoldo still managed to organise a few meetings of the Accademia, but eventually, no more experiments were reported in the diary, bringing the Cimento to an end after only ten years in operation.

This does not mean that nothing was happening during those last five years between 1662 and 1667. On the contrary, Magalotti, along with three members of the Cimento, and several linguistic and ecclesiastical authorities were called upon by the Prince to work on the completion of a text narrating the experiments performed by the Cimento. Clearly then, since 1662, they had taken on a new direction inside the Accademia, one which was now aimed towards ensuring that they obtained a presentation of their work that the Prince would be happy to show to his friends and rival courts. Actual experimenting was pushed aside.

Indeed, as the diary entry cited earlier indicates, Leopoldo took an interest in the experiments that were chosen for publication.[5] In fact, we will find that the entire publication process occurred under the watchful eye of the Prince and proceeded according to his own political aims and ambitions. The publication of the Cimento's work was undoubtedly his idea and it provided the Tuscan Court with exposure all over Europe. According to Middleton, Leopoldo already had this

[2] Middleton, *The Experimenters*, 316–317.
[3] *Ibid.*, 324–325.
[4] BNCF, Ms. Gal. 282, ff. 150r–v. It is easy to assume that Leopoldo's appointment as cardinal was a political manoeuvre by the Pope, designed to either close down the Cimento, or perhaps reward the Prince for not creating any religious conflict through the *Saggi*. However, as Middleton argues, the correspondence between Tuscany and Rome during July and August 1667 reveals that Leopoldo's appointment was simply a family matter. The Medici regularly appointed a member of the family to represent them in Rome, and following Cardinal Carlo de' Medici's death, Leopoldo was the most suitable of the remaining Medici princes to receive the honour. Meanwhile, Pope Clement IX demonstrated no particular interest in which of the Medici brothers was to be made a cardinal. W.E.K. Middleton, 'A Cardinalate for Prince Leopoldo de' Medici', *Studi Secenteschi* (1970), **xi**, 167–180.
[5] See note 1, above.

aim in mind when he founded the Cimento, and was pushing for a publication possibly from as early as 1657, soon after the Cimento's first meeting. This seems like a plausible possibility, but there is little evidence to support Middleton's suggestion.[6]

Nevertheless, in December 1660, Leopoldo was undoubtedly receiving calls for the publication of the Cimento's work,[7] and by 1662, the rest of Europe was eagerly awaiting a publication from Leopoldo's academy. It was not only mentioned in a letter from Dutch philologist Nicholas Heinsius to Huygens in March of that year,[8] but in November, Ricci also informed Leopoldo that 'Huygens, and the gentlemen in Paris and England, are waiting very impatiently for the book of the experiments made in your Highness' Academy'.[9] It would be another five years before the text was completed and the rest of Europe could read about the Italians' exploits, but by the time Ricci had written to Leopoldo in November 1662, the Cimento's secretary had already compiled the first draft. Indeed, it is likely that this first version of the *Saggi* was written even before the academicians met on 31 July of that year to repeat the experiments intended for publication – we must remember here that the diary account of that meeting reports their aim to *complete* the text, not to begin work on it, suggesting that Magalotti had already compiled some material. Furthermore, as Middleton points out, many of the experiments carried out after that date were suggested as a result of the editorial comments to the draft submitted by Borelli, Viviani, and Rinaldini.[10] All this meant that by 31 July 1662, Magalotti had probably already compiled the first draft of the *Saggi* and had already submitted that draft to his fellow academicians for editing.

Clearly by early 1662, the Accademia del Cimento shifted its focus from constructing and interpreting experiments, to presenting its work to the rest of Europe. Additionally, this change in direction seems to have been due to the Prince's desire to advertise the exploits of his academy to his learned correspondents and rival Courts in Europe. Therefore, our aim in this chapter is to analyse exactly in what light the Cimento intended to present its work and why Leopoldo and his courtiers made the decisions regarding the *Saggi* that they did, especially concerning their wish to exclude their accomplishments in astronomy from publication. In combination with our earlier analysis of the Cimento's cognitive concerns, this will provide us with a complete and in-depth picture of the processes

6 Such conjecture is based on a letter Leopoldo wrote to Ismael Boulliau on 13 October, promising the French astronomer that he would soon have a publication 'in hand, which (if I succeed in bringing into a conclusion, and if I am not mistaken) ought in some ways to be of no small use to the Republic of Letters'. Bibliothèque Nationale, fonds français, 13039, f. 71r. As cited by Middleton, *The Experimenters*, 66. Middleton claims that Leopoldo could only have been referring to a publication by the Cimento. However, there is no evidence that such a work was in progress at that time. Furthermore, it would be curious why Leopoldo should wish to advertise the forthcoming Cimento publication to an astronomer when, as we shall see in Chapter Eight, the Prince never had any intention of publishing the Accademia's observations in astronomy.

7 Middleton, *The Experimenters*, 66.

8 *Ibid.*, 67.

9 Fabroni (ed.), *Lettere inedite*, ii, 111. See also Middleton, *The Experimenters*, 67.

10 Middleton, *The Experimenters*, 69.

of making natural knowledge during this institution's ten years of existence. In other words, this look at the presentational techniques of the Accademia will provide us with the final layer of the cultural complexities that came to make up the so-called experimental life in mid to late seventeenth-century Tuscany.

1. WRITING AND EDITING THE *SAGGI*

In Magalotti's first draft of the *Saggi*, published by Abetti and Pagnini, the same rhetorical stance is evident as in the final version of the text.[11] For example, in a proposed preface to the *Saggi*, which was never actually used, Magalotti referred to the aim of presenting the Cimento's work by maintaining some distance from any conceptual framing of their experiments. Much as in the preface that was actually published, Magalotti claimed that such speculations were the opinions of certain academicians, 'but never of the whole Accademia, whose only purpose is to experiment and narrate'.[12] Similarly, reading through the 1662 version of the text, we find the same non-emotive narrative of the experiments that is in the final publication. On the odd occasion, Magalotti made the mistake of attributing some experiments to certain academicians, but all such references to individuals were cancelled out in the same manuscript, suggesting that the intention was certainly never to reveal who inside the Tuscan Court was constructing each specific experiment. The anonymity of the members responsible for each experiment was considered crucial to maintaining the uncontroversial and unbiased image of the Cimento, since it suggested that no individual academician put his theoretical aims and interests ahead of the collective, atheoretical experimental philosophy of the group.[13] So, given that all this was certainly written sometime approaching the date 31 July 1662, five years before the text was finally published, it would seem certain that not only was the Cimento taking on a new direction with the presentation of their work, but Magalotti was from the outset of this publication process, adhering to some type of policy, one that was to exclude all the natural philosophical concern and contention that we have seen in the case studies and that existed throughout the careers of each of the academicians.

In other words, only when they decided to establish the façade that they were going to present of the Cimento to their European colleagues, did the academicians begin to talk about an atheoretical experimental practice. The experimental

[11] Abetti and Pagnini (eds.), *Le Opere*, 280–322.

[12] '... ma non già mai dell'Accademia tutta, della quale unico istituto si è di sperimentare e narrare.' *Ibid.*, 275.

[13] Those references that were cancelled out were to Viviani's suggestion to measure the compression of air, as well as the weight of liquids. Candido del Buono was also mentioned by Magalotti for suggesting an experiment also testing the weight of different liquids. And finally Magalotti made it known in his draft of the *Saggi* that Prince Leopoldo was the first to perform the experiment measuring the freezing process. A reader of the final draft of the *Saggi*, who might also have been familiar with the individual interests of each academician, may have been able to infer the authors of each experiment. In any case, the anonymity of the text still provided the Cimento with the experimentalist façade that the Prince was seeking. *Ibid.*, 313.

rhetoric was the policy that the Cimento, under the insistence of the Prince, was to adopt in the presentation of its work, as reflected in Magalotti's first draft. Now we shall see to what extent it can also be traced in the editorial notes to the draft written by Borelli, Viviani, and Rinaldini.[14]

We saw from the case studies in Part Two that in his notes to Magalotti's draft of the *Saggi*, Borelli occasionally wished that the text provide stronger indications of his mechanistic ideals behind the construction of some experiments. At times, he insisted that it at least show how he conceived of certain ways of carrying out observations. However, he was also often very cautious about how much the text should reveal of the academicians' natural philosophical opinions. For example, when it came to the presentation of one of the many experiments testing the pressure of air, Borelli noted that much of the discussion that Magalotti included in the draft could have been omitted, since 'it does not produce experiments, but opinions and counter-arguments of the things that could be observed against the pressure of liquids'.[15] One could suggest that Borelli was simply arguing the exclusion of any Aristotelian comments against the pressure of the air. This may well be the case, but we cannot ignore the more specific comments that follow. In response to Magalotti's description of another air pressure experiment, Borelli stated: 'I should think that it would be more suitable for our aim to write historically rather than in the form of a debate.'[16] On a separate occasion, Borelli mentioned once again that the experiments should be narrated 'historically without showing partiality or any opinion'.[17]

In a moment we shall see what Borelli meant by a 'historical narration', and in what way, or for whom such a narration is supposedly 'suitable'. In the meantime we can make the following summation. Borelli was well aware that he was not going to be publicly credited for his contributions to the Cimento, but clearly he was still concerned about how the Accademia's work should be presented; as the accumulation of knowledge claims through rigorous experimenting and free of any controversial theorising.

Viviani gave no such recommendations in his brief editorial notes to Magalotti, preferring instead to make minor suggestions regarding the clarity of the wording of the narrative. However, at this point Viviani was also quite aware of the atheoretical and experimentalist rhetoric required in the presentation of the Cimento's work. This is evident in a letter he wrote on 17 July 1663, to Ricci in Rome. By this time Leopoldo had also called Ricci into the editing process of Magalotti's work, so Viviani was simply advising Ricci of the issues that preoccupied the Prince with regard to the style of presentation of the *Saggi*. These issues included how the academicians wished to report their work on the

[14] These notes were also published by Abetti and Pagnini. *Ibid.*, 323–348.

[15] 'Pongo in considerazione se tutto il seguente discorso sia bene tralasciarlo non arrecandosi esperienze, ma opinioni e risposte alle cose che si potrebbero osservare contro la pressione de'fluidi.' *Ibid.*, 329. See also Middleton, *The Experimenters*, 68.

[16] 'Crederei che fosse più conveniente al nostro instituto scrivere istoricamente che in forma di disputa.' Abetti and Pagnini (eds.), *Le Opere*, 330.

[17] 'Istoricamente senza mostrar parzialità ad alcuna opinione.' *Ibid.*, 331.

Torricellian tube, since, Viviani admitted, 'amongst the academicians here, the opinions about the reasons for the effects experimented on, particularly regarding the vacuum and the movements of the air, could be diverse'.[18] For this reason, continued Viviani, the Prince and his academicians had decided to publish merely a simple narration of the experiments carried out with the Torricellian tube, 'not to incur a reputation of little unity amongst the academicians'.[19] In other words, Viviani was well aware that an atheoretical and experimental rhetoric in the *Saggi* was essential in order to avoid giving the impression to readers that the academicians were divided on this issue on the basis of contrasting natural philosophical opinions. Revealing such divisions would harm the efficacy of their work, and in the process, the image of the Cimento, its patron, and its academicians.

This would also explain why Viviani compiled his own list of all the experiments he had suggested for the Cimento, as was mentioned in Chapter Two.[20] Viviani was well aware that the *Saggi* was never going to include the authors of the experiments or the natural philosophical reasoning behind their construction, so he ensured through this list that his contributions to the Accademia would not be completely forgotten.

The intention to omit any overtly controversial opinions from the text is particularly obvious in Rinaldini's comments about Magalotti's draft. Although Rinaldini too occasionally insisted on the use of words more sympathetic to his own Aristotelian beliefs, his stance on the objective appearance of the Cimento in the publication was made clear when he insisted that there is no need to enter into disputes on certain topics when they simply wish 'to narrate the story' of how their experiments were carried out.[21] Similarly, when referring to Magalotti's descriptions of the observations of the effects of magnetism in the void, Rinaldini stated that the experiments must be given in detail to the reader so that the observation was reduced to nothing more than the 'narration of a fact'.[22]

Finally, regarding their work on the pressure of the air, Rinaldini wrote that the impression the academicians must give is that 'these experiments were set up to see the effects of nature without any passion, and not to defend one opinion or another ... with metaphysical speculations'.[23] This stance may well have been partly because Rinaldini was continuing to resist the expression of any

[18] 'Tra gli accademici di qua potranno esser diversi i pareri intorno alle cagioni dell'effetti fin qui sperimentati e particolarmente circa al vacuo et al modo dell'operare dell'aria.' BNCF, Ms. Gal. 234, ff. 234–235r. See also Galluzzi, 'L'Accademia del Cimento', 809.

[19] 'Non incorrere in qualche taccia di poca unione tra gli accademici.' BNCF, Ms. Gal. 234, ff. 234r-235r. See also Galluzzi, 'L'Accademia del Cimento', 809.

[20] Nelli, *Saggio*, 110–111.

[21] 'Seguitarei quel ch'è disteso nè aggiungerei altro perchè s'entra in disputa di molte cose da non trattarsi da chi vuol semplicemente raccontare l'istoria.' Abetti and Pagnini (eds.), *Le Opere*, 338.

[22] 'Si dovrebbe anco ridurre in memoria al lettore che l'ambra riscaldata con la confricazione o in qualunque modo tiri il foglio, e benchè talvolta queste cose venghino toccate nel discorso, niente di meno l'ordine del dire richiede che questi si dichino prima, in ciò consistendo molto la lode della narrazione d'un fatto.' *Ibid.*, 343.

[23] '... queste esperienze furono instituite per vedere come stanno gli effetti della natura senza passione alcuna, e non per difendere un'opinione o un'altra ... con speculazioni metafisiche.' *Ibid.*, 339.

mechanistic beliefs. But from the sentiments that both Borelli and Rinaldini shared about how the experiments should be generally reported, we can be certain that there was a wider policy in place that these academicians were adhering to during the publication process.

According to Middleton, Borelli and Rinaldini were simply keeping 'a watchful eye open for violations of the unwritten laws by which the Academy was supposed to be governed'.[24] Now, it is rather ambiguous what Middleton means by 'unwritten laws', but from his reference to how the 'Academy was supposed to be governed', it would seem that he is suggesting that the academicians were following some type of general code of conduct for accumulating knowledge claims. Historians seeking the origins of modern 'experimental science' could even take this to be a reference to the type of gentlemanly conduct for obtaining matters of fact that Steven Shapin and Simon Schaffer believe existed in the early Royal Society of London, and that was mentioned in Chapter One. However, given that from their foundation in 1657 until their last months as active experimentalists in 1662, the academicians were never too timid to debate their natural philosophical concerns, it would seem unlikely that any type of ambiguous 'unwritten laws' or code of conduct existed for the production of atheoretical matters of fact. It would instead be more accurate to claim that from 1662 on, the academicians were applying a certain policy simply for the uncontroversial presentation of their experiments. In other words, when they began to devise how they were to present their work to the public, they were aware that they could not expose, and must censor, the natural philosophical contention that they went through during the construction and interpretation of their experiments. This censorship policy is evident in the following slightly more subtle comments that Borelli made in his editorial remarks to the first draft of the *Saggi*. These comments reveal what Borelli meant by a 'historical narration' and especially what he believed was 'suitable' for presentation to the public – what he believed would give the Cimento a more 'honourable' appearance.

When providing the mathematical basis for his claims regarding the force needed to overcome the resistance of a very thick metal container,[25] Borelli referred to the public image of the Cimento and the impressions passed on to the readers of the *Saggi*. On this occasion he mentioned that the description of the experiments and the effects of nature should be presented with such finesse that the curiosity of the reader would be easily satisfied. 'It seems to me that these experiments, although beautiful, should be handled in a way which, in their sequence and historical narration, they may present such diligent and fine judgement that the curiosity of the reader should have nothing to desire.'[26] To this he added that the experiments should, therefore, be presented with the precise

[24] Middleton, *The Experimenters*, 68.

[25] See Chapter Six.

[26] 'Parmi che queste sperienze benchè belle debbono essere maneggiate in modo che, nella serie e racconto storico de esse, so scorga industria e finezza tali di giudizio che la curiosità del lettore non vi abbia nulla da desiderare.' Abetti and Pagnini (eds.), *Le Opere*, 333.

measurements of the thickness of the containers and the force needed to overcome their resistance to breaking, that we have already examined in Chapter Six.

Once again we see a mention of the 'historical narration' that Borelli and Rinaldini insisted was so important to the presentation of the Cimento's work. More specifically, we see that this referred to fulfilling the curiosity of the reader. However, according to these editors of the *Saggi* and leading members of the Cimento, this certainly did not mean that they should have recourse to openly expressing the natural philosophical reasoning that for contending parties lurked behind the construction of the experiments and framed their different explanations of the outcomes, although we have seen continually in the case studies that such reasoning is hinted at in parts of the *Saggi*. Instead, they are insisting that the performance of an experiment and the observation of a natural phenomenon be described for the ideal reader with such detail, so as to leave no room for speculations or opinions. This is also evident on a separate occasion, when Borelli made further mention of this type of presentation. When once again referring to the wording Magalotti used in the description of an experiment, Borelli not only mentioned the need to apply a 'historical narration', but claimed that this was even a method. 'Wishing to comply with the historical method used in all of this writing, I believe that it could be said as such: ....'[27] This reference to 'method' is not to be confused with the inductive matter-of-fact producing 'experimental method', that is the preferred topic for many traditional and some 'cultural' historians of Italian science. Instead, Borelli was clearly referring to method in the well-known contemporary sense of a 'presentational technique' that, in this case, included an experimental rhetoric that happened to provide authority and credibility to the academicians' work.

Such talk about policies and techniques for presentation that involve careful and detailed descriptions about the performance of an experiment and the carrying out of observations, can also be identified in the early Royal Society of London. As Peter Dear points out, when the Royal Society was founded in 1660, its fellows were seeking new presentational techniques for strengthening the credibility of their work. Since the Society faced many criticisms, even from Charles II, it was imperative for their survival that they should establish an authoritative and persuasive style of presentation for their knowledge claims. They found in the example set by Robert Boyle's work in the early 1660s, that such authority and persuasiveness could be established upon detailed descriptions of experiments and natural phenomena.[28]

[27] 'Volendo osservare il metodo istorico usato in tutta questa scrittura, crederei che si potesse dir così: ....' *Ibid.*, 330.

[28] P. Dear, '*Totius in verba*: Rhetoric and Authority in the Early Royal Society', *Isis* (1985), **76**, 145–161. This is not to suggest that experiments were not in use in England prior to the formation of the Royal Society in 1660. Indeed, they were used extensively during the 12 years in which some of the future members of the Society met informally at Wadham College in Oxford. T. Sprat, *History of the Royal Society* (ed. J.I. Cope and H.W. Jones), London, 1966, 66–67. However, during this period experiments were not used as a tool of presentation. The secretive group at Wadham College did not need to persuade patrons of the validity of their work.

Indeed, Steven Shapin also identifies how the circumstantial detail in Boyle's published narratives of his experiments, extending also to his accompanying figures, was aimed towards turning the reader into a 'virtual witness'. Boyle's 'literary technology' was such that the reader would be convinced that the experiments were carried out effectively and achieved the only possible results. In other words, the experiments were described with such detail and diligence that there was nothing left for the reader to do to confirm the results, except of course replicate the narrated experiment, if he so wished.[29] In what seems almost like a copy of the Cimento academicians' rhetoric in the *Saggi*, in 1667 Thomas Sprat narrated how the Royal Society simply relied on 'diligent and laborious' observations, and not speculative interests, in order to construct so-called matters of fact.[30] As Dear argues, this type of rhetoric helped to establish the image of the Royal Society as provider of unbiased and objective observations of natural events.[31]

So the credibility of knowledge claims depended on the suitable presentation of an observation of nature. For this reason, experimental reports contained no speculations that could be misconstrued as having led the experimenter astray. Instead, they simply contained historical narrations of how and when experiments were carried out. As Dear puts it:

> The resulting style of presentation allowed no clear distinction to be made between a 'natural historical' and an 'experimental' report; each was, in the same way, given as an experience defined in space and time by an actor, the observer. The credentials that established the actuality of the event were provided by surrounding the description by a wealth of circumstantial detail. This detail generally included information regarding time, place, and participants, together with additional extraneous remarks about the experience, all serving to add verisimilitude.[32]

This reliance on experiments and their presentation in such a historical narrative, added authority to the Royal Society's work. According to Shapin and Dear, the status of the Royal Society rose as a result of excluding all speculative material from the presentation of their work, and by appealing to the persuasiveness and authority of experiments.[33] Therefore, for the Cimento, as for the Royal Society, the use of experimental historical narratives to describe their work in the *Saggi*, became a policy, or method, as Borelli referred to it, for the presentation of their work, since to expose their natural philosophical arguments would have drained their publication of the authority and persuasiveness they were seeking.

This is not to suggest that the academicians were deviously trying to compose a rhetoric that would deceive their readers about what was really happening inside the Cimento. Rather, as we have seen from Borelli's, Viviani's, and even Leopoldo's faith in their experimental work, they did indeed believe that they were compiling irrefutable knowledge claims. Neither is it my intention, as was

[29] S. Shapin, 'Pump and Circumstance: Robert Boyle's Literary Technology', *Social Studies of Science* (1984), **14**, 481–494.
[30] T. Sprat, *History of the Royal Society*, London, 1667. As cited by Dear, '*Totius in verba*', 152.
[31] Dear, '*Totius in verba*', 154.
[32] *Ibid.*, 154.
[33] *Ibid.*, 158–159.

mentioned in Chapter One, to argue that the academicians were breaking away from the speculative and theory-driven work carried out by natural philosophers in the early to mid seventeenth century. The academicians were instead continuing to debate the natural philosophical issues that had occupied their predecessors, including of course, Galileo, but they were also concerned with the most persuasive way of *presenting* their work to their colleagues.

So what *is* being said here is simply that considering the natural philosophical concern and contention that was behind the construction and interpretation of these claims made between 1657 and 1662 inside the Cimento, Magalotti and the editors of the *Saggi* now faced a difficult presentational task. They needed to present their experiments in a manner which would not reveal the natural philosophical controversy inside the Accademia and would instead only show supposed experimentally based factual knowledge, or the 'historical narration of a fact', as Rinaldini and Borelli referred to it. This was the only way to maintain the status of the Cimento, and especially of the Court that protected them.

This leaves us then to analyse exactly what Borelli meant by the 'suitability' of such a 'historical narration'. Here we may begin by returning to Borelli's wish, written in his editorial notes to Magalotti's draft and mentioned earlier in Chapter Six, to tell the readers of the Cimento's publication about how he and Leopoldo devised the experiment on the effects of heat and cold on a solid ring. This is where he made the following curious statement: 'I would believe that the narration of this fact as it followed would not dishonour the Accademia.'[34] From this statement, it would appear that when Borelli earlier mentioned the suitability of the policy of natural philosophical censorship in the *Saggi*, he was referring to the 'honour' of the Cimento. In other words, the academicians still had to give the air of authority and credibility to their work so that their status would be preserved across Europe as trustworthy practitioners of knowledge.

Furthermore, what this means is that the Medici were also relying quite heavily on what resulted from the patronage they afforded these academicians. The publication of the Accademia's work seemed to be Leopoldo's idea and well before the text was even published he was promoting it to his friends and correspondents in other parts of Europe, who were eagerly awaiting the work. Therefore, the Prince's and his family's status, reputation, and honour as worthy protectors of natural knowledge-making, was at stake with the success of this publication, and that success obviously depended upon a suitable style of presentation, a rhetoric that would convince the reader of the reliability of the academicians' knowledge claims.

This is why, on 27 July 1661, Borelli wrote to his friend Alessandro Marchetti about the image that his Medici patrons wished to portray to the public of the work carried out in their Court. 'These princes try to avoid a clamorous

[34] 'Crederei che il racoontar questo fatto come seguì non arrechi disonore all'Accademia, quando i lettori possono scorgere che le persone che ci travagliano sono tali che, da un cenno di sperienze eccitati, seppero trovare le vere cause di esse.' Abetti and Pagnini (eds.), *Le Opere*, 335.

appearance that might arouse malevolence and clamour, and in short [they see] that true philosophy spreads in a pleasant way and soft manner.'[35] This statement provides a neat illustration of how greatly the Medici valued the presentation of the Cimento's work. The 'appearance' they were eager to obtain with the *Saggi* was that of a purely experimentalist academy. So in order to give the impression to readers of the *Saggi* that this was an academy producing reliable knowledge claims, Leopoldo ordered his courtiers to work on a presentational technique that would not expose the natural philosophical concern and contention that had occurred inside the Cimento from 1657 to 1662. Therefore, there was indeed a policy during the publication process from the year 1662 until 1667 that was aimed at censoring the natural philosophical commitments the academicians had expressed during the first five years, and at preserving the status and reputation of the Cimento and the Tuscan Court.

2. LEOPOLDO'S RELIGIOUS CONCERNS AND THE REST OF THE *SAGGI*'S EDITING PROCESS

Aside from the thorough revision provided by Borelli, Rinaldini, and Viviani of the first draft of the *Saggi*, Magalotti's work continued to undergo many changes as it went through the hands of various other editors. Those editors included linguistic experts and ecclesiastical censors, all called upon by the Prince to give their recommendations on the suitability of the text. It is important, then, that we look at this editing process since it reveals quite a bit about what Leopoldo was trying to achieve from the publication and whom he was aiming to please. This will provide us with the adequate cultural and political framework needed to analyse the academicians' work in astronomy, and their refusal to publish any of their astronomical observations in the *Saggi*.

After Magalotti considered the editorial comments from his fellow academicians, and after they retested many of the experiments during the second half of 1662, a second draft was written. By the end of that year, Leopoldo sent Magalotti to Rome with the manuscript of this new version of the text. The Prince desired that Ricci, a well-respected literary expert, especially when it came to scientific texts, should critique Magalotti's latest draft.[36] Ricci was to have much to say about the style of writing appropriate for a text published by an

[35] T. Derenzini, 'Alcune lettere di Giovanni Alfonso Borelli ad Alessandro Marchetti', *Physis* (1959), i, 233. See also Segre, *In the Wake*, 140.

[36] See letter from Ricci to Leopoldo dated 25 June 1663. BNCF, Ms. Gal. 276, f. 204r. Leopoldo also called upon Ricci to work in this editorial capacity on other occasions, demonstrating the faith the Prince had in his Roman correspondent to critique important works produced by Tuscan courtiers. For example, in December 1664, Leopoldo asked Ricci to give his opinion of a treatise written by an engineer for the Tuscan Court, Famiano Michelini, and published in Florence. Entitled *Trattato della direzione dei fiumi*, this was an important work for Italian engineers and as such was heavily promoted by the Prince who sent it to many of his friends and correspondents. He therefore entrusted Ricci to ensure that Michelini's work was well written. See BNCF, Ms. Gal. 277, ff. 56r–56v.

academy working in the Medici Court and proclaiming to be followers of Galileo. His criticisms were expected to address language, argument structure, and the type of phrasing that would be required for a book containing some potentially controversial material. Following the religious controversy that continued to surround Galilean astronomy and natural philosophy, Leopoldo would have been wary of the scrutiny his book and his academy would receive from ecclesiastical authorities and so would have valued the opinion of someone as experienced in this field as Ricci, a Roman mathematician and member of the Papal Court.[37]

Indeed, Leopoldo's concern with the approval of the Catholic Church was made clear in a letter written by him to Magalotti on 25 March 1664: 'I prefer that the manuscript be sent on through Cardinal Ranucci. I warn you that nothing will be printed against his wishes.'[38] This is the first clear indication that Leopoldo was concerned about what ecclesiastical authorities thought about the *Saggi*. That is to say, once again, that only after turning their attention to the presentation of their work, were the academicians, including the Prince, concerned with formulating a type of rhetoric that would not be offensive to the Church, but that would, nonetheless, show-off the talents of the Cimento. Religion was, therefore, quite important to the publication and editing process and this was made even more evident in May 1664. The last page of the manuscript believed to be the second draft compiled by Magalotti, contains signed statements by four ecclesiastical censors, including the vicar general of Florence and the Chancellor of the Holy Office for Florence. Leopoldo had sent the manuscript to be read and analysed by these representatives of the Church, all of whom approved of the suitability of the text. According to eighteenth-century author, Angelo Fabroni, Leopoldo had even sent several pages of Magalotti's work to the Pope himself, who also gave his approval of the text.[39] On 31 July 1664, it was given the Imprimatur.[40]

Following this lengthy process of editing and rewriting that the Prince had insisted on carrying out, it would seem that Magalotti's work was finally ready for the printer. However, Leopoldo once again sought an expert opinion of Magalotti's style. This time, an expert in Tuscan language, Cardinal Sforza Pallavicini made some minor criticisms. Unfortunately, they were not taken well by Magalotti who left the manuscript unchanged and unprinted for the next 18 months.[41] When he finally turned his attention once more to this work in January 1666, he progressed slowly towards its completion. By July 1667, he finished writing the dedication to the Grand Duke and the *Saggi* was finally published. But before going to the printer, the manuscript once again had to pass through the hands of the ecclesiastical censors. Such was Leopoldo's concern

[37] As was mentioned in Chapter One, the Catholic Church was still so concerned with the prohibition of Galileo's astronomical claims that during the 1650s they insisted that the *Dialogue* could not be published amongst a Bolognese collection of Galileo's works.

[38] Bibl. Laurenziana, Ashburn 1818, f. 16r. As cited by Segre, *In the Wake*, 72.

[39] Fabroni (ed.) *Delle lettere*, i, xx.

[40] Abetti and Pagnini (eds.), *Le Opere*, 276; Middleton, *The Experimenters*, 73.

[41] Fermi, *Lorenzo Magalotti*, 91–92; Middleton, *The Experimenters*, 73.

with the approval of the Church that despite the delay in publishing the Accademia's work, he would not leave anything unchecked. This left Magalotti extremely frustrated, but as the following letter written during that time to his friend in Rome, Ottavio Falconieri, shows, he was also quite aware of the caution being taken due to religious scrutiny.

> I should not like to be obliged by Signor Michelangelo Ricci to recast my thoughts; for I would rather have it printed in Geneva. Good heavens! This man ruins my temper; either the difficulty arises from the thing itself, or from ignorance, or from the subtlety of the censor .... Now let them do almost anything they want and make a mess of it in the end.[42]

Therefore, during the final stage of the so-called experimental life of the Accademia del Cimento, we find once again that there were far wider-reaching issues at stake than simply the practice of a supposedly inductive and empirical method of research, or even an adherence to a courtly culture of gentlemanly conduct that supposedly inspired the accumulation of factual knowledge. During their first five years as a somewhat formal society, the academicians were actually preoccupied with constructing and interpreting experiments that fitted in with the natural philosophical concerns of seventeenth-century Europe. During their last five years, they wanted to present their work to the rest of Europe, but not as an example of the controversy that surrounds speculation based on cosmological and ontological opinions. Rather, for the sake of keeping onside with the volatile Catholic Church and ensuring that the Medici Court gained a status and reputation as patrons of reliable knowledge-making, they had to present an atheoretical experimental rhetoric. These are the political and religious issues explaining why Leopoldo was so cautious with the publication and editing process. The following chapter will reveal that this policy for presenting their work, what Paolo Galluzzi referred to as 'self-censorship', is especially evident when the academicians were carrying out observations in astronomy and dealing with the particularly sensitive topic of Copernicanism.

[42] As translated by Middleton, *The Experimenters*, 76. 'Io non vorrei, che il Sig. Michel Agnolo Ricci mi obbligasse a rimutare il pensiero, perchè più tosto voglio farlo stampare in Ginevra. Capperi quest'uomo mi riesce stitico: o la difficoltà nasce dalla cosa in se, o da ignoranza, o sottigliezza del revisore: se nasce di qui, perchè non si può render capace? or facciano un po quel ch'e' vogliono, e la finiscano in tanta malora.' Fabroni (ed.) *Delle lettere familiari*, i, 176.

CHAPTER EIGHT

THE SATURN PROBLEM AND THE PATH OF COMETS: AN ANALYSIS OF THE ACADEMICIANS' THEORETICAL AND OBSERVATIONAL ASTRONOMY

There is no indication in the Cimento's only publication, *Saggi di naturali esperienze*, that the academicians were at all interested in making astronomical observations. Additionally, in the official Cimento diary, the only work in astronomy that the academicians recorded was their involvement in the dispute between Christian Huygens and Father Honorè Fabri (1607–1688) in 1660 regarding the appearance of Saturn. This dispute was mentioned briefly in the diary on only five occasions. From this evidence, it would appear, therefore, that the Cimento's interests in astronomy were only marginal, or as Middleton argues, a mere 'digression' from the academicians' work in other areas, such as the vacuum, air pressure, and the effects of heat and cold.[1] However, while the Cimento's publication and diary provides little indication that the academicians were interested in astronomy, unpublished letters and manuscripts again reveal much more about the amount of work they carried out in this field and the intellectual aims and interests that work encompassed. More specifically, although they registered very little activity in the field of astronomy, some of the academicians still wrote reports and letters about the anti-Aristotelian natural philosophical skills, commitments, and agendas that they carried into their investigations of some celestial phenomena. In fact, the Galilean manuscripts held at the Biblioteca Nazionale Centrale in Florence, consist of several folders containing letters and manuscripts written to and by Leopoldo, and concerned with the natural philosophical interpretation of various observations of Saturn, Jupiter, the moon, eclipses, and comets.[2]

This chapter will examine two of these cases. The first concerns the academicians' interest in Saturn's ring, while the second case considers their interpretation of comet sightings in 1664 and 1665. While it is undoubtedly true that these topics

[1] Middleton, *The Experimenters*, 256. This argument is also made by Segre, *In the Wake of Galileo*, 133; and Abetti and Pagnini (eds.), *Le Opere*, 54.
[2] Those folders include Mss. Gal. 271, 272, 273, 276, and 282.

were a digression from the Cimento's regular research interests, they were still eagerly pursued by most of the academicians, and involved natural philosophical questions to do with the structure and movements of the celestial realm. In both cases, the members of the Accademia, particularly Borelli, constructed and interpreted their knowledge claims according to their commitments to Copernicanism. This was, therefore, a topic that was obviously also sensitive to the same religious issues that saw the condemnation of Galileo by the Catholic Church, an episode in Tuscany's history that was fresh in the minds of Galileo's students and followers in the Accademia del Cimento. For this reason, the academicians were careful about how they approached these cases, the experiments, and observations they were willing to carry out and the claims they were prepared to make, especially since, from as early as 1660, they were probably considering the possible publication of their work, and the reputation that publication would bring them and their patron.

So, in addition to the anti-Aristotelian beliefs involved in their work on Saturn and the comets, we shall see that Leopoldo and his academicians faced some serious political and religious pressures when working in the field of astronomy. The Cimento concerned itself heavily in resolving issues that were laden with natural philosophical implications, but they were carefully trying to avoid any controversy with the Catholic Church by not allowing the natural philosophical opinions of the academicians to be published in their own text, or in any other. This attempt to avoid controversy also assisted Leopoldo in his endeavour to create an image of the Cimento as an unbiased and uncontroversial institution.

1. THE SATURN PROBLEM

In July 1610, Galileo made his first telescopic observations of Saturn.[3] Following his discovery of the Medicean stars in 1609, he believed that he had uncovered another astounding celestial phenomenon with his telescope: that Saturn is not just another planet of ordinary appearance, but is in fact, the composite of three spherical bodies. Galileo believed that he had observed two small stars sitting very close to either side of the much larger central globe. He may have even thought that he was observing two satellites of Saturn, much like the Medicean stars moving around Jupiter.[4] However, after having observed no changes at all in the positions

[3] The following account of the Saturn controversy, from Galileo's observations in 1610 until the debate between Huygens and Fabri in 1660, is based largely on the work carried out on this topic by Albert Van Helden. Van Helden's publications from the early 1970s include: 'Eustachio Divini versus Christian Huygens: a reappraisal', *Physis* (1970), **xii**, 36–50; 'The Accademia del Cimento and Saturn's Ring', *Physis* (1973), **xv**, 237–259; 'Saturn and his Anses', *Journal for the History of Astronomy* (1974), **v**, 105–121; '"Annulo Cingitur": the solution of the problem of Saturn', *Journal for the History of Astronomy* (1974), **v**, 155–174. Van Helden's historiographical position with regard to his study of the Saturn problem is analysed in the Conclusion. One other more recent secondary source concerned with the observations of Saturn inside the Cimento, is P. del Santo and G. Strano, 'Il Cimento degli astri', in *Scienziati a Corte: l'arte della sperimentazione nell'Accademia Galileiana del Cimento (1657–1667)* (ed. P. Galluzzi), Livorno, 2001, 29–35.

[4] Van Helden, 'Saturn and his Anses', 106; S. Drake (ed.), *Discoveries and Opinions of Galileo*, New York, 1957, 74.

of the smaller bodies for the following two years, it seems highly unlikely that Galileo would have felt comfortable with such a hypothesis. In fact, as he expressed in his first letter on sunspots, dated 4 May 1612, Galileo was quite certain that these two bodies on either side of Saturn were not like any other stars.[5]

By the end of 1612, a sudden and dramatic change in Saturn's appearance justified Galileo's suspicion that these were not ordinary satellites. On 1 December of that year, when he wrote the third of his published letters on sunspots, he mentioned that Saturn was apparently no longer triple-bodied, but could instead be seen as a single perfectly spherical planet, 'without its customary supporting stars'.[6] Regardless of this sudden and strange change in the planet's appearance, Galileo predicted, with some degree of caution and reservation, that the two small bodies on either side of Saturn that he had earlier observed with his telescope, would return to sight by the northern hemisphere summer of 1613 for only two months, before reappearing in the winter solstice of 1614 for another brief period, and finally again in the summer solstice of 1615. According to Galileo, they would then remain in view for many years.[7]

It is not clear what type of model Galileo was using as the basis of his predictions or how he believed the three apparent spheres were moving.[8] In any case, those predictions were fairly accurate. At the very least, the three-bodied appearance of Saturn was confirmed to have returned briefly in 1613, and again in 1615.[9] Despite these predictions, Galileo could not have possibly expected the observations he was to make of Saturn later, in 1616. Around August that year, in a letter addressed to Federico Cesi in Rome, Galileo stated that Saturn no longer appeared to consist so clearly of three separate bodies, or even one body on its own. Instead, now only one middle globe could be observed, with 'two half eclipses', or 'handles', on either side (Figure 17).[10] Any confidence that Galileo may have gained from accurately predicting some of Saturn's phases could have possibly been destroyed by this latest observation.

Nevertheless, he continued to observe Saturn's movements until he lost his sight towards the end of the 1630s. During this time, as Van Helden states, Galileo's observations of Saturn's appearance, and his early predictions about its phases, became the most trusted available authority on the subject. Few other astronomers made any effort during this time to improve on Galileo's work. Only

[5] Favaro (ed.), *Le Opere*, v, 110.

[6] 'senza l'assistenza delle consuete stelle'. Favaro (ed.), *Le Opere*, v, 237.

[7] *Ibid.*

[8] Van Helden, 'Saturn and his Anses', 107.

[9] A letter from a correspondent of Galileo in Rome, Giovanni Battista Agucchi, confirmed the return of Saturn's satellites in July 1613. Agucchi congratulated Galileo on the accuracy of his predictions thus far. Favaro (ed.), *Le Opere*, xi, 532.

[10] 'Non voglio restare di significare a V.E. un nuovo et stravagante fenomeno osservato da me da alcuni giorni in qua nella stella di Saturno, li due compagni del quale non sono più due piccoli globi perfettamente rotondi, come erano già, ma sono di presente corpi molto maggiori, et di figura non più rotonda, ma come vede nella figura appresso, cioè due mezze ecclissi con due triangoletti oscurissimi nel mezzo di dette figure, et contigui al globo di mezzo di Saturno'. As cited by Giovanni Faber, an associate of the Accademia dei Lincei, in a letter to his friend Federigo Borromeo in Milan, dated 3 September 1616. Favaro (ed.), *Le Opere*, xii, 276.

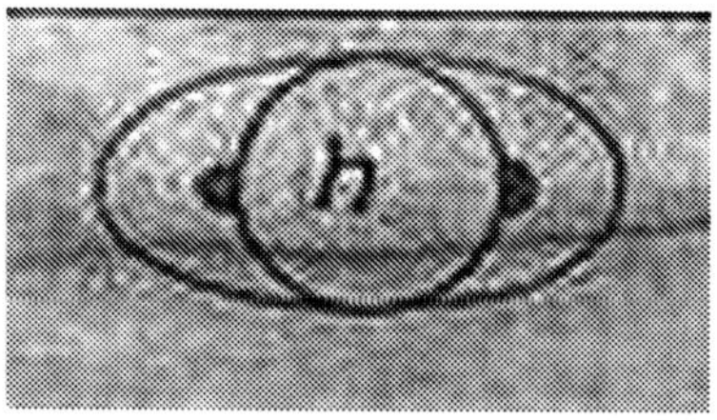

Figure 17. Galileo's depiction of his observation of Saturn. G. Galilei, *Il Saggiatore,* Rome, 1623, 217.

Gassendi and Francesco Fontana (1585–1656), a Neopolitan instrument maker, recorded several more observations of Saturn during the 1630s. While Gassendi observed the 'handled' appearance in 1633 and described it in the posthumously published *Opera Omnia* (1658), Fontana used his own telescopes, more powerful than Galileo's, to view the handles and to depict them in a manuscript in 1638.[11] But it was not until August 1642 that interest in the Saturn problem began to increase across Europe. At this point Gassendi observed the planet without its 'handles' and discussed this apparent change in Saturn's appearance with colleagues. Following this, several other astronomers around Europe began to make further observations of the planet and to contribute to the resolution of the problem regarding its strange phases. So within the following two decades, several publications were released on the topic and various theories were proposed.[12] However, despite this sudden increase in interest in Saturn's strange appearance, nobody devised a hypothesis that could be agreed upon by most astronomers, to account plausibly for each of the planet's phases. Christopher Wren summarised the situation in his 1658 treatise about the planet:

> For Saturn alone stands apart from the pattern of the remaining celestial bodies, and shows so many discrepant phases, that hitherto it has been doubted whether it is a globe connected to two smaller globes or whether it is a spheroid provided with two conspicuous cavities or, if you like spots, or whether it represents a kind of vessel with handles on both sides, or finally, whether it is some other shape.[13]

It was at this point that Huygens put forward his hypothesis of a ring in his *Systema Saturnium* in 1659. In this text, dedicated to Leopoldo , Huygens confidently claimed that what he could see through his telescope was a solid, rigid, and thick ring surrounding Saturn, and at the same time, completely detached from the planet. He believed that this ring created the illusion of handles

[11] Van Helden, 'Saturn and his Anses', 112–113.

[12] These publications include: F. Fontana, *Novae coelestium terrestriumque rerum observationes*, Naples, 1646; Johannes Hevelius, *Selenographia*,Gandsk, 1647; Pierre Gassendi, *Animadversiones in decimum librum Diogenis Laertii*, Lyons, 1649; Giovanni Riccioli, *Amalgestum novum*, Bologna, 1651; and Christopher Wren, *De Corpore Saturni*, London, 1651. For more information on the opinions of Fontana, Hevelius, Gassendi, Riccioli, and others regarding Saturn, see Van Helden, 'Saturn and his Anses', 105–121.

[13] As cited by Van Helden, 'Saturn and his Anses', 105.

or satellites that had been observed by other astronomers during the previous 50 years. Such illusions, claimed Huygens, were assisted by the use of telescopes inferior in quality to his own.[14] In fact, in what he believed to be a demonstration of the superiority of his telescope, which he had himself constructed, over those used by his colleagues in other parts of Europe, Huygens also claimed that in 1656 he had seen a small satellite of Saturn, never before spotted by anyone else.

So from the combination of these two observations, the new satellite and the ring, Huygens was convinced that not only had he constructed the best telescope in Europe, but that he had also solved the puzzle regarding Saturn's various appearances. However, the situation was hardly that simple and the problem of Saturn was not so easily resolved. Huygens' work carried some serious natural philosophical, religious, and political implications for the traditional Aristotelian astronomers still dominating the Jesuit schools, especially in Rome, and who were assessing the validity of Huygens' ring theory.

2. HUYGENS VERSUS FABRI AND DIVINI: RELIGION, REPUTATIONS, AND NATURAL PHILOSOPHICAL COMMITMENTS ON THE LINE

The implications arising from Huygens' work were numerous. In the first place he was rejecting Galileo's claim that Saturn is the composite of three spherical bodies. Second, and more important, Huygens was providing a definitive claim against the Aristotelian notion that planets, since they belong to the celestial realm, are perfectly spherical and incorruptible. Furthermore, Huygens' confidence in the validity of his hypothesis was based on his openly expressed Corpernican commitments. In a clearly illustrated diagram used to explain Saturn's phases, Huygens was suggesting that the appearance of Saturn from Earth depended upon the illumination of the ring from the centrally located Sun (Figure 18). In fact, Huygens openly expressed his support for Copernicanism by clearly stating in the dedicatory letter of *Systema Saturnium* that Saturn, like Earth, orbits the Sun.[15] He even suggested that Saturn and Earth were quite alike, contrary to scholastic belief in the uniqueness of the Earth: he claimed that both have only one satellite, and both have the same degree of inclination.[16]

So for our Tuscan academicians, the Medici Court, and traditional scholastic astronomers, the same issues that saw the condemnation of Galileo by the

[14] Van Helden, 'Eustachio Divini versus Christian Huygens', 37.

[15] 'Unum hoc inanimadversum eos praeterire nolim; nempe quam non leve argumentum ad astruendum pulcherrimum illud mundi universi ordinem qui a Copernico nomen habet Saturnius hic mundus adferat'. C. Huygens, *Oeuvres complètes de Christian Huygens*, 22 vols., The Hague, 1888–1950, iii, 433. As cited by Galluzzi, 'L'Accademia del Cimento', 826.

[16] Van Helden, 'Annulo Cingitur', 163.

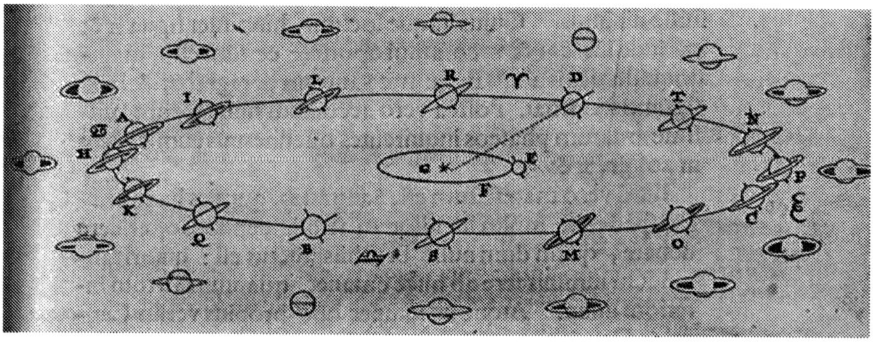

Figure 18. Huygens' diagram of Saturn's trajectory around the Sun. C. Huygens, *Oeuvres completes de Christiaan Huygens, 1888–1950*, xv, 309.

Catholic Church in 1633 regarding the teaching of Copernican and anti-Aristotelian cosmology as truth, were threatening to re-emerge in public debate during the 1660s. Rather than resolve the problem with Saturn, Huygens' claims actually helped to increase the contention surrounding its phases and appearances. In particular, since Huygens dedicated his text to Leopoldo, the issue was especially fraught with political, religious, and natural philosophical dangers for the Tuscan Court and its members, including of course, the Accademia del Cimento. This is, therefore, where we begin to examine the reasons why the academicians' first became involved in this issue, and why their astronomical work was omitted from the *Saggi*.

After Huygens published and distributed his text to his friends and colleagues across Europe, criticisms of the ring theory were immediately raised. Some astronomers were sceptical of the validity of Huygens' hypothesis, because Saturn sometimes appeared to be unaccompanied by any ring or the illusion of satellites. That is, if the ring were as thick as Huygens proposed, then it should be visible all the time.[17] These types of criticisms were additionally aimed against Huygens' claim that he was using a telescope superior in power and quality to anyone else's in Europe. This statement obviously did not sit well with other highly respected instrument makers. Johannes Hevelius (1611–1687), as an example, in a letter to Ismael Boulliau in December 1659, insisted that his telescopes were not inferior to Huygens'. Hevelius stated that he had made the same observations as Huygens ten years earlier, only he was too 'careless', as he put it, to speculate upon whether the new satellite of Saturn that Huygens claimed to have discovered, could be anything other than a fixed star. As for Huygens' proposed ring, Hevelius was also sceptical of this notion on the basis of its supposed thickness. In any case, he claimed to have already carefully annotated each of Saturn's apparent phases

[17] Ismael Boulliau and Christopher Wren were particularly critical of Huygens' theory on these grounds. Van Helden, 'Annulo Cingitur', 163.

before Huygens even suggested the existence of a ring. Hevelius concluded to Boulliau: 'Thus, on this point I don't concede anything to him.'[18]

Meanwhile, Eustachio Divini (1610–1685), a Roman manufacturer of telescopes was also quite annoyed by Huygens' claims. In a letter he wrote to Leopoldo on 10 July 1660, in which he mentioned how he had only recently received and read the *Systema Saturnium*, Divini gave his estimation of Huygens' work: 'I found that he placed too much faith in some things, in himself, and in his lenses.'[19] As Van Helden points out, it was to be expected that Divini, whose livelihood depended on his reputation as a worthy manufacturer of telescopes, would be critical of an opponent who claimed to be making the best telescopes in Europe, and believed that he was carrying out far more accurate observations of Saturn than anyone else.[20] Divini was therefore unimpressed by Huygens' claims and set about trying to discredit them.[21]

Divini's letter to Leopoldo accompanied a text, *Brevis annotatio in systema Saturnium*, which was also dedicated to the Tuscan Prince, and which was intended to be a public response to Huygens' claims. Not surprisingly, this writing also contained criticism of Huygens' overwhelming faith in his telescopes. But the discrediting of Huygens as an instrument maker was not the only aim of this text. The author of *Brevis annotatio* also suggested replacing Huygens' theory with a hypothesis that was more of a compromise between Galileo's earliest observations of a triple-bodied Saturn, and traditional Ptolemaic astronomy.

This theory suggested that there were four stars near Saturn, apart from the satellite discovered by Huygens. These stars did not orbit Saturn, but rather two points behind that planet. Since two of the stars reflected light and the other two did not, an illusion was supposedly created in which, when the light-reflecting stars were partially obscured by the dark stars, two half eclipses could be observed from Earth. In addition, at their greatest elongation, the light-reflecting satellites could be seen in their entirety sitting closely by the sides of Saturn, but

[18] Bibiotheque Nationale, Mss. Collection Boulliau', ff. 89v–90r. As cited by Van Helden, 'Eustachio Divini versus Christian Huygens', 38.

[19] '... trovai ch'in qualche cosa troppo egli si sia fidato, e di sé, e delli suoi occhiali'. BNCF, Ms. Gal. 276, f. 33r.

[20] Van Helden, 'Eustachio Divini versus Christian Huygens', 38.

[21] Both Divini's and Hevelius' objections to Huygens' observations and claims illustrate some of the sociological issues involved in the historiography of this case study. The significance of an observation was being challenged and negotiated, yet the task for historians is not to discuss who had the 'best' instrument and who was making the 'right' observations. Indeed, rather than make 'whiggish' statements about this case, we should come to understand that knowledge claims were being debated by rival telescope makers with social and political concerns extending well beyond who had the best theory. The efficacy of one's instrument was grounds upon which to be critical of a rival theory, and to be supportive of one's own intellectual, political, and religious commitments. So these astronomers were challenging each other's observations on the basis of their own social and natural philosophical agendas. We will find that these were concerns that extended to the Cimento in their involvement in this topic in 1660, and their decisions regarding the presentation of their work. This discussion recalls the sociological analysis of scientific knowledge pioneered by Harry Collins and Trevor Pinch mentioned in Chapters Five and Six (see note 424, Ch. 5; and note 472, Ch. 6). Collins, 'The Seven Sexes', 205–224; Pinch, 'Towards an Analysis of Scientific Observation', 3–36.

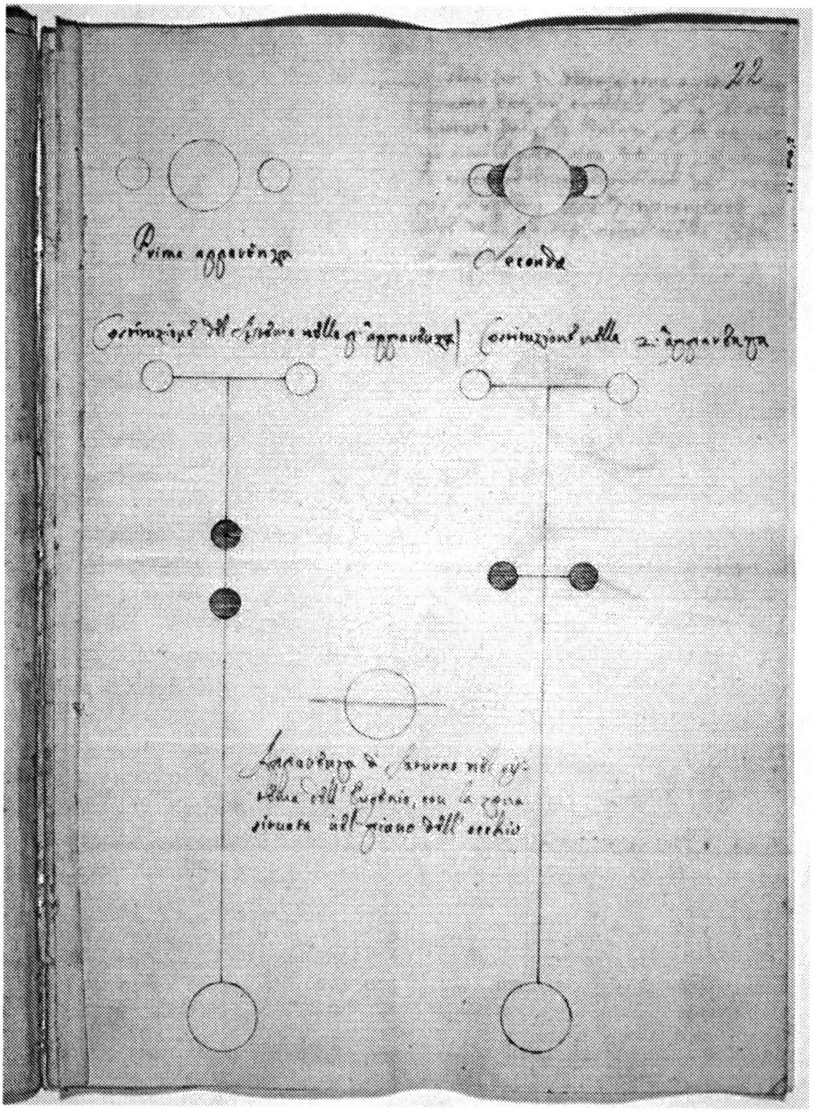

Figure 19. Saturn's appearance according to the hypothesis in Divini's *Brevis annotatio*. This illustration is in a manuscript in Lorenzo Magalotti's handwriting, dated July 1660 and entitled Osservazioni delle stele di Saturno. Note that Saturn is also drawn here according to Huygens' system. BNCF, Ms. Gal. 271, f. 22r. Courtesy of the Ministero per i Beni e le Attività Culturali / Biblioteca Nazionale Centrale di Firenze. Protected by Copyright.

when they were hidden from view behind the planet, obviously only the sphere of Saturn could be seen (Figure 19). So despite making the same telescopic observations as Huygens, Divini's publication proposed a theory very distant from the suggestion of the existence of a ring around Saturn. Significantly, it did not include any references to Copernicanism. In fact, it insisted that the phases of Saturn were as viewed from the centrally located Earth, and it rejected the notion that any imperfect solid could be suspended around Saturn. This theory even validated Galileo's observations of separate spherical bodies surrounding Saturn, and framed those observations within an acceptable geocentric Tychonic model of the universe. As the first of the 23 propositions outlined in *Brevis annotatio* states, the heliocentric system could not be acceptably applied to the movements of any planet, since, according to Aristotle, '[T]he earth is immobile at the centre of the world and ... the celestial spheres turn around it.' This is an opinion, continues the text that the author defended 'with tenacity, judging it to conform at the same time to the Catholic decrees, to the Sacred Scriptures, to the phenomena observed, and to sane reason'.[22]

Let us be clear, this was not a theory of Saturn that was strongly and logically tied to a geocentric universe or traditional natural philosophy, and it is even possible that any such ties were being overplayed by its author, but it was still designed to put forward an alternative hypothesis that rejected Huygens' Copernican-based model, and retained scholastic astronomy. Therefore, should this theory be sanctioned by a natural philosophical authority such as the patron of the Cimento, Divini's reputation would not only be restored, but traditional cosmological and astronomical observations and values would also be kept intact. Leopoldo and his academicians were being led into a minefield of political, religious, and natural philosophical issues.

One other point remains to be mentioned before we analyse exactly how the Cimento entered into this debate and what decisions they made regarding the two competing theories. Although *Brevis annotatio* was published under Divini's name, in the above-mentioned letter to Leopoldo from 10 July 1660, Divini hinted at the possibility that the text was actually written by a correspondent of the Cimento, Father Honorè Fabri (1607–1688), a French Inquisitor with the Holy Office and Jesuit mathematician in the Roman College. Divini stated that he called upon Fabri to assist him in making this response to Huygens' work available in Latin, since Divini was himself only experienced in writing in Italian, and as he confessed to Leopoldo, a tract in Italian 'would only be of service to a few'.[23] This would seem to suggest that Fabri simply translated Divini's manuscript. But given that this public response to Huygens was littered with the type of religious and anti-Copernican rhetoric that we could expect from a Jesuit astronomer such as Fabri, it is likely that Fabri's contribution went much further than a mere translation. In fact, from the moment the tract arrived in Italy, it was

[22] C. Huygens, *Oeuvres complètes de Christian Huygens*, 22 vols., The Hague, 1888–1950, xv, 422. As cited by Van Helden, 'The Accademia del Cimento and Saturn's Ring', 242. See also Galluzzi, 'L'Accademia del Cimento', 826–827.

[23] 'ad alchuni pochi serviriano'. BNCF, Ms. Gal. 276, f. 33r.

considered to be Fabri's work.[24] According to Van Helden, Divini probably provided Fabri with some doubts about Huygens' telescopic observations and Fabri, with his own agenda to disprove the heretical arguments put forward by the Dutch Protestant, Huygens, compiled the *Brevis annotatio* himself.[25]

Further evidence suggesting that Fabri composed this text lies in the fact that all the subsequent references in letters and manuscripts to the theory opposing Huygens' ring hypothesis mentioned Fabri as the innovator. For example, Michelangelo Ricci, who was regularly dispatching news to the Tuscan Court from Rome, made mention on several occasions of his discussions with Fabri, 'talking to him about his system of Saturn'.[26] So it is likely that Fabri was the central figure behind the scholastic objections from Rome regarding Huygens' claims, and as we have already seen, his aim was to defend traditional Aristotelian cosmology and the Scriptures, 'with tenacity'.[27]

The involvement of Fabri and his use of such scholastic rhetoric against the Protestant astronomer, Huygens, certainly elevated the stakes in these seventeenth-century studies of Saturn. Supporting Copernicanism did not seem to be the central focus of Huygens' *Systema Saturnium*. Instead, as we have just seen, he was far more concerned with boasting about the superiority of his telescope over all others and advancing his ring theory. Copernican astronomy was, nevertheless, the basis of his description of Saturn's phases and this left him open to criticism from Catholic authorities and scholars still determined to have Copernicanism taught as nothing more than hypothetical. Therefore, Huygens could not avoid having to defend the anti-Aristotelian implications in his work. Once he received Fabri's and Divini's tract against the ring theory in August 1660, he immediately composed a reply, *Brevis assertio systematis Saturni*, once again dedicated to Leopoldo.

This work was eagerly anticipated in Rome and probably also in Florence, where Leopoldo was continually receiving news from Ricci about his conversations with Fabri. But the anticipation surrounded not so much the technicalities of Huygens' argument, such as his beliefs regarding the inclination of Saturn or the thickness of the ring, as the cosmological framing of his work. In a letter from Rome dated 13 September 1660, Ricci mentioned to Leopoldo his expectations of Huygens' reply to the publication made in Divini's name and against the ring theory. Ricci revealed the dangers that he believed Huygens faced and the restraint and caution that Huygens should practise when compiling his response to the criticisms of the *Systema Saturnium*.

> A friend of mine sent Eustachio's book to Huygens. I said to him that Huygens should write carefully without insulting anyone, or touching on the motion of the Earth or anything else that could give the Congregation in Rome reason to prohibit him, impeding the book from being seen and also prejudicing the reputation of the cause.[28]

[24] When summarising the accompanying letter from Divini to the Prince, Magalotti wrote: 'Eustachio Divini manda a S.A. il suo libro contro l'Ugenio. Dice essere stato disteso dal Padre Fabri col fondamento d'alcune poche particolarità notate da lui nel libro dell'Ugenio'. BNCF, Ms. Gal. 276, f. 33v.

[25] Van Helden, 'Eustachio Divini versus Christian Huygens', 39.

[26] 'Parlandogli io di quel suo sistema di Saturno'. 26 July 1660. BNCF, Ms. Gal. 276, ff. 42r–42v. See also Van Helden, 'The Accademia del Cimento and Saturn's Ring', 244.

[27] See page 203.

[28] Fabroni, *Lettere inedite*, ii, 97. See also Galluzzi, 'L'Accademia del Cimento', 827.

This was not a warning about Huygens' safety, but more so about the threat his theory was posing for scholastics in the Jesuit colleges, the Courts close to Rome, and in the Catholic Church, those who remained determined to uphold traditional cosmology and Aristotelian natural philosophical beliefs. The 'cause', in all probability, refers to the acceptance of Huygens' ring theory, and was probably Ricci's motivation behind having sanitised versions of Huygens' work available in Italy. In any case, if Huygens actually received this warning from Ricci, he completely disregarded it. In his *Brevis assertio*, Huygens was critical of Fabri's use of Aristotelianism and defended the Copernican basis of his work by maintaining that Copernicus' model was closer to the truth than Ptolemy's, or even Tycho's. This, so Huygens believed, was even widely accepted by many Catholic astronomers.[29] These statements compounded the practical problems that Huygens adduced in Fabri's four-satellite theory. According to Huygens, the apparent 'handles' were not circular, as should be the case in the theory about the dark stars eclipsing the light-reflecting satellites. Instead, they were clearly elliptical. Furthermore, Huygens criticised Fabri and Divini for failing to provide a model for predicting the phases of these satellites moving behind Saturn.[30]

Clearly then, aside from Divini's own concern about his reputation, there was a strong natural philosophical and religious perspective at stake in these three treatises mentioned so far: Huygens' first work on Saturn, Fabri's and Divini's criticism, and Huygens' response to that criticism, all published in 1659–1660. As is revealed in these writings, as well as in the unpublished letters between the central figures in this debate, natural philosophical and religious beliefs were crucial to the acceptance or rejection of the opposing theories and the instrumentation used by the rival astronomers. So this was certainly not about who was using a correct method of observation, but rather how the rival instrument makers and astronomers protected their careers, and pursued their social and political concerns. In other words, what religious, political, and natural philosophical aims they were each trying to achieve.

So how did the Cimento, who we have seen were acutely aware of maintaining a distance from such controversy, become involved in such a potentially volatile situation? When a copy of the *Systema Saturnium* arrived in Tuscany with the dedication to the Medici Prince in July 1659, Leopoldo delayed his reply to Huygens for over one year. The reason for this, as Van Helden suggests, may have been partly due to Huygens' failure to send Leopoldo an accompanying letter with the text.[31] But we may be willing to believe that Leopoldo was concerned

[29] Galluzzi, 'L'Accademia del Cimento', 827.

[30] Van Helden, 'Annulo Cingitur', 165.

[31] It is not even clear why Huygens decided to dedicate this work to the Tuscan Prince since he had never met or even written to Leopoldo prior to August 1660. It is a possibility that Hevelius or Boulliau, both mutual acquaintances of Huygens and the Prince may have given Huygens the idea to write the dedication to Leopoldo, given the Prince's interest in natural philosophy. As Van Helden suggests, Huygens obviously followed this advice, if it was indeed given. He did not include an accompanying letter, perhaps for the reason that he did not wish to give Leopoldo the impression that he was seeking patronage from the Tuscan Court. Van Helden, 'The Accademia del Cimento and Saturn's Ring', 240.

with far more than a mere lack of communication from the Dutch astronomer. As Galluzzi claims, such a long delay was quite out of the ordinary and reflected the Prince's extreme caution when facing the possibility of giving his approval to a Protestant astronomer with openly Copernican beliefs.[32] Furthermore, not only did Huygens' Copernicanism count against him, but Leopoldo may have also been aware of the practical concerns with the ring theory that Boulliau and others had been expressing, especially Huygens' proposed thickness of the ring. So the Medici Prince, being so close to Rome and so eager to avoid any controversy, understandably hesitated in providing his approval of a theory that lacked credibility according to many astronomers, and more importantly, had the potential of being highly controversial because of its Copernican content. If this was indeed the reason for Leopoldo's seemingly cautious response to having *Systema Saturnium* dedicated to him, his judgement was not misplaced considering Divini's and Fabri's objection to Huygens' claims.

It was in fact, Fabri's and Divini's public criticism of Huygens that forced Leopoldo to take some action regarding the Saturn problem. Forming an assessment of Huygens' work was unavoidable once Fabri and Divini also dedicated their publication to the Prince. This was not only because Leopoldo was the recipient of both dedications by the opposing astronomers, but also because Divini even specifically pleaded with the Prince to act as mediator in the debate between him and Huygens. Divini asked Leopoldo 'to inquire into which of us got it right and if the glasswork from Holland is more perfect than ours'.[33] Furthermore, in *Brevis annotatio* Fabri and Divini again appealed to Leopoldo's 'very good censure' and his 'enlightened judgement' to adjudicate between Huygens' theory and the hypothesis from Rome.[34]

3. LEOPOLDO TAKES CONTROL

In July and August 1660, Leopoldo called upon his Cimento academicians to assist him in the resolution of this controversy between Huygens and Fabri. The following actions taken by the Cimento were indicative of the natural philosophical concerns pursued by the academicians, led once again by Borelli, and their social, political, and religious concerns when presenting their work. They maintained an anti-Aristotelian agenda that reflected upon their arguments in favour of Copernicanism. In this respect, their work in astronomy followed the same patterns of natural philosophical speculation as that which we have seen in the two case studies in Part Two. However, the Cimento's experiments concerned with the Saturn problem carried far greater political and religious pressures than the work they had performed earlier on pneumatics and the effects of heat and

[32] Galluzzi, 'L'Accademia del Cimento', 826.
[33] This appeal to Leopoldo was made by Divini in his previously mentioned letter on 10 July: '... esplorare chi di noi habbia accertato e se li vetri d'Ollanda siano più perfetti della nostra Italia'. BNCF, Ms. Gal. 276, f. 33r.
[34] Huygens, *Oeuvres*, xv, 436.

cold. Leopoldo was forced to choose between a Dutch Protestant who was not afraid to express the same views that saw the condemnation of Galileo by the Catholic Church, and a Jesuit mathematician in Rome who was also an Inquisitor for the Holy Office, no less. As a result, the academicians were aware that their work on this topic was going to be anticipated in several parts of Europe, particularly Rome, where scholastics may have felt the most threatened by the Copernican content of Huygens' work. This was to have an impact upon the way in which they would decide to carry out their observations, how they would choose to discuss their natural philosophical concerns, and how they would present their work to their colleagues. In other words, they knew about the controversial nature of this case when they approached it, and this influenced how they carried out their work. We can look forward, therefore, to the decisions and actions that Leopoldo and his academicians were to take, keeping in mind the deep concern the Prince had for the public image of the academy under his control and his relationship with ecclesiastical authorities.

As they began to investigate the contrasting theories regarding Saturn's appearance, the situation was not looking favourable for Fabri and Divini. On 17 July 1660, when the academicians had only just read through the work that had been published under Divini's name, that is, before they had even agreed on a course of action, Magalotti recorded in the official Cimento diary that Borelli was already making some remarks in Huygens' defence.[35] Furthermore, during late July, while the academicians were planning their approach to the problem, the correspondence to Leopoldo from Rome and France regarding the choice that Leopoldo had to make between the competing theories, was also quite favourable for Huygens.

First, on 26 July 1660, Ricci wrote to Leopoldo giving his own judgement of Fabri's theory. According to Ricci, Fabri's work was certainly worthy of praise. However, upon closer investigation of his theory, Ricci claimed that Fabri did not provide a satisfactory account of the phases and movements of the planet and its apparent satellites. In fact, so sceptical was Ricci of the validity of Fabri's hypothesis, that he believed the Jesuit astronomer was only concerned with doing everything possible to defend traditional scholastic beliefs: 'It only seems so far that the Father introduces many changes in order to save only one ancient opinion of Saturn moving around the Earth.'[36] This reinforces the idea mentioned earlier that the tenuous ties between Fabri's and Divini's hypothesis and their broader natural philosophical concerns about geocentricity, were being emphasised and even exaggerated in order to counter Huygens' Copernican-based theory.

Several days later, on 9 August 1660, as he was awaiting news from Florence about the Cimento's observations, Ricci again wrote to the Prince expressing

[35] 'Si lesse tutto il libro del Divini scritto contro il sistema Saturnico di Christiano Eugenio, et in esso quello, che ha inventato il Padre Fabri Gesuita. Si sentirono alcune annotazioni fatte dal Sig. Borelli sopra a detto Libro in difesa dell'Eugenio, e si stabilirono alcune esperienze in questo istesso proposito'. BNCF, Ms. Gal. 262, ff. 93r–93v.

[36] 'Quel che appare fin ora è che'l Padre introduce molte novità per salvare una sola antica opinione di Saturno mosso intorno la terra'. BNCF, Ms. Gal. 276, f. 42v.

more criticism about Fabri's hypothesis and expecting a similar sceptical report from the academicians. Ricci wrote that although Fabri had heard several criticisms of his theory, including Ricci's own thoughts, Fabri was convinced that he would be able to formulate a suitable response. Ricci concluded: 'I doubt that perhaps things will not come out for him as easily as what he believes.'[37]

In the meantime, Huygens' first letters to the Prince were arriving from France. One of those, written on 16 August, stated Huygens' eager anticipation of the Prince's judgement. In particular, Huygens expressed his confidence that the academicians would decide in his favour, especially since observations from England, such as those made by Wren, could be used to support the ring theory, or at least falsify Fabri's hypothesis, even though Wren himself argued that Huygens' theory was problematic.[38] By this time, the academicians had already performed an experiment to assist them in their assessments of the opposing hypotheses. But we may see from the above-mentioned correspondence from Rome and Paris addressed to Leopoldo, that Fabri's and Divini's work was not gaining a great deal of favourable publicity in Florence, despite its conformity with scholastic beliefs, and despite also the criticisms that were aimed against the credibility of Huygens' theory. In addition to this, we must not forget that most of the academicians, including the Prince, were supportive of Galileo and his work on Copernican astronomy. All this reflected the unlikely advantage that Huygens held over his Roman colleagues. Indeed, as we shall now see, although they did not dare to express openly any anti-scholastic sentiments, especially in astronomy, the Cimento's work on Saturn was far from favourable for Aristotelian natural philosophers such as Fabri. We shall also see that the judgement reached by his academicians placed Leopoldo in a difficult position with regard to the Catholic Church. That is, he now had to try to negotiate the credibility of his judgement against Fabri in the face of traditional religious and natural philosophical pressures. This means that Leopoldo was also fighting to preserve the credibility of the academy under his control, and its reputation as an institution producing reliable and uncontroversial knowledge.

4. MODEL EXPERIMENTING USED TO RESOLVE THE SATURN PROBLEM

The Cimento diary entry for 20 July 1660, mentions how the academicians discussed their options for investigating the appearance of Saturn. Since, they agreed, observing Saturn's phases would require years of telescopic observations, they would have to devise other ways of arriving at a quick resolution for this

[37] '... dubbito forse non sia per riuscirli così facilmente come si crede'. BNCF, Ms. Gal. 276, f. 49r.

[38] BNCF, Ms. Gal. 276, ff. 51r–51v. Wren did not believe that what he saw around Saturn was a ring unattached from the planet, but rather an elliptical corona that touched Saturn at two opposite ends. Although not agreeing with Huygens' theory, it was certainly far from being supportive of Fabri's hypothesis about four satellites of different light-reflecting capabilities. For a summary of Wren's corona theory, see Van Helden, 'Annulo Cingitur', 160.

topic, including the construction of models.[39] So, during the weeks that followed this meeting, the academicians made two models of Saturn, one with Huygens' ring, and the other with Fabri's satellites (Figure 20).[40] How the first of those models was set up, and how the following observations were carried out, was described in a letter Borelli wrote to Leopoldo in August, 1660.

Borelli described how the academicians set up their model of Huygens' system of Saturn in a long gallery, probably in the Pitti Palace, at a distance of about 37 braccia (75 m) away from the two telescopes, a powerful and large one, and

Figure 20. Illustration of the model used by the Cimento to test Huygens' ring theory. BNCF, Ms. Gal. 289 f. 81r. Courtesy of the Ministero per i Beni e le Attività Culturali / Biblioteca Nazionale Centrale di Firenze. Protected by Copyright.

[39] 'Si consultò il modo, e il tempo da farvi le osservazioni di Saturno con l'occhiale del Divini, perciò si discorsero diverse maniere di macchine per addopperare con facilità il Telescopio'. BNCF, Ms. Gal. 262, f. 93v. As Borelli stated in his letter to Leopoldo cited below, and published by Targioni Tozzetti, regular observations of Saturn's phases would have taken 'eight or nine years'. For this reason they decided to make the models and record their conclusions as soon as possible. BNCF, Ms. Gal. 271, f. 3r.; Targioni Tozzetti, *Notizie*, ii, 740.

[40] A sketch of the ring-theory model was sent to Huygens and was published in: Huygens, *Oeuvres*, iii, 154–155. Rough sketches of the model, as well as drawings of the ring around Saturn, and even of Fabri's proposed satellites can also be found in BNCF, Ms. Gal. 271, ff. 34r–47r.

another one smaller and inferior to the first. Four torches were set up to illuminate the model, but were hidden from the observer's field of vision. The initial observations of these models through the telescopes were favourable for Huygens' theory, since the powerful telescope observed the ring clearly, while the inferior instrument created the illusion of two small satellites on either side of Saturn. This strengthened the suggestion that the ring theory explained the strange appearances and phases of Saturn better than any other hypothesis produced in the seventeenth century. But having constructed the model themselves, the academicians recognised that they were of course already aware of its dimensions before observing it through the telescopes, and were therefore not unbiased viewers. Not trusting their own senses, they called upon neutral observers, 'who had not seen the shape of this device from nearby', to look through the smaller telescope.[41] Besides some men who, for several possible reasons, recorded rather odd observations, '[I]t was obvious that the appearance that they almost all drew was the disc of Saturn in the middle of two little round balls and separated from it by a sensible distance.'[42] That is to say that Galileo's first observation of Saturn was recreated through the Cimento's experiment showing that the observations of the supposed satellites were simply illusions created by the ring and by the imperfections of those telescopes inferior in quality and power to the Huygens' instruments. So the experiment was a resounding success from Huygens' point of view. Indeed, Borelli reported in his letter to the Prince that he could even observe Saturn's shadow being cast upon the ring. With this final observation, '[I]t would seem that a very efficacious argument', so Borelli claimed, 'could be deduced in Sig. Huygens' favour.'[43]

The only doubt that remained for the academicians was the same one for which Huygens was criticised by other French and English astronomers: in their observations of the model through the strongest of the two telescopes, it seemed that some trace of the ring was always apparent, meaning that Huygens' theory could not explain how in reality Saturn occasionally appeared only as a single sphere with no accompanying 'handles'. The only argument that Huygens made here in defence of his theory was that the edge of the ring was made out of a material that did not reflect light and was therefore sometimes invisible from Earth. This type of ad hoc claim was understandably not very well received, since it also meant that the academicians would have to accept Fabri's and Divini's

[41] Borelli did not actually state what telescopes, if any, these observers used. But judging from their observations and Borelli's comments about the results, cited below, it would seem that these independent participants could have been asked to observe the model only either with the inferior of the two instruments, or even with no telescope at all.

[42] BNCF, Ms. Gal. 271, f. 7v.; Targioni Tozzetti, *Notizie*, ii, 742. 'Per chiarire adunque la verità di questa apparenza furono chiamati molti, fra quale anche delle persone idiote, e che non avessero veduta da presso la struttura di quella macchina, ad osservarla e fatta gliela vedere dalla detta distanza di 37 braccia, e disegnare ciascuno a parte ciò che se gli appresentasse, fu così patente l'apparenza che disegnarono quasi tutti il disco di Saturno in mezzo a due palline rotonde, e distaccate per sensibile spazio di essa'. As translated by Van Helden, 'The Accademia del Cimento and Saturn's Ring', 245.

[43] '... pare che possa dedursene argumento molto efficace a favore del Sig. Ugenio'. BNCF, Ms. Gal. 271, f. 13r.; Targioni-Tozzetti, *Notizie*, ii, 745.

suggestion that two of the hypothetical satellites behind Saturn could also be made out of this material.[44] In any case, the academicians' experiments proved to be far more successful for Huygens than for Fabri. A model of Fabri's theory with the satellites provided the three-bodied appearance of Saturn and the single sphere, but the 'handled' appearance was never achieved.[45] So while both Huygens and Fabri had practical problems with their hypotheses, according to the Cimento academicians, the observations carried out by Huygens were far more acceptable and less obviously flawed than those made by Fabri.

This signalled the end of the academicians' observational work on the topic, but it only marked the beginning of their religious and political concerns when presenting their results – the type of concerns that guided the Cimento's publication process. The Tuscan Court faced a problem when Huygens' Copernican-based work was dedicated to the Prince, but now that problem intensified as Leopoldo was hearing recommendations from his friends and courtiers in favour of Huygens' controversial theory. Therefore, Leopoldo had to decide how to present the Cimento's work to the public. Now that Huygens' work was not rejected, but in fact supported by the academicians, there was still plenty of room for the type of controversy with ecclesiastical authorities that Leopoldo was obviously anxious to avoid for the sake of preserving the uncontroversial status and reputation of his Court and his academy.

On 17 August 1660, two reports on the experiment and its outcomes, one written by Borelli, and the other by Carlo Dati, were sent to Rome. They were addressed to Ricci, but the accompanying letters by Leopoldo and Magalotti were both intended for Fabri.[46] As we saw from Borelli's earlier letter to Leopoldo describing the experiment, the Cimento's leading contributor was quite supportive of the ring theory, despite Huygens' doubtful claims about the ring's thickness. Borelli's official report reflected similar sentiments against Fabri, and in favour of Huygens and his Copernican-based theory of Saturn. Meanwhile, Dati was seemingly more impartial in his assessment, suggesting that there could be grounds to dismiss either of the competing theories. Nevertheless, he agreed that from the observations performed by the Accademia, Fabri's hypothesis was the less likely to be true.

In the meantime, a different report, also written by Borelli, was sent to Huygens via Heinsius. A letter from Dati to Heinsius also accompanied the report. After presenting the same arguments supporting Huygens' and based on the Cimento's observations of the models, Borelli praised the Dutch astronomer for his observations and interpretation of Saturn's phases. This was finally, as Van Helden points out, the approval of his *Systema Saturnium* that he had been seeking from the Tuscan Court when he dedicated his publication to the Prince in 1659.[47]

[44] See BNCF, Ms. Gal. 271, f. 10v.

[45] This observation was not mentioned in the above-cited letter from Borelli to Leopoldo. But it was described in the final report Borelli wrote in August, 1660, one of two reports written by the Cimento. BNCF, Ms. Gal. 289, ff. 15r–19v.

[46] Copies of the letters and the reports are preserved in BNCF, Ms. Gal. 289, ff. 6r–9r.; 15r–21v. A summary of these reports and the accompanying letters by Leopoldo and Magalotti can be found in Van Helden, 'The Accademia del Cimento and Saturn's Ring', 248–249.

[47] Van Helden, 'The Accademia del Cimento and Saturn's Ring', 250–251.

Fabri's response to the academicians' conclusions can be gauged first from Ricci's 22 August letter to Leopoldo. Ricci himself was delighted with the Cimento's 'ingenious' experiment and suggested that Huygens' theory was clearly shown to be the more accurate of the two. However, Ricci warned that Fabri was far from convinced and would seek Divini's collaboration to analyse the Cimento's claims and to compile a defence of their theory.[48] Indeed, on 30 August, Ricci sent Leopoldo a manuscript, again written under Divini's name, which continued to defend the quality of Divini's telescopes.[49] In this apologia, entitled *Pro sua annotatione in Systema Saturnium*, Fabri also defended his theory by adding two more light-reflecting satellites to his system of Saturn that could create the elliptical shape of the 'handles' (Figure 21). Nevertheless, Fabri was still careful not to dismiss Huygens' claims completely. He conceded, for example, that while he disagreed with the ring theory, Saturn could appear to have a ring surrounding it. As Van Helden suggests, Fabri was allowing himself the opportunity to retreat gracefully should any more criticisms be aimed against him.[50] Once again this demonstrates how observations of natural phenomena, including the instrumentation used, were laden with social, political, and natural philosophical commitments. Fabri and Divini were willing to go to any length to defend scholastic principles in which they had been trained and that formed a cornerstone of traditional intellectual endeavours, and in the process, to show that Divini's telescopes were not inferior to those of Huygens. Yet, realising that their claims were attracting criticism from highly respected sources, such as the Cimento, they still manoeuvred to present their work in a way that could offer them an escape and save their own careers and reputations.

In the meantime, since the Cimento was now overwhelmingly in favour of the validity of Huygens' theory, and had even made their support clear in the reports they sent to the competing astronomers, Leopoldo had to ensure that no accusations of heresy could possibly be made against him or his academy. For this reason, first he made a request to Fabri and to Huygens that they should not refer to the Cimento's work on this topic in their writings. In what is probably the clearest demonstration of the academicians' intention to keep well away from any type of conflict that could harm their reputation and relations with other courts, especially the Papal Court, Dati wrote the following message to Heinsius, intended for Huygens, in August 1660:

> For the moment it is desired that no public mention is made of it. For one thing this is because these men are very cautious in affirming anything, not wishing to commit themselves without much consideration and repeated trials ..., and for another thing

[48] 'Non vedo però che fin ora si possa dir altro se non che'l Sig. Ugenio non sia convinto dal P.re Fabbri di falsità, ma che ne meno ci costi esser vero il di lui sistema, restandovi pur assai da smaltire. Gran diletto ha poi recato all'animo mio l'esperienza che mostra la fascia intorno il globo formato a simiglianza di Saturno, ora in forma di due globi separati, ora nella sua natural figura: pensiero de' più ingegnosi e pellegrini ch'i'udissi mai. Lo dissi al P.re Fabbri prima di consegnargli il piego del Sig.r Lorenzo Magalotti, e mi rispose che 'l Divini avrebbe voluto provar tutto questo e per quel che m'imagino ambidue s'armano alla difesa'. BNCF, Ms. Gal. 267, ff. 55r–55v.

[49] Fabroni (ed.), *Lettere inedite*, ii, 94–95.

[50] Van Helden, 'The Accademia del Cimento and Saturn's Ring', 256–258.

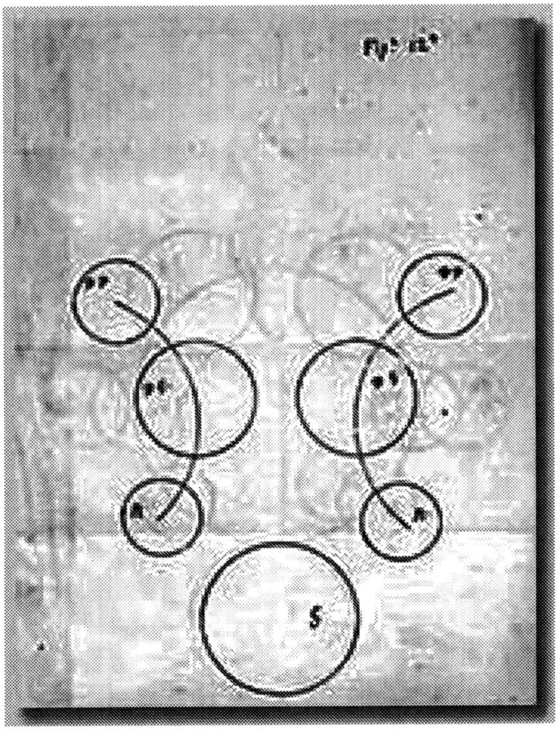

Figure 21. A drawing of Fabri's hypothesis with six satellites. The light absorbing satellites are the larger spheres in the centre. BNCF, Ms. Gal. 283, f. 106r. Courtesy of the Ministero per i Beni e le Attività Culturali / Biblioteca Nazionale Centrale di Firenze. Protected by Copyright.

> because, having written some rather severe censures against Father Fabri, they would not wish to commit themselves and to be held by the world to be impassioned and partial .... In this I commit myself to your prudence.[51]

This appeal to Huygens' discretion would certainly save the academicians from being exposed by the Dutchman, but they still needed to convince Fabri that they did not intend to promote Copernicanism as the truth. This would require some subtle diplomatic manoeuvring. Leopoldo decided to publish, in Florence, Huygens' *Brevis Assertio*, the Dutch astronomer's reply to Fabri's and Divini's critical analysis of the *Systema Saturnium*. This text, as was mentioned, defended the ring theory largely on the basis that Copernicus' system was true, and criticised its Roman detractors for their scholastic beliefs. Obviously the complete publication of this tract, including Huygens' anti-Aristotelian rhetoric, would hardly have aided the Prince's reputation in Rome, so in order to avoid any

[51] Huygens, *Oevres*, iii, 149–150. As cited by Van Helden, 'The Accademia del Cimento and Saturn's Ring', 250. This message was in the letter Dati sent to Heinsius accompanying Borelli's report to Huygens of the academicians' Saturn experiment.

controversy and appease Fabri's natural philosophical and religious concerns, Leopoldo simply omitted Huygens' references to Copernicanism. This piece of editing did not detract greatly from Huygens' ring hypothesis, nor did it deny the Copernican basis of his work, which most astronomers outside Rome probably still would have inferred from having read, or heard about, Huygens' previous publications and utterances on the topic. But it did maintain the academicians' impartial reputation by showing that they were willing to listen to the varying claims and make an assessment without becoming participants in controversial speculations and debates about theory.

Furthermore, the experiment with a model, although an unusual approach to an astronomical question, provided Leopoldo and his courtiers with an opportunity to maintain a purely experimental, and therefore seemingly factual and atheoretical, approach. This awareness of the value of the academicians' experimental work was reflected in Borelli's report of the experiment to Leopoldo:

> In this matter too we have inviolably observed the custom of the Academy of Your Highness, which is to search out the truth through many experimental proofs, to a degree, however, in which it can be adapted to things so far removed from our senses, and we have fully and dispassionately examined the opinions of Mr. Huygens and those of the adversaries who oppose him, in the meeting before Your Serene Highness.[52]

This is the type of rhetoric that appealed to the Prince's political aims and interests. Through the construction of a model, the academicians managed to give the impression that they were avoiding the natural philosophical controversy surrounding the issue. This was an experimental approach and rhetoric that was intended to build their reputations as reliable producers of natural knowledge. In fact, this was the reputation that Leopoldo advertised to Huygens on 14 September 1660. The Prince wrote to Huygens about the unbiased quality of the Cimento's work and their ability to carry out an impartial judgement of the competing theories of Saturn. In this letter, Leopoldo first praised Huygens for his 'great desire to recognise the truth in everything' and then claimed that this same search for truth was also 'the most important maxim of an academy of many virtuous men, who gather together before me almost every day without impassioning themselves to the opinions of others, or even to their own'.[53]

By the time Leopoldo had written this letter, the academicians' reports about their observations of the models had already been sent to Huygens. Dati had also

[52] BNCF, Ms. Gal. 271, ff. 3v–4r.; Targioni Tozzetti, *Notizie*, ii, 740. 'Noi però altrimenti, secondo il costume dell'Accademia di Vostra A.S., che è d'investigare il vero per via di riprove sperimentali, l'abbiamo inviolabilmente osservato anche in questo affare, per quella parte però che può ridursi ad Esperienza di cose tanto remote da' nostri sensi, et esaminando per ultimo nei Congressi tenuti d'avanti all'A.V, disappassionatamente i Concetti dell'Ugenio, e quei degl'Avversari che gli oppongono, vi sono cadute alcune Riflessioni'. As translated by Van Helden, 'The Accademia del Cimento and Saturn's Ring', 244.

[53] 'un desiderio grande di riconoscere la verità in ciascheduna cosa, come ho determinato che sia la principal massima di un'Accademia di molti Virtuosi, che quasi ogno giorno si radunano avanti di me, senza appassionarsi non solo alle opinioni altrui, ma nemmeno alle proprie'. Targioni Tozzetti, *Notizie*, i, 382.

already appealed to the Dutch astronomer's discretion to preserve the Cimento's uncontroversial image. Now Leopoldo was personally advertising this image to Huygens by insisting that his courtiers were only searching for the truth and did not purposefully set out to support or reject any opinions or speculations. The academicians' reputation as unbiased knowledge makers was, therefore, undoubtedly quite important to Leopoldo and reflected the Cimento's censorship policy when it came to the presentation of their work.

Soon afterwards, Magalotti suggested to Leopoldo that since a censored copy of Huygens' work was to be published in Florence, the academicians could also publish censored versions, without Copernican theorising, of their own reports that they had sent to Huygens and Fabri. In a single-page memorandum to the Prince, Magalotti proposed that the academicians' observations of Saturn should be recorded, but that any statements by Borelli made in support of Copernicanism should also be omitted, in order 'to avoid difficulties'.[54]

Magalotti's proposal was never accepted, but through the Prince's other efforts to avoid controversy, by the end of this debate the Cimento came out with its reputation high amongst European astronomers and intact amongst ecclesiastical authorities and Jesuit thinkers such as Fabri.[55] Although the work carried out by his academicians was in favour of the Dutch Protestant instead of the Jesuit Inquisitor, remarkably Leopoldo had managed to keep himself and his academy away from any controversy, and even reinforced their image as unbiased experimentalists. In other words, by refusing to publish or make public the natural philosophical skills, commitments, and agendas of the academicians, as they were expressed in the reports and in letters, they could not possibly have been condemned by the Catholic Church. They could support Huygens without publicly acknowledging his belief in Copernicanism. This then relieved them of being threatened by any accusations from Fabri that the ring theory was heretical. In the meantime, the Cimento was to boost its reputation across Europe for producing reliable experimental knowledge claims, creating the status which the Medici Prince longed for as a protector of truth and knowledge in the Tuscan Court, and strengthening the reputations and careers of the academicians.

The Cimento's rhetoric in this case study was crucial to the political and religious concerns of its patron and members. As Galluzzi points out, there is little evidence that the Prince was implementing a formal policy of censorship of natural philosophical expression. Nevertheless, Leopoldo seems to have adopted 'self-censorship' for the sake of attaining respectable status and reputation for his academy.[56] While they could get away with publishing their work on the vacuum, air pressure, and the effects of heat and cold with clear natural philosophical underpinnings, astronomy was a different story. Leopoldo could dare to publish

[54] 'Si è pensato di metter in sicuro tutto quello che l'anno 1660 si speculò, e si operò nell'Accademia di V.A. intorno a Saturno .... Bisognerà però che il Sig. Borelli, si contenti di ridurre fuori del sistema Copernicano quelle sue dimostrazioni per sfuggir difficultà'. BNCF, Ms. Gal. 271, f. 16r.; Targioni Tozzetti, *Notizie*, i, 385.

[55] Van Helden, 'The Accademia del Cimento and Saturn's Ring', 254.

[56] Galluzzi, 'L'Accademia del Cimento', 823–832.

experiments that hinted at a belief in microstructures such as atomism and the existence of the vacuum, as long as such beliefs were never openly supported. But astronomy was still based on the analysis of the macrostructures that not only formed the basis of Aristotelian cosmology, but were also used to support Catholic biblical teachings. To make such immensely controversial anti-Aristotelian statements in seventeenth-century Italy, the academicians would have been exposing themselves to the same type of religious scrutiny that resulted in the condemnation of their hero, Galileo.

The academicians' reputation of impartiality could not have been achieved if they had made public their work on Saturn and risked exposing the controversial Copernican interests of some of its members. This is why the Prince did not allow that he or his Court members should be publicly seen as participants in this debate. It is also the reason why Magalotti's proposal to publish the academicians' work was not acceptable, and why not a single word of the Cimento's role in this controversy was mentioned in the *Saggi*. In fact, so effective was Leopoldo's strategy to maintain this reputation in this case study, that even twentieth-century historians have found themselves marvelling at the academicians' approach to the Saturn problem, including the construction of models for experimenting, and extolling the virtues of the 'Experimental Method'. This point shall be discussed further in the Conclusion. In the meantime, we shall now see that the academicians' work on Saturn was not their only interest in astronomy. In 1664, they engaged in a new debate on comets that proved to be just as controversial, if not more so, than the Saturn problem. How Leopoldo handled these comet observations is further reflection of how the academicians were determined not to create a controversial image of the Cimento, for the sake of their patron's reputation and their own careers.

5. COMETS

The observations of comets during the 1660s also provided Europe's astronomers, including members of the Cimento, with some social, political, and natural philosophical concerns. Some of the academicians were again involved in providing Copernican and anti-scholastic interpretations of their observations of comets in 1664–1665. Much as we have already seen with the academicians' arguments about Saturn, these natural philosophical interpretations of comets were never presented to the public. But unlike the Saturn controversy, by the time the academicians became interested in the debate regarding the movements of comets, it had already been entangled in natural philosophical controversy for three-quarters of a century since Tycho Brahe's first comet observation in 1577. For this reason, Leopoldo ensured that the comet observations, made mostly by Borelli, were never to be published in the *Saggi*.

According to Aristotle, because of the ephemeral and independent nature of comets, far removed from the perfectly circular and consistent motions of all celestial bodies, they could only possibly exist inside the corrupt and imperfect sublunary realm. Aristotle believed that comets were nothing more than the hot and dry exhalations from Earth being carried around by the sky until they burnt up and died. In contrast, Tycho claimed that the comets he had observed in 1577, 1580, 1582, 1585, and 1590, were travelling at great distances from the Earth, somewhere outside the orbit of Venus, around the Sun. This suggested that comets, with their vague trajectories and their inconsistent lifespans, could actually be moving in the celestial superlunary realm, believed by Aristotelians to be perfect and incorruptible. Tycho also calculated the trajectories of the comets and estimated that they ran across the paths of the planets, meaning that the crystalline spheres, said by scholastics and Copernicans to be keeping the planets in their orbits, could not possibly exist. Contrary to scholastic thought, Tycho had both the planets and comets moving independently.[57]

In the beginning of the seventeenth century, Galileo used Tycho's work regarding the existence of comets beyond the sphere of the moon, in order to assist him in his argument for the mutability of the celestial realm. In the second letter on sunspots, published in 1613, Galileo discussed the mountainous surface of the moon and the apparent spots on the sun in order to prove that there is no distinction between celestial and terrestrial realms as Aristotle believed. In addition to his own evidence, he referred to Tycho's observation of the comets and his estimation that they moved beyond the terrestrial region: '[A]s if to remove all doubt from our minds, a host of observations come to teach us that comets are generated in the celestial regions.'[58]

However, this was the last and only tribute Galileo gave to Tycho's comet observations, since Tycho's system eventually began to pose a much bigger threat to the acceptance of Copernicanism, than what Galileo may have initially believed.[59] The reason for this was that while unfavourable, in some minor respects, for Aristotelian cosmology, Tycho's work on comets was equally critical of Copernicus' heliocentric system. Although Tycho and Copernicus agreed that comets orbit the Sun, Tycho argued that comets do not display any regular motion corresponding with the supposed annual motion of the Earth. In other words, both planets and comets retrogress, but while planets display regular patterns of retrograde motion, supposedly due to the Earth's movement according to Copernicus, cometary motion was irregular and far less predictable.[60] So Tycho's claims provided a convincing argument against both Ptolemaic and Copernican astronomers, and in support of his own compromise system. In fact, as William Shea shows, even some of Galileo's correspondents during the 1610s were expressing their concerns that Copernicus' heliocentric claims may have been

[57] While Tycho questioned certain aspects of traditional astronomy, he was still committed to preserving Aristotelian cosmology. He had compromised between Ptolemy and Copernicus and suggested a geocentric universe, with the Sun and the moon orbiting Earth, and the planets orbiting the Sun. With this theory, Tycho defended the Aristotelian distinction between the matter, structure, organisation and movements of bodies in the celestial and terrestrial realms.

[58] 'Ecco, da virtù superiore, per rimuoverci ogni ambiguità, vengono inspirati ad alcuno metodo necessarii, onde s'intenda, la generazion delle comete esser nella regione celeste'. Favaro (ed.), *Le Opere*, v, 139–140. As translated by Drake (ed.), *Discoveries and Opinions*, 118–119.

[59] Drake (ed.), *Discoveries and Opinions*, 119–120 n. 11.

[60] C.J. Schofield, *Tychonic and Semi-Tychonic World Systems*, New York, 1981, 74; W.R. Shea, *Galileo's Intellectual Revolution*, London, 1972, 86–87.

refuted and replaced by Tycho's system.[61] From the point of view of scholastics, Tycho still retained fundamental aspects of Aristotelian cosmology while refuting modern followers of Copernicus. For this reason, when a new comet appeared across the skies of Europe in 1618, scholastics adopted Tycho in their astronomical work to counter Galileo's arguments in defence of Copernicus. In other words, following Galileo's first confrontation with the Catholic Church in 1616, the emphasis of all astronomical observations fell squarely within the field of natural philosophy: Tycho was being used by scholastics to defeat anti-Aristotelians such as Galileo and their commitments to Copernicus.

From late November 1618 until early January 1619, a Jesuit mathematician at the Collegio Romano, Orazio Grassi, made careful observations of three bright comets. Like most Jesuit scholars at this time, Grassi favoured the Tychonic world system ahead of Ptolemy's. His observations and claims regarding the 1618–1619 comet were therefore favourable for Tycho and in the process, did not seek to upset any of the entrenched Aristotelian beliefs. Grassi reported his observations and conclusions in a lecture that he delivered at the Collegio Romano in 1619, and in a text that he published anonymously that same year.[62] In this publication, Grassi insisted, as did Tycho, that the only way of determining whether the comet belonged to either the celestial or terrestrial realms, is through its parallax. If the comet were sublunary, then when observed from different places on the earth, it would appear to be in different parts of the sky because its proximity would allow parallax to be detected. Meanwhile, if it were superlunary and therefore far away, the parallactic effect would be diminished considerably or be totally undetectable.

Grassi cited observations of the 1618–1619 comet made from Antwerp, Rome, Parma, Innsbruck, Arcturus, and Cologne and demonstrated that there appeared to be little or no change in the position of the comet when observed from each of these cities. 'Therefore, you have it from parallax, however observed, that our comet was not sublunar but clearly celestial.'[63] Furthermore, claimed Grassi, while the telescope improved observations of nearby planets and satellites, its power of magnification seemed barely perceptible on the comet, much like on the stars that are at such a great distance from the earth. This could only mean that the comet too was far beyond the distance of the moon. According to Drake and O'Malley, these claims in themselves were not offensive to Copernicans. Indeed, Grassi did not mention Copernicanism in his pamphlet on the 1618 comet and was concerned primarily with the position of comets in the celestial realm, an argument more damaging to Aristotelian cosmology than heliocentricism. Drake and O'Malley suggest that it was the Jesuit reaction to Grassi's work that might have stirred Galileo to write a response. For Jesuit astronomers at the Collegio Romano, such reports seemingly compiled in support of Tycho's model, were continuing to discredit Copernicus' heliocentric system.[64]

[61] Shea, 86.

[62] *De Tibus Cometis Anni MDCXVIII: Disputato Astronomico Publice Habita in Collegio Romano*, Rome, 1619. See Favaro (ed.), *Le Opere*, vi, 20–34; S. Drake and C.D. O'Malley (eds.) *The Controversy on the Comet of 1618*, Philadelphia, 1960, 3–19.

[63] 'Habetis igitur ex parallaxi utcunque observata, non sublunarem, sed plane caelestem, fuisse cometam nostrum'. Favaro (ed.), *Le Opere*, vi, 31. As translated by Drake and O'Malley (eds.), 14.

[64] Drake and O'Malley (eds.), xv. See also Shea, 74.

Serious illness had prevented Galileo from carrying out extensive observations of the 1618–1619 comets himself, but as he revealed in *The Assayer* in 1623, while he was bedridden he had received visits from several friends with whom he had discussed the movements of these comets. Galileo claimed that in those conversations he had 'cast doubt upon the doctrines that have been previously held on this matter', including Tycho's.[65] One of Galileo's companions who was present at these discussions, Mario Guiducci (1584–1646), used Galileo's arguments to deliver two lectures on comets in 1619 to the Florentine Academy. These lectures, soon after published in Florence, were based almost entirely on the arguments Galileo had expressed to Guiducci about the properties and movements of comets, and responded directly to the claims made by Grassi.[66] They begin to reveal the rather strange claims that Galileo made with regard to comets, but that were, nonetheless, aimed to discredit Aristotelianism. Guiducci argued on behalf of his friend that comets were probably nothing more than an illusion, the image created by the reflections of light on an accumulation of vapours just above the surface of the Earth. This being the case according to Galileo and Guiducci, the search for parallax was irrelevant since such observations are only valid for real and permanent objects, not for mere illusions. So while they acknowledged parallax as a valid tool for astronomers, Galileo and Guiducci denied that it was useful for determining the distances of comets. Anyone who used parallax in this situation, therefore, had to prove that comets were nothing more than illusions: 'I shall not believe that parallax has really any place in comets until it is first proved that comets are not reflections of light, but are unique, fixed, real, and permanent.'[67]

Furthermore, wrote Guiducci, these accumulations of vapours emanate from the earth and move in a straight line towards the heavens, perpendicular to the Earth's surface, only becoming visible after rising 'beyond the cone of the earth's shadow'.[68] That is why they do not exhibit much change in their position and why they appear to become dimmer until they finally fade away. In addition, this perpendicular and rectilinear motion, would explain why comets appear to slow down from the point of view of the observer standing on Earth; the increasing angle between the comet and the observer would provide the appearance that the comet is slowing (Figure 22).[69]

[65] 'Per tutto il tempo che si vide la cometa, io mi ritrovai in letto indisposto, dove, sendo frequentemente visitato da amici, cadde più volte ragionamento delle comete, onde m'accorse'dire alcuno de' miei p'nsieri, che rendevano peina di dubbi la dottrina datone sin qui'. Favaro (ed.), *Le Opere*, vi, 225; Drake and O'Malley (eds.), 236.

[66] M. Guiducci, *Discorso delle Comete*, Florence, 1619. Published in Favaro (ed.), *Le Opere*, vi, 36–108; Drake and O'Malley (eds.), 20–55.

[67] Favaro (ed.), *Le Opere*, vi, 71. 'Io credo che ella veramente non sia per aver efficacia nelle comete, se prima non vien determinato ch'elle non sieno di queste cotali reflessioni di lume, ma oggetti uni, fissi, reali e permanenti'. As translated by Drake and O'Malley (eds.), 39.

[68] Favaro (ed.), *Le Opere*, vi, 94. '... abbia sormentato il cono dell'ombra terrestre'. As translated by Drake and O'Malley (eds.), 50.

[69] The problem with this argument was that if the comet possessed a rectilinear motion perpendicular to the Earth's surface, then that would mean that it should be moving towards the zenith. This was not confirmed by observational reports. According to Shea, Galileo recognised this problem but countered it by suggesting that observations of the comet's movements were distorted by the refraction of light through vapours. This was equally problematic, since it meant that such vapours would distort the observation of all celestial phenomena. Shea, 82.

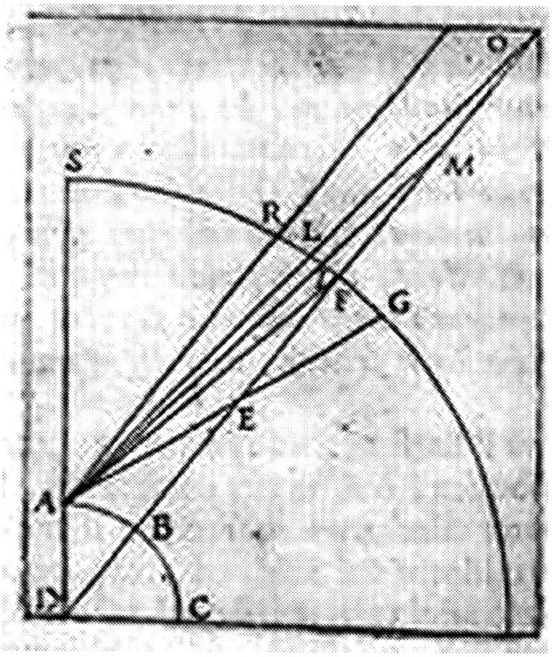

Figure 22. Galileo's drawing of the movement of comets in a straight line away from the Earth's surface, showing the increasing angle between the observer, A, and the comet, DO. G. Galilei, *Il Saggiatore*, Rome, 1623.

As for Grassi's proclaimed difficulties in viewing comets with the telescope, Guiducci also argued that this was simply an erroneous claim made by Grassi since distant stars are indeed magnified by the telescope.

This reflects how observations and instruments were laden with natural philosophical skills, commitments, and agendas. Galileo was using his limited observations of the 1618–1619 comet to propose a theory constructed on the basis of his anti-Aristotelian agenda. In fact, Guiducci only made these arguments about the vaporous nature of comets in his lectures after he took great care in undermining the notion that comets originate in the extremities of the terrestrial region because of the combustion of fire and gaseous materials. In the conclusion, Guiducci returned to this point, as if to emphasise the natural philosophical significance of Galileo's theory, and argued that the scholastic distinction between terrestrial and celestial regions was inadequate for explaining the movements of comets.[70] Therefore, Aristotelian philosophy of nature remained Galileo's main

[70] Favaro (ed.), *Le Opere*, vi, 93. 'A me, al quale non ha nel pensiero avuto mai luogo quella vana distinzione, anzi contrarietà, tra gli elementi ed i cieli, niun fastidio o difficultà arreca che la materia in cui s'è formata la cometa avesse tal volta ingombrate queste nostre basse regioni, e quindi sublimatasi avesse sormontato l'aria e quello che oltre di quella si diffonde per gl'immensi spazi dell'universo'. As translated by Drake and O'Malley (eds.), 53.

target of criticism. But since Tycho was now being used by scholastics to undermine Copernicus and his followers, the Danish astronomer became Galileo's opposition. Despite the fact that Galileo had earlier mentioned Tycho's observations of comets in the sixteenth century in support of his anti-Aristotelian claims, the use of Tycho by scholastics in the early seventeenth century to attempt to undermine Copernicanism, forced Galileo to alter his position.

Grassi immediately responded with another publication, this time under the pseudonym of Lothario Sarsi.[71] Written in the pretence that the author was a student of Grassi, this text did not even bother to mention Guiducci, but instead made its claims directly against Galileo. Grassi was then concealing his own identity, but at the same time revealing the name of his opponent. Galileo was so annoyed at this that he compiled *The Assayer* in order to undermine not only Grassi's and Tycho's comet observations, but also their very methods of investigation. Here Galileo defended his observations and claims regarding comets and maintained that Sarsi relied far too much on his senses when dealing with such a difficult topic as the position and movements of comets. Instead he believed that observation had to be combined with mathematical data in order to produce credible knowledge claims. This, according to Drake and O'Malley, demonstrates that Galileo was far from a pure empiricist.[72] To this statement we may add that this also reflects how Galileo's experiments were subordinate to his investment in the emerging mathematical traditions of the seventeenth century.

Following the favourable reception of *The Assayer* in Rome, Galileo focused his attention on compiling the *Dialogue*. Meanwhile, Kepler contributed to the debate about comets in an appendix to his 1625 publication, *Tychonis Brahei Dani Hyperaspistes*. Kepler insisted that comets do indeed move in straight lines. But instead of emanating from the earth as a vaporous illusion, as Galileo believed, Kepler claimed that they are simply celestial objects often travelling straight through the earth's field of vision, 'approaching through one zone and receding from the earth into another'.[73] Kepler made some further remarks in support of Copernicus, but he refused to be too critical of his former teacher, Tycho. Later, Descartes wrote about his hypothesis regarding comets in his posthumously published *Le Monde* (1658), and in *Principia Philosophiae* (1644).

According to Descartes, the largest and heaviest planets revolve around the outermost circumference of a vortex. Such a large and heavy planet could gain so much centrifugal force as to be able to spin out of its vortex and into another, becoming what is known as a comet, travelling continually in and out of vortices. A comet would thus only travel momentarily inside any vortex, it would be bright at first as it falls into the circular orbit of the central body, in our case the Sun, and would move slower and fade away as it moves out of the vortex. Descartes therefore suggested that comets are carried around by the Sun beyond the sphere

[71] L. Sarsi, *Libra astranomica ac philosophica*, Perugia, 1619. The pseudonym, according to Shea, was to avoid being caught up in a public debate that would not be in the best interests of Grassi's religious superiors. Shea, 83.

[72] Drake and O'Malley (eds.), xxiv.

[73] As cited by Drake and O'Malley (eds.), 347.

of Saturn. This theory rejected the Galilean notion of a rectilinear path for comets emanating from earthly vapours, as well as the Aristotelian belief that comets were carried around by the sky within the terrestrial realm. But more importantly, what harmed the chances of Descartes' theory being accepted by scholastics, and the reason why he pulled out of publishing *Le monde* in 1633, after hearing about Galileo's condemnation, was that he even rejected Tycho's belief that comets travel amongst the planets, and he insisted upon a Copernican heliocentric system within an infinite universe.

By 1664 the Cimento academicians still wished to determine whether comets belong beyond the sublunary region, and whether their trajectories are rectilinear or circular. We shall see once again from the academicians' letters and manuscripts that just like Galileo, the leading members of the Cimento, including Leopoldo, were eager to strengthen their Copernican and anti-Aristotelian beliefs. In fact, Borelli, Viviani, and the Prince did a much better job of this than their deceased mentor, Galileo, but once again none of their work in this field was to be published under the Cimento's name.

6. THE ACCADEMIA DEL CIMENTO AND THE COMET OF 1664

On 18 December 1664, Borelli excitedly wrote to Leopoldo from Pisa about a comet he had observed that morning.[74] Borelli recounted the position of the comet and its apparent trajectory, that is, in a straight line, but he also conceded that these first observations were very limited. On 19 December he again wrote to the Prince with some more news about the comet's movements, claiming that the comet no longer appeared to be moving in the same direction, but had in fact gone through 'an oblique retrograde motion in a straight line'.[75] Borelli claimed in this second letter to the Prince that the most important aspect of his observations regarded the comet's parallax.

> If the comet continues along this path and is observed as such from other countries, we will have a certain and secure argument of the comet's lack of parallax. So I implore Your Serene Highness to inform me of the observations received from other locations, because if at the same hour of the night, that is, if at the same time we find that other observers see it at the same place, this controversy would be quite over.[76]

We can safely assume that the 'controversy' to which Borelli referred was the same one that Tycho Brahe had instigated decades earlier regarding the location of comets, whether they travel in the celestial or terrestrial realms. In other words, Borelli was concerned about what his observations of the comet's parallax implied for the con-

[74] BNCF, Ms. Gal. 277, f. 58r.
[75] '... il viaggio suo viene ad essere retrogrado obliquo per una linea retta'. BNCF, Ms. Gal. 277, f. 60v.
[76] 'Se ella seguita questo stesso viaggio e così sarà osservata in altri paesi averemo un argomento certo e sicuro della poca parallasse di detta cometa, che però supplio V.A.S. che mi faccia partecipe dell'osservazioni che ricerverà da altri luoghi, perchè se nella stessa ora della notte cioè nello stesso tempo ci abbatteremo con altri osservatori a vederla nello stesso sito sarebbe bell'e finita questa controversia'. BNCF, Ms. Gal. 277, f. 60v.

tention between Aristotelian, Tychonic, and Copernican astronomers concerning the mutability of the heavens. Indeed, Borelli made clear his natural philosophical stance against traditional astronomy in his next letter to Leopoldo, dated 22 December 1664. Here Borelli was critical of comments made by Jesuit astronomer Giambattista Riccioli, who claimed that it was impossible to demonstrate the lack of parallax of comets, since two observers from different parts of the globe could not be certain that they were observing the comet at exactly the same instant.[77]

Despite his scepticism about the ability to properly observe (or rather, not observe) parallax, Riccioli was still of the opinion that comets belonged in the celestial regions, and that this impacted on the debate between Tychonic and Copernican followers. Riccioli was an ardent anti-Copernican. In his *Amalgestum novem*, published in Bologna in 1651, he made his admiration for Copernicus' mathematical work quite clear, but insisted that the heliocentric system could not be considered as anything more than hypothetical because of its clash with Aristotelian and biblical teachings. Furthermore, he claimed that a modified version of Tycho's system, with Saturn and Jupiter still orbiting the Earth, was far more acceptable and credible than the suggestion of a sun-centred universe.[78]

In the meantime, while Borelli assessed some other observations of the comet made from Rome,[79] and while he continued to observe the comet's movements, more claims in favour of Tychonic astronomy and Aristotelian cosmology were being made by Gian Domenico Cassini, professor of astronomy and mathematics at the University of Bologna. During the early 1660s, Cassini often found himself in Rome overseeing engineering projects for the Papal court, where he took the opportunity in 1664 to make observations of the movements of the comet and publish those observations along with some of his astronomical beliefs.[80] In a letter addressed to Leopoldo, Ottavio Falconieri wrote about the arguments Cassini was hoping to establish in this text. According to Falconieri, Cassini's theory concerned with the motion of comets was aimed at demonstrating that they 'did not move in a straight line perpendicular to the surface of the earth, but along the plane of the greatest circle [beyond the orbit of Saturn]' around the sun, which is itself orbiting the stationary earth.[81] More specifically, as he described in his published works on the

[77] BNCF, Ms. Gal. 277, ff. 61r–63v.

[78] J.L. Heilbron, *The Sun in the Church: cathedrals as solar observatories*, London, 2001, 183–184.

[79] By 27 December, 1664, Borelli had received other observations of the comet from Rome, but as he mentioned to Leopoldo, since those observations were not carried out with 'the accuracy normally applied by astronomers', they were inconclusive. Therefore, Borelli continued to observe the comet himself while awaiting news from other parts. BNCF, Ms. Gal. 277, f. 72r.

[80] G.D. Cassini, *Lettere astronomiche di Gio: Domenico Cassini al signor abbate Ottavio Falconieri sopra il confronto di alcune osservazioni delle comete di quest'anno*, Rome, 1665; and *Ephemeris prima motus cometae novissimi*, Rome, 1665.

[81] 'Il D. Cassini va mettendo in ordine il suo Discorso sopra la Cometa, e la Teoria del moto di essa, con la quale spera di poter dimostrare che la cometa non si è mossa per una linea retta perpenidoclare alla superficie della terra, ma per il piano d'un cerchio massimo'. BNCF, Ms. Gal. 277, f. 1r. This letter is dated 10 January 1664. That is, more than 12 months before anyone saw the 1664–1665 comet. A possible explanation for this is that Cassini may have been compiling his treatise on the movements of comets since as early as October 1663, after the appearance of a smaller comet that month. He then would have decided to delay printing his letters to Falconieri and a treatise on comets dedicated to Queen Chrisitna of Sweden until the appearance of the 1664 comet in December and another comet in April 1665.

topic, Cassini believed that the 1664 comet travelled in epicycles around the distant bright star of Sirius, while that star orbits the earth. In other words, he believed the comet to be moving around the earth, not the sun, as Falconieri had intimated in his letter to Leopoldo.[82] Additionally, he clearly denied that the rapid movement of the comet when in opposition to the sun could be used by Copernicans as proof of the mobility of the earth, since, according to Cassini, such motion could also be explained within a Tychonic geocentric system.[83]

Therefore, Cassini was proposing a theory that dismissed Galileo's claim about the rectilinear path of comets emanating from vapours in the earth's atmosphere. Furthermore, by placing the comet amongst the sphere of stars, Cassini was proposing a radical departure from most theories since the late sixteenth century on the location and movements of comets. Nevertheless, he still maintained a finite geocentric and geostatic model with circular motion, consistent with Tychonic astronomy.[84] Despite some anomalies in his calculations, he even compiled an ephemerides of the comet based on this theory.[85]

Borelli's views about Cassini's work were apparent from as early as March 1664, when he wrote the following words to Leopoldo: 'With regard to the theory of the comet that he claims to have discovered, it seems to me that it is very exhausting and confusing, from which in the end few results and benefits can be extracted.'[86] Borelli was so concerned about such arguments and the claims Tycho's followers were continuing to make against Copernicus, that he compiled a tract on the movements of comets in 1665, entitled *Del movimento della cometa apparsa nel mese di dicembre 1664*. It is interesting to note that this work was published under the pseudonym of P.A. Mutoli, giving the impression that Borelli may have been concerned about the controversy into which he could have been leading his academy and his patron.[87] While the discussion over comets did not find its way into the Cimento's meetings, which were held only on rare occasions

[82] For a succinct summary of Cassini's cometary theory, see Donald K. Yeomans, *Comets: A Chronological History of Observation, Science, Myth and Folklore*, Toronto, 1991, 70–72.

[83] According to Cassini, Copernicans such as Adrien Auzout (whose theory I discuss below) believed that the change in the comet's speed when in opposition to the sun, is comparable to the retrograde motion of the superior planets when in a similar position. G.D. Cassini, *Lettere astronomiche di Gio: Domenco Cassini al signor abate Ottavio Falconieri sopra il confronto di alcune osservazioni delle comete di quest'anno,* Rome, 1665, 6–7.

[84] As Heilbron points out, Cassini was quite content to admire the mathematical quality of Copernicus' work, but much like Riccioli, he could never accept that a heliocentric system could be anything more than hypothetical. Heilbron, 185.

[85] Gian Domenico Cassini, *Ephemeris prima motus cometae novissimi*, Rome, 1665. For a succinct summary of Cassini's cometary theory, see Donald K. Yeomans, *Comets: A Chronological History of Observation, Science, Myth, and Folklore*, Toronto, 1991, 70–72.

[86] 'Circa la teoria della cometa ch'egli pretende aver ritrovata mi pare che sia una cosa molto faticosa et imbrogliata, dalla quale alla fine poco frutto et utile se ne cava'. BNCF, Ms. Gal. 277, f. 4r.

[87] It is likely that Borelli was wary of how his work would be received in Rome, since Pope Alexander VII, sensitive to the growth of heretical movements such as Jansenism, wished to crack down on Copernicanism. The 1664 papal Bull, 'Spies in the House of Isreal', is rather forceful in its assertion that all followers of the Church should adhere to the doctrines that frame the Holy Office's decree against certain books, such as those teaching 'the mobility of the earth, and the immobility of the sun'. William Roberts, *The Pontifical Decrees Against the Doctrine of the Earth's Movement and the Ultramontane Defence of Them*, London, 1885, 47.

at this point, Leopoldo was still acting on behalf of one of his courtiers and academicians when he asked his colleagues around Europe for their observations of the 1664 comet. There is little doubt, therefore, that Leopoldo would have felt that his Court and his academy could come under scrutiny because of Borelli's Copernican opinions. A pseudonymous publication was probably believed to be in the Cimento's, and Leopoldo's, best interests.

So by late 1664, when Leopoldo began asking astronomers in other parts of Europe whether they had observed the same comet as Borelli, natural philosophical and religious arguments continued to be raised by some prominent Italian astronomers. Just as Orazio Grassi was determined to defend scholastics against Galileo's Copernican and anti-Aristotelian claims, so Riccioli and Cassini were intent on strengthening the acceptance of Tycho amongst European astronomers, as the only true system. The same natural philosophical and religious controversy surrounding comets that had existed during the first two decades of the seventeenth century, continued to play an important role in how theories and claims in astronomy were constructed and negotiated during the 1660s, when the cautious Cimento academicians became involved in the topic. There was no chance, therefore, that Borelli's and Leopoldo's interests in the movements of comets could be published in the *Saggi.*

7. BORELLI VERSUS ADRIEN AUZOUT

While Borelli was writing to Leopoldo about his observations of the comet in December 1664, a French natural philosopher, Adrien Auzout (1622–1691), was also observing the same comet and preparing a document that contained a table of its nightly position. Auzout claimed to have been able to predict the comet's path, and as a result, to have settled any arguments regarding the position and movements of comets. In January 1665, Auzout sent this document, *L'ephêmérides du nouveaus comete*, to some of his fellow astronomers around Europe. Copies were eventually received by the Royal Society of London and the Tuscan Court. Auzout's ephemerides were well received in England as his observations and predictive ability were confirmed by others who claimed to have followed the comet's trajectory.[88] Some dispute emerged about one significant difference between Auzout's observations, and those of another continental astronomer, Hevelius. But according to Shapin, the fellows of the Royal Society resolved the dispute in Auzout's favour by drawing on issues of trust and gentlemanly civility, as well as the search for 'matters of fact'.[89] These issues may well have been crucial to the process of investigating and presenting the analysis of comets in London, but Auzout's work also carried some broader natural philosophical concerns that Shapin does not consider important to his narration of events in England.[90] The situation in Florence at the time reveals how such

[88] Shapin, *A Social History of Truth*, 269.

[89] *Ibid.*, 266–291.

[90] Shapin restricts his commentary about the theoretical significance of the observations of comets in the seventeenth century, to a general and brief footnote. *Ibid.*, 270, n. 70.

concerns were indeed at stake in Auzout's writing, and shaped the way that Leopoldo and Borelli presented their own claims about cometary movement.

In reaction to Borelli's enthusiastic letters about the comet he had begun to observe in December 1664, Leopoldo asked Boulliau, one of his correspondents in Paris, if he had known about any further arguments regarding the distance of the comet from Earth. Boulliau replied on 20 January 1665, enclosing some of the observations of the 1664–1665 comet carried out by Auzout and published in *L'ephémérides*.[91] As Borelli revealed in his commentary on Auzout's work, rather than concern himself with establishing the exact path of the comet, he was interested in what Auzout offered to the debate over the role of comets in a Copernican universe. Auzout had contended that the comet had an apparent retrograde motion, this time meaning a movement in the opposite direction to the planets, and that this was explained by its trajectory in a straight line as well as by the movement of the Earth. That is to say that according to Auzout, and contrary to Cassini and Riccioli, the Copernican system could easily accommodate the apparent movement of comets. Furthermore, Auzout wrote that Galileo and Kepler were correct in their calculations that the comet moved along rectilinear paths, although for Auzout, comets were solid bodies and not made from vapours in the terrestrial realm, as Galileo had proposed. In other words, Auzout was supporting some of Galileo's claims in order to strengthen his own anti-Aristotelian position in the field of astronomy. After all, Auzout revealed in his 1665 publication, *Lettre à M. l'abbé Charles*, that he regarded Copernicus' system to be the truth and he hoped that the scandal surrounding it would eventually be dismissed.[92]

This was also precisely Borelli's aim when examining the path of the 1664–1665 comet; but instead of simply complying with Galileo's claims, he rejected them in the belief that he could devise a stronger argument in favour of Copernicus and against Aristotelians. Borelli wrote two commentary pieces about Auzout's ephemerides, in which he was, nevertheless, critical of several claims made by the French astronomer.[93] Those criticisms referred first to Auzout's statements regarding what past astronomers have said about the regularity of the movement of comets. Borelli was particularly annoyed at Auzout's rather vague and inaccurate comment that '[U]ntil now, the whole world has persuaded itself that the movements of comets are irregular.'[94] Borelli contested what Auzout meant by this statement, considering that since antiquity, comets had often been

[91] BNCF, Ms. Gal. 277, ff. 94r–94v. From the date recorded on the letter from Leopoldo to Boulliau, it would appear that it was written as early as on 16 January 1664, months before Borelli observed the comet. Although this is undoubtedly the date on this letter (BNCF, Ms. Gal. 282, f. 84r.), Middleton claims that Leopoldo actually wrote to Boulliau in January 1665. This would indeed seem more likely considering that the Prince would have made the request for more observations after reading Borelli's letters about the comet from Pisa during late December, 1664. See Middleton, *The Experimenters*, 257.

[92] See Heilbron, 186.

[93] Auzout also replied to Borelli's criticisms in a letter to Leopoldo in April, 1665: BNCF, Ms. Gal. 272, ff. 146r–150r.

[94] 'tutto il mondo s'è persuaso fino al presente, che i movimenti delle comete sono irregolari'. BNCF, Ms. Gal. 272, f. 177r.

considered to be moving according to regular patterns that fitted in with astronomical and cosmological beliefs. For example, argued Borelli, Tychonic astronomers claimed to have identified a regularity in the movement of comets by stating that they travel around the Sun.[95] Borelli also questioned Auzout's predictive ability, since, so Borelli thought, anyone could plot out the comet's path after observing it for just the first few nights. According to Borelli, predictions about cometary movement could only be legitimate if they were to be made before the comet even appeared.[96] Finally, and more importantly for Borelli's astronomical and natural philosophical commitments, he denied Auzout's claim that the comet was moving in a straight line.[97] Furthermore, he did not think it plausible to deduce that because of the apparent retrograde movement of the comet, the Earth must be in motion.

In contrast to all the theories so far put forward concerned with the path of the comets (aside from Descartes' suggestion based on the centrifugal force of heavy bodies moving through vortices), Borelli contended that the comet he had observed did not move in a straight line, as Galileo and Auzout believed, and certainly not in a circle as Tycho's followers claimed. Instead, as a variety of drawings, found amongst the Galilean manuscripts, also reveal, Borelli proposed that the comet traced a curved path.[98] Borelli stated: 'I am more than certain that ... it is necessary to make the motion of the comet along a curved line, and this I think can even be demonstrated against Kepler himself.'[99] So Borelli traced the small section of the comet's path that is observable from Earth, and estimated that it followed a curved trajectory. This was the perfect opportunity for Borelli to fit his work on comets into the theory of planetary motion that he was developing for his 1666 publication, *Theoricae*. He believed that comets, just like the planets, travel around the Sun because they are subjected to the physical centripetal and centrifugal tendencies that keep all the heavenly bodies in orbit. This would explain the curved paths that he believed the comet followed; it was just like the elliptical orbits of the planets, only on a much grander scale. As usual, this was supported by the type of geometrical demonstrations that Borelli had learned to use throughout his career, particularly the properties of conic sections upon which he based his theory of elliptical planetary motion.[100] This is what we have come to expect from Borelli and his fellow Galilean followers when they talk about demonstrating their natural philosophical opinions.

[95] BNCF, Ms. Gal. 272, f. 186r. According to Shapin, Auzout's statement was accepted by the fellows of the Royal Society and was even reiterated in the *Philosophical Transactions*. Auzout was, therefore, considered by the English to be doing ground-breaking work by assigning a regular trajectory for comets. Shapin, *A Social History of Truth*, 269.

[96] BNCF, Ms. Gal. 272, f. 186v–187r.

[97] Evidently Borelli had shifted from his position, cited earlier, that the comet had gone through 'an oblique retrograde motion in a straight line'. See page 222. This change in opinion may have come after further observations of the comet and a consideration of the possible interpretations of its movements.

[98] BNCF, Ms. Gal. 272, ff. 72v, 74r, 77r.

[99] 'io son più che sicuro che ... il moto della cometa è necessario farlo per una linea curva, e questo stimo potersi dimostrar anche contro il Keplero stesso'. BNCF, Ms. Gal. 272, f. 183v.

[100] See Chapter Three.

Therefore, while Riccioli, Cassini, and Auzout were either using or denying Galileo's work in 1618 in order to promote their own natural philosophical concerns, Borelli too was arguing against the validity of many of Galileo's claims in order to promote his physico-mathematical and natural philosophical cause. Regardless of whether Galileo was correct in his assumption of the vaporous illusion of the comets, Borelli still insisted that a lack of parallax indicated the comet's celestial existence, and that comets did not travel in a straight line emanating from the Earth. However, this did not mean that he also opposed the validity of Copernicus' system, as Cassini and Riccioli proposed, but simply that Galileo's observations could be improved upon in order to preserve the efficacy of contemporary anti-Aristotelian arguments. This is why Borelli's proposed curved trajectory of comets was so important, since it could be demonstrated with the same geometrical principles that he was using to describe the elliptical orbits of planets. By fitting cometary movement neatly into his celestial mechanics, Borelli was proposing a complete heliocentric system with mathematical and mechanical causes, accounting for all celestial bodies. This went beyond the geocentric claims of his Tychonic rivals who struggled to combine the movements of planets with those of comets. Therefore, as we have seen with his work concerning Jupiter's moons in Chapter Three, Borelli was constructing an argument that played upon previous astronomical claims, especially those made by Galileo, as well as upon his own background in geometry, in order to preserve his anti-Aristotelian and pro-mechanist agenda.

8. MAINTAINING LEOPOLDO'S POLICY OF SELF-CENSORSHIP AND CONCLUDING THE ACADEMICIANS' WORK IN ASTRONOMY

Similar speculations surrounded the observations of another comet in April 1665. On this occasion, Borelli seemingly made no observations, but his fellow academician, Viviani, as well as the Cimento's correspondents, Cassini, Levera, Riccioli, and Auzout, all assisted Leopoldo to compile a table of the comet's movements and parallax.[101] The topic was therefore not restricted to Borelli amongst the academicians, but well and truly seized the interest of Viviani and especially Leopoldo.

In any case, despite this commitment of the Prince and his leading academicians to examine the movements of comets, none of their efforts in astronomy were published in the *Saggi*. Following our earlier analysis of Leopoldo's suppression of the academicians' investigations of Saturn's ring, it is not difficult to imagine that the highly controversial speculations concerned with comets, would have also fallen victim to the Prince's policy of self-censorship. This is not only evident in the *Saggi*'s lack of references to the academicians' heavy involvement

[101] BNCF, Ms. Gal. 272, ff. 115v–116r. Reports about the comet's movement from different cities and from different people are also contained in this folder: BNCF, Ms. Gal. 272, ff. 117r–142r.

in the Saturn and comet controversies, but also in the caution that surrounded the two other publications by Borelli during this time.

We should not forget that Borelli's 1665 treatise, *Del movimento della cometa*, outlining a theory on the movements of comets that denied scholastic beliefs and attempted to strengthen the Copernican system that was also controversially supported by Galileo, was published under a pseudonym. During the height of the comet controversy in the 1660s, and while the Cimento was in the process of compiling its own collection of observations, Leopoldo would not have been too eager to expose a high-profile member of his academy to the scrutiny of ecclesiastical authorities. Similarly, Borelli's analysis of the elliptical orbits of Jupiter's moons, also observed during this period in collaboration with the Prince and other members and correspondents of the Cimento, was quite carefully presented in his 1666 publication, *Theoricae*.[102] Borelli veiled his Copernican commitments by avoiding any mention of a heliocentric system, instead giving the impression that he might be applying his theories according to Tycho's system, which was of course far more acceptable to the Catholic Church.[103]

However, according to Domenico Bertoloni Meli, the *Theoricae* passed through the hands of the ecclesiastical censors within only two weeks. Furthermore, Meli claims that Leopoldo intervened in the Roman inquisitor's assessment of Borelli's work by sending Francesco Redi and Antonio Uliva to learn if there was any problem with the treatise and possibly to hasten the publication process. Meli claims that the ecclesiastical authorities were, therefore, not Borelli's main concern, but rather, as some of Borelli's letters indicate, he was simply attempting to have his work made public before Fabri could release any of his own observations of Jupiter's satellites.[104]

Meli provides a convincing argument for the publication process behind Borelli's *Theoricae* that goes beyond the usual discussions about religious censorship. But while Borelli may well have been eager to offset any criticisms of his work made by Jesuit rivals such as Fabri, this does not mean that he was any less concerned about the religious implications behind his claims regarding Jupiter's moons and comets. Indeed, as Paolo Galluzzi argues, the Medici Prince had no choice but to control the content of the writings published in Florence and pertaining to issues in natural philosophy.[105] Leopoldo himself was clearly interested in constructing corpuscularian, mechanistic arguments, and supporting the anti-Aristotelian speculations put forward by his courtiers, such as Borelli. But from the censorship that Borelli faced in compiling his arguments and from the

[102] Once again, although Borelli was the leading figure in the observations of Jupiter and the only academician to publish anything regarding that planet, this did not mean that he did not receive any support from his patron and colleagues in the Tuscan Court. Viviani and Leopoldo also carried out observations of Jupiter and its moons and assessed the arguments made by Boulliau, Huygens, Fabri, Cassini, Campani, and other European astronomers and telescope makers.

[103] Koyré, *The Astronomical Revolution*, 471.

[104] D.B. Meli, 'Shadows and Deception', 389–391. Michael Segre provides a similar argument, suggesting that since Leopoldo was willing to publish two texts by Borelli that touched on controversial Copernican themes, the Prince was probably not as worried about ecclesiastical authorities as we may believe. Segre, *In the Wake*, 139–140.

[105] Galluzzi, 'L'Accademia del Cimento', 825.

restrictions that we have seen the academicians face when compiling the narration of their observations, especially those to do with astronomy, Leopoldo clearly didn't wish 'to ignite clamorous controversies'.[106]

So to conclude this analysis of the Cimento's work in astronomy and their publication process, we may now argue that we can understand the religious and political interests behind the experimental rhetoric of the *Saggi*. The atheoretical style of the text assisted the Medici Court and its academy to build a reputation as uncontroversial and unbiased knowledge makers, thus increasing the credibility of the Cimento and its academicians and the status of the Court. Additionally, Leopoldo and his academicians needed to avert any religious or political controversy if they wished to preserve that very reputation and allow their work to be read throughout Europe without having to endure the type of condemnation by the Catholic Church that Galileo faced.

The importance Leopoldo and his academicians placed on the Court's status and reputation, including their uncontroversial rhetoric, is reflected in a letter Borelli wrote to the Prince on 30 May 1670. After he had returned to Messina, Borelli heard that some of his former colleagues and students who had been working in the field of physiology at the University of Pisa, were engaged in some controversial discussions with scholastics who refused to accept the type of iatrophysiology that Borelli had developed while working in Tuscany. Borelli confessed to Leopoldo that he reproached his friends in Pisa 'because they were not following the modest style happily used by me for twelve years so as to not irritate or to vilify the overly devoted followers of the common School'.[107] Furthermore, Borelli pleaded with Leopoldo to protect the Pisan physiologists from any accusations made against the validity of their knowledge claims by scholastics. Borelli praised Prince Leopoldo for the support he had always shown to Galilean followers, and insisted that supporting Borelli's friends in Pisa would also 'result in profit and reputation not only for the University of Pisa, but even for our Italy'.[108]

This shows that although the Cimento had no formal regulations in place for maintaining a rhetorical style in their writings that would ensure that the academicians stick solely to the narration of experiments, Leopoldo was still maintaining an informal policy of self-censorship. In order to avoid the same type of controversy that Galileo had encountered with the Catholic Church, and to advance the reputation of Tuscan natural philosophy, as well as their own concerns inside the Court, the academicians were careful to veil the contentious arguments made in the *Saggi* behind an experimental rhetoric. Those arguments

[106] *Ibid.*

[107] '... io ripresi i miei amici perché non seguitavano lo stile modesto usato da me felicemente per dodeci anni di non irritar né vilipendere i troppo affettionati seguaci della comune Schuola'. BNCF, Ms. Gal. 279, f. 18r.

[108] '... io parlo con un Principe di tale e tanta dottrina che non occorre esagerare qual'utile rechino i comunali filosofi, o pure i seguaci del Galileo. Però la supplico humilmente quanto posso, e voglio che si compiaccia proteggere e favorire la giustizia della nostra causa, il che poi, se non m'inganno, risultarà in utile e reputatione non solo dello Studio di Pisa, ma anco della nostra Italia'. BNCF, Ms. Gal. 279, f. 18v.

might have been inferred by readers familiar with the topic and the thinkers involved, but it was clear that the academicians were attempting to avoid any public demonstrations of having contributed to controversial speculations and natural philosophical theorising. Since they were so eager to protect the status and reputation of the Medici Court and the Accademia del Cimento, the academicians did not dare to disrupt the traditional established cultural settings in natural philosophy, politics, and religion by openly arguing for corpuscularian, mechanist and above all, Copernican, beliefs.

CONCLUSION

The last recorded meeting of the Accademia del Cimento was on 5 March 1667, when the academicians performed an experiment fittingly related to the pressure of the air.[1] On 18 March, Borelli accepted an offer to return to Messina, Sicily. Weeks later, Rinaldini was offered a position at the University of Padua. Additionally, Uliva, as was mentioned in Chapter Four, mysteriously decided to return to Rome where he was jailed by the Holy Office and eventually committed suicide. The departure of these three from Tuscany left the Cimento with only half of its founding members still available, and without at least two of its biggest contributors in Borelli and Rinaldini. Despite Leopoldo's efforts to find suitable replacements, the Cimento did not resume its meetings after the publication of the *Saggi* in 1667.[2] This, however, did not stop Leopoldo from continuing to use the Cimento to promote the status and reputation of his Court.

The *Saggi* was finally published in October 1667 and offered the Prince and the Grand Duke plenty of opportunity to advertise the experimental exploits undertaken under their protection and patronage. For this reason, copies of the text were not sold to the public, but were rather distributed to friends and correspondents of the Cimento academicians and the Medici Court. Magalotti was sent on a European tour with the purpose of personally distributing specially printed and bound copies of the *Saggi*. His visit to the Royal Society of London was of particular importance, since it showcased the Cimento's work to arguably

[1] BNCF, Ms. Gal. 262, ff. 177r–178v.

[2] Several theories exist about the reasons why these three academicians decided to leave Tuscany and the Accademia. Middleton suggests that the rivalry between Borelli and Viviani had become unbearable by 1665 and by the end of 1666, Borelli was already planning his departure. Middleton, *The Experimenters*, 315–317. Similarly, according to eighteenth-century historian, Riguccio Galluzzi, Borelli and Rinaldini were probably extremely upset with the news that Viviani and Dati had been awarded a pension by King Louis XIV in 1664, and perhaps thought that they were not receiving enough merit for their work in the Tuscan Court. R. Galluzzi, *Istoria del Granducato di Toscana*, iv, 170. In the meantime, in their correspondence to the Grand Duke, both Borelli and Rinaldini claimed that the reason for their departure was that the climate in Pisa was unsuitable for their age and health. Despite all these possibilities, it is more likely that they simply wished to take advantage of far more lucrative offers being made by rival universities. In any case, each academician was looking to profit from a career inside the Tuscan Court, demonstrating that natural philosophical skills and commitments aside, they were still participants in a culture of religious and political pressures of seventeenth-century courts, as was discussed in Chapters Seven and Eight.

the premier natural philosophical institution of the time.[3] Unfortunately for the Cimento, the Royal Society's reaction to the work that the academicians' presented in the *Saggi* was not enthusiastic. Several fellows had the opportunity to review the text, but they found that most subjects discussed in it had already been considered by the Royal Society. Despite this criticism, the report that Oldenburg eventually sent Magalotti in May 1668 was, according to Middleton, probably quite tactful and careful not to be too critical of the *Saggi*.[4]

In any case, we may still argue that the *Saggi* had the desired effect for Leopoldo and his academy. Oldenburg's report, judging from Magalotti's reply, did not mention the natural philosophical interests and opinions that formed the theoretical basis of the academicians' experiments. Furthermore, despite the topics discussed in the *Saggi* being outdated, from the point of view of some English readers, the text was still representative, through its rhetoric, of the Cimento's achievements as an experimentalist academy, and the Prince's ability to protect such practices considered valuable to the production of natural knowledge. In fact, these were precisely the sentiments expressed by the Royal Society's president when he received the *Saggi* from Magalotti. The Cimento's secretary described the occasion in a letter to Prince Leopoldo on 13 March 1668:

> The president, taking off his hat, replied that these matters were among the most essential and the most difficult in the order of natural phenomena, and that having been examined in the presence and under the protection of a prince so great, so splendid, and so wise, they could not be otherwise than extremely well determined and illuminated.[5]

Evidently, the Royal Society assumed the reliability of the experiments carried out by the academicians simply because they were constructed under the protection of a Prince reputed to be 'so great, so splendid and so wise'. Such praise for the Cimento, its princely patron, and the group's experimentalist approach to producing natural knowledge, was echoed by reviewers of the *Saggi* in Rome and Paris. Early in 1668, the *Giornale de Letterati* in Rome, published a review of the text that, again, did not mention the natural philosophical issues underpinning the Cimento's experiments, and instead praised Leopoldo for protecting a highly esteemed and learned experimentalist institution.

> Nor must we fail to consider the merit of the virtuosi who have contributed to this work their sublime intellect and mature judgement, and much skill, industry, and diligence in making such experiments; and finally that which gives the most singular excellence to the whole, the contribution made by Prince Leopoldo, today a cardinal of the Holy Church, with his authority and protection, his sublime judgement and profound intelligence.[6]

[3] Magalotti's letters to Leopoldo regarding his visits to the Royal Society and the arrival of the *Saggi* in England are in: BNCF, Ms. Gal. 278, ff. 145r–158r. Extracts from these letters have been translated and published by Middleton, *The Experimenters*, 291–296.

[4] Middleton, *The Experimenters*, 1971, 295. I could not find Oldenburg's letter to Magalotti, but Middleton makes this judgement, a fair one in my opinion, on the basis of Magalotti's reply to Oldenburg on 25 May 1668, in which he was apologetic for the delay in publishing the Cimento's work, inferring that this might have been the reason for the *Saggi*'s failure to make a deep impression. See A.R. Hall and M.B. Hall (eds. and trs.), *The Correspondence of Henry Oldenburg*, 5 vols., Madison, 1965, iv, 410–412.

[5] BNCF, Ms. Gal. 278, ff. 154r–155r. As cited by Middleton, *The Experimenters*, 294.

[6] *Giornale de Letterati* (1668), i, 4–6. As cited by by Middleton, *The Experimenters*, 334–335.

At around the same time, Hubert de Montmor wrote to Leopoldo from Paris expressing his gratitude for having received a copy of the *Saggi*, as well as his admiration for the Prince and his academy:

> I have also just received ... the gift that it has pleased you to make me, of the first samples of the experiments of your illustrious Academy, where everything is excellent, magnificent, and worthy of your genius and that of those great personages who assemble under the protection of your Sublime Excellency. I can assure you that this work has received general approval, and all the scholars and interested people to whom I have shown it have praised it highly to me.[7]

Such public proclamations from London, Rome, and Paris, of admiration for the work undertaken under the Medici prince's patronage, might have been predictable formulas of diplomacy. But they still reflected the status and reputation Leopoldo was achieving for himself and his courtiers by compiling a text that avoided philosophical contention and demonstrated that his academicians were practitioners of uncontroversial inquiry and producers of reliable knowledge. In summary, the way the *Saggi* was publicly received in some of the main centres of learning in Europe boosted the reputation of the academicians who had produced this volume of supposedly uncontroversial and reliable experimental knowledge claims in the reputedly learned environment of the Tuscan Court. At the same time, Leopoldo also became recognised for his own learning and his wise judgement to protect such an institution as the Cimento. This justified the time and energy Leopoldo spent in ensuring that the *Saggi*, effectively the Accademia's public façade, consist only of a narrative of the Cimento's experiments thus creating an image of an authoritative and uncontroversial experimentalist institution. This was, of course, precisely the type of institution that in their own public presentations, the Royal Society and the Parisian Academy of Sciences valued so highly and sought to represent as having been established in London and Paris.

It is also important for our understanding of the historiography of the Cimento that we appreciate that this is precisely the argument that has been made by 'cultural' historians such as Biagioli, Tribby, and Findlen. As we saw in Chapter One, in recent years these three authors have enlightened us about the social and political nature of the organisational and institutional practices that helped to shape the production of the first examples of experimentally based knowledge in seventeenth-century Europe. In other words, these historians have helped us to understand the social and political aims and interests of the Medici Court in the seventeenth century and the constant efforts of the Grand Duke to establish his reputation and status across Europe. Additionally, they point to the cautious reporting of experiments in the *Saggi* as an example of how the Cimento elevated the credibility of its work and how Leopoldo increased his status as a patron of reliable and objective knowledge-making.[8]

In addition to such analyses of the social, political, and religious circumstances in seventeenth-century Italy, we have noted the views put forward by

[7] BNCF, Ms. Gal. 314, ff. 968r–969r. As translated by Middleton, *The Experimenters*, 336. This passage also reveals that the *Saggi* was read by several of Montmor's colleagues, probably those who were also members of his informal meetings.

[8] Biagioli, 'Scientific Revolution', 27–28; Tribby, 'Dante's Restaurant', 328.

Steven Shapin and Simon Schaffer regarding the 'experimental life' inside the Royal Society of London. These authors skilfully identify the codes of gentlemanly civility and trust that existed in organised, English natural philosophical pursuits. In particular, they claim that those codes helped the Royal Society to avoid theoretical discussions based on controversial natural philosophical concerns, and instead to search for experimental, atheoretical, matters of fact. At one point, Shapin and Schaffer even enrolled the Cimento into their analysis of early modern decorum for producing trustworthy experimental facts.

However, while Shapin and Schaffer's analysis of the Royal Society's 'experimental life' is admirable, such stories about the origins of experimental science at the expense of natural philosophical theorising, have had some serious implications for 'cultural' historians and other writers looking to describe the activities of the Accademia del Cimento. Since 'cultural' historians have attempted to examine the social situations that encouraged the experimental practices of the Accademia del Cimento, they have also sometimes slipped into general discussions about the origins of experimental science. In the process, just like Shapin and Schaffer, they have overlooked the more wide-reaching intellectual skills, commitments, and agendas that existed across Europe, including in the Cimento's actual day-to-day discussions and experiments.

Marco Beretta, as an example, provides a simplistic look at the experimental rhetoric in the *Saggi*, without an analysis of the natural philosophical concern and contention behind the group's work, and arrives at a conclusion that describes the birth of modern science in the Tuscan Court. In other words, he ends up adopting the traditional account of the origins of atheoretical, inductivist, modern experimental science. Beretta stated that the Accademia was simply the first scientific academy to use an experimental method, a practice for knowledge-making that is tantamount to so-called modern experimental science, free from any natural philosophical theorising and based on unbiased and uncontroversial experimental fact-making.[9]

The Saturn episode is perhaps the clearest example of the essentialist simplicity of this historiographical approach. The academicians were intent on building models of the rival theories of Saturn's appearance and movements and seemingly relied purely on this astronomical experiment in order to resolve the controversy. Indeed, not trusting their own bias, they called upon naive passers-by to describe what they saw through the telescope to ensure that the results were achieved objectively. For historians searching for evidence of an experimental method, this is a perfect example of how the academicians were supposedly avoiding controversial theorising by basing their work simply on their experiments and observations. Indeed, even Albert Van Helden cannot resist the temptation to provide a sweeping generalisation in conclusion to this case study. Van Helden provides an excellent account of the theoretical issues that tormented the academicians as they prepared to solve the dispute between Fabri and Huygens. But such is Van Helden's admiration for the academicians' skills in

[9] For a summary and review of Beretta's work, see Chapter One.

observation and experimenting that he still believes that the most important part of this case study is the Cimento's demonstration of their 'mastery of the experimental method'.[10] So despite Leopoldo's political manoeuvring to suppress the publication of the natural philosophical opinions of his courtiers regarding this issue, Van Helden cannot resist concluding that the academicians' experiments were simply an 'illustration of the height of sophistication to which the experimental method had risen in Florence by 1660'.[11]

It is not difficult to understand how Van Helden arrives at such a conclusion when considering the academicians' efforts to use so-called objective and independent observers, and their intention to examine the merits of both competing theories about Saturn's movements. In his report to Leopoldo about the Saturn observations carried out by the Cimento, Borelli even stated that they simply searched for the truth 'through many experimental proofs', and that they had 'fully and dispassionately examined the opinions of Mr. Huygens and those of the adversaries who oppose him'.[12] This also led Middleton to claim, in his brief analysis of the Saturn problem, that the academicians' observations were an example of the Cimento's 'experimental psychology'.[13]

There is no doubt that the academicians were ardent and at times talented experimentalists, but despite the rhetoric in the *Saggi* and Borelli's statement earlier, is it fair to describe the mere use of experiments as a belief in a modern experimental method? If by 'experimental method' we are to understand the type of fact-gathering and inductive reasoning that so many other traditional historians imply existed in the seventeenth century, then Middleton's and Van Helden's statement could not be further from the truth. The observations of Saturn during the late 1650s and in 1660, especially the Cimento's work on the topic, were laden with natural philosophical concern and contention. Borelli in particular took a special interest in supporting the Copernican view represented by Huygens against the scholastic beliefs defended by Fabri. In fact, these observations were so heavily laden with natural philosophical contention that Leopoldo refused to have them published in the *Saggi*, or anywhere else for that matter, in order to maintain the Cimento's uncontroversial and unbiased image. So, while some historians evidently take this episode to be an example of some type of atheoretical experimental programme, equating with some purported modern scientific method, upon closer examination we find that the academicians never relied on such a practice when actually carrying out and interpreting their observations. Experiments were a crucial part of the culture of natural philosophising in mid to late seventeenth-century Tuscany by adding authority and persuasiveness to the academicians' work, but there is no indication that the Cimento was following an atheoretical experimental method that is tantamount to the so-called modern experimental science.

[10] Van Helden, 'The Accademia del Cimento and Saturn's Ring', 247.
[11] *Ibid.*, 259.
[12] '... per via di riprovi esperimentali'; '... esaminando per ultimo ... dispassionatamente i concetti del Sig. Ugenio e quie degli avversari che se gl'oppongono.' BNCF, Ms. Gal. 271, ff. 3v–4r.
[13] Middleton, *The Experimenters*, 262.

In fact, in November 1658, in a letter to Leopoldo, Borelli clearly stated that his concerns as a member of the Cimento, were not with the practice of experiments, but rather the theoretical speculations that formed the basis of any of the group's experimental work. Upon receiving news that Montmor's academy in Paris was seeking to communicate with the Cimento on a number of topics, Borelli wrote to Leopoldo about his

> ... doubts and suspicions that ... the foreigners will make themselves the authors and discoverers of the inventions and speculations of our masters, and of those that we ourselves have found. This fear makes me go slowly in beginning this correspondence with those gentlemen of the Parisian academy, since in writing, one cannot do less than communicate something or other, and I fear that this may give those foreign minds an opportunity to rediscover the things; I am speaking of the causes, not the experiments.[14]

Here Borelli was not discussing the academicians' need to maintain an unbiased, atheoretical experimental method for accumulating matters of fact and avoiding controversies fuelled by natural philosophical concerns. Instead, he was stating his opinion about what he believed was the most valuable aspect of the Cimento's work, their concerns with causal knowledge, a cornerstone of natural philosophising in the seventeenth century.

Method talk, therefore, provides us with no explanation of the conceptual complexities and negotiations that were at work throughout the history of the Accademia del Cimento and that the academicians' considered so valuable to their work. With this in mind, I have presented an alternative scenario based on the following three historiographical presuppositions: that there is no attempt here to establish a misleading origin story, that the Cimento's work should be examined within the context of seventeenth-century natural philosophy, and finally that we rely on manuscript evidence as an insight into the Accademia's daily discussions and activities.

In other words, there is no attempt here to narrate an origin story, based on modern day concepts of what it means to conduct proper scientific research. Instead, I have attempted to find the continuity of natural philosophical concern and contention in the seventeenth century, how the varying contemporary views about Aristotelianism, corpuscularianism, mechanism, and the role of the mathematical sciences in investigating nature's organisation, movements, and causes, helped to shape the work carried out by the Accademia del Cimento. While experiments were a part of natural philosophising, this did not mean that the Cimento was using an inductive and theory-free experimental method.

[14] BNCF, Ms. Gal. 275, ff.126v. 'Ora io godo sommamente che da quei Sig.ri in Francia si vada con nuove esperienze e speculazioni promovendo la natural filosofia, ma ho anche qualche sospetto e gelosia che dell'inventioni de' nostri maestri e di quelle ch'abbiamo trovate noi se n'habbino, secondo l'usanza vecchia, far autori e ritrovatori gli stranieri. Questo rispetto mi fa andar ritenuto ad attaccar questo commercio con quei Sig.ri dell'Accademia Parigina perché non si puol far di meno nello scrivere di non communicarli qualche cosa e l'istesso dubitare dà campo a quegl'ingegni pellegrini di ritrovar le cose tratto delle raggioni, non dell'esperienze.' As translated by Middleton, *The Experimenters*, 300.

Experiments were a valuable authoritative tool that helped to persuade the academicians' colleagues that they were presenting natural knowledge that was non-speculative and free of controversial theorising. Therefore, experiments were critical to how natural philosophical knowledge was constructed and presented, but they were still subsidiary to the academicians' natural philosophical concerns.

With this point in mind, we have seen the need to provide a far more contextual account of the Cimento's practices. This has been done by looking at the academicians' biographies and seeking clues about what skills and commitments each academician brought to the Tuscan Court before the Cimento opened in 1657, through their education and training. This is what has allowed us to identify the physico-mathematical practices and mechanistic skills and commitments of the majority of the academicians, as well as the scholastic opinions of a minority within the group. Finally, the accuracy of the arguments made in this book lies in the use, and careful interpretation, of manuscript evidence. The Galilean manuscripts in particular, provide us with an insight into the day-to-day activities of the Cimento between 1657 and 1662 when they were not afraid to openly debate their natural philosophical concerns when constructing and interpreting their experiments.

These presuppositions lie at the foundation of my argument. They attempt to dispose of, once and for all, the origin stories surrounding the Cimento and to create a contextual understanding of the natural philosophical climate across Europe, in which Tuscany was only a participant. As we position ourselves to launch further investigations into the academicians' work, such as their studies of magnetism and electricity, as well as the natural philosophical interests that emerged among the next generation of Italian thinkers, we may appreciate that the Cimento's façade should not be mistaken for the group's actual structure and workings.

BIBLIOGRAPHY

Abetti, G. and Pagnini, P. (eds.) *Le opere dei discepoli di Galileo Galilei. Edizione Nazionale. I. L'Accademia del Cimento. Parte Prima*, Florence, S.A.G. Barbèra Editore, 1942

Aristotle, *Generation of Animals* (tr. Peck, A.L.), London, William Heinemann, 1963

Armitage, A. 'Borelli's Hypothesis and the Rise of Celestial Mechanics', *Annals of Science*, Vol. 6, 1950, 268–282

Baldini, U. 'Giovanni Alfonso Borelli e la rivoluzione scientifica', *Physis*, Vol. 16, 1974, 97–128

Baldini, U. 'Giovanni Alfonso Borelli biologo e fisico negli studi recenti', *Physis*, Vol. 16, 1974, 234–266

Baldini, U. 'Un libertino accademico del Cimento. Antonio Oliva', *Annali dell'Istituto e Museo di Storia della Scienza*, Monografia 1, 1977

Barbensi, G. *Borelli*, Trieste, Zigiotti Editore, 1947

Barnes, B. *T.S. Kuhn and Social Science*, London, Macmillan, 1982

Bennett, J.A. 'The Mechanics' Philosophy and the Mechanical Philosophy', *History of Science*, Vol. 24, 1986, 1–28

Bennett, J.A. 'The Challenge of Practical Mathematics', in Pumfrey, S. Rossi, P.L. and Slawinski, M. (eds.) *Science, Culture and Popular Belief in Renaissance Europe*, Manchester, Manchester University Press, 1991

Beretta, M. 'At the Source of Western Science: The Organisation of Experimentalism at the Accademia del Cimento (1657–1667)', *Notes and Records of the Royal Society of London*, Vol. 54, No. 2, 2000, 131–151

Biagioli, M. 'Galileo's System of Patronage', *History of Science*, Vol. 28, 1990, 1–62

Biagioli, M. 'Scientific Revolution, Social Bricolage, and Etiquette', in Porter, R. and Teich, M. (eds.) *The Scientific Revolution in National Context*, Cambridge, Cambridge University Press, 1992, 11–54

Bonelli, M.L. 'L'ultimo discepolo: Vincenzio Viviani', in Maccagni, C. (ed.) *Saggi su Galileo* (2 vols.), Florence, G. Barbèra Editore, Vol. 2, 1972, 656–688

Bonera, G. *Galileo Oggi*, Pavia, Università degli Studi di Pavia, 1995

Borelli, G.A. *Theoricae mediceorum*, Florence, 1666

Borelli, G.A. *De motionibus naturalibus*, Bologna, 1670

Borelli, G.A. *On the Movement of Animals* (tr. Marquet, P.), Berlin, Springer, 1989

Brundell, B. *Pierre Gassendi: From Aristotelianism to a New Natural Philosophy*, Dordrecht, D. Reidel, 1987

Caverni, R. *Storia del Metodo Sperimentale in Italia* (6 vols.), Florence, 1891–1900 (reprinted: Bologna, Forni Editore, 1970)

Clagett, M. *The Science of Mechanics in the Middle Ages*, Madison, University of Wisconsin Press, 1961

Clagett, M. *Archimedes in the Middle Ages* (4 vols.), Madison, University of Wisconsin Press, 1980

Clarke, D.M. *Occult Powers and Hypotheses: Cartesian Natural Philosophy under Louis XIV*, Oxford, Clarendon Press, 1989

Clavelin, M. *The Natural Philosophy of Galileo: Essay on the Origins and Formation of Classical Mechanics* (tr. Pomerans, A.J.), Cambridge, MA, MIT Press, 1974

Cochrane, E. *Florence in the Forgotten Centuries, 1527–1800*, Chicago, University of Chicago Press, 1973

Collins, H.M. 'The Seven Sexes: A Study in the Sociology of a Phenomenon or the Replication of Experiments in Physics', *Sociology*, Vol. 9, 1975, 205–224

Collins, H.M. 'Son of Seven Sexes: The Social Destruction of a Physical Phenomenon', *Social Studies of Science*, Vol. 11, 1981, 33–62

Collins, H.M. *Changing Order: Replication and Induction in Scientific Practice*, London, Sage, 1985

Dear, P. 'Totius in verba: Rhetoric and Authority in the Early Royal Society', *Isis*, Vol. 76, 1985, 145–161

Dear, P. *Discipline and Experience: The Mathematical Way in the Scientific Revolution*, Chicago, University of Chicago Press, 1995

Dear, P. *Revolutionising the Sciences: European Knowledge and its Ambitions, 1500–1700*, Basingstoke, Palgrave, 2001

Del Gaizo, M. *Una lettera di G.A. Borelli ed alcune indagini di pneumatica da lui compiute*, Tipografia della R. Accademia delle Scienze, Naples, 1886

Del Santo, P. and Strano, G. 'Il Cimento degli astri', in Galluzzi, P. (ed.) *Scienziati a Corte: l'arte della sperimentazione nell'Accademia Galileiana del Cimento, 1657–1667*, Livorno, Sillabe, 2001, 29–35

Derenzini, T. 'Giovanni Alfonso Borelli, Fisico', in *Celebrazione della Accademia del Cimento nel Tricentenario della Fondazione*, Pisa, Domus Galileina, 1957

Derenzini, T. 'Alcune lettere di Giovanni Alfonso Borelli ad Alessandro Marchetti', *Physis*, Vol. 1, 1959, 224–243

Dijksterhuis, E.J. *The Mechanisation of the World Picture* (tr. Dikshourn, C.), Oxford, Oxford University Press, 1969

Drake, S. (ed.) *Discoveries and Opinions of Galileo*, New York, Doubleday, 1957

Drake, S. 'Galileo Gleanings – XXIII Velocity and Eudoxian Proportion Theory', *Physis*, Vol. 15, 1973, 49–64

Drake, S. 'Euclid Book V from Eudoxus to Dedekind', in Grattan-Guiness, I (ed.) *History in Mathematics Education*, n.s. 21, 1987, 52–64. Reprinted: S. Drake, *Essays on Galileo and the History and Philosophy of Science* (3 vols.), Toronto, Toronto University Press, 1999, iii, 61–75.

Drake, S. 'Ptolemy, Galileo, and Scientific Method', *Studies in the History of the Philosophy of Science*, Vol. 9, 1978, 99–115. Reprinted: S. Drake, *Essays on Galileo and the History and Philosophy of Science* (3 vols.), Toronto, Toronto University Press, 1999, i, 273–292.

Drake, S. *Galileo at Work: His Scientific Biography*, Chicago, University of Chicago Press, 1978

Drake, S. *Cause, Experiment and Science: A Galilean Dialogue Incorporating a New English Translation of Galileo's 'Bodies That Stay Atop Water or Move in It'*, Chicago, University of Chicago Press, 1981

Drake, S. and O'Malley C.D. (eds.) *The Controversy on the Comet of 1618*, Philadelphia, University of Philadelphia Press, 1960

Eamon, W. 'Court, Academy, and Printing House: Patronage and Scientific Careers in Late-Renaissance Italy', in Moran, B.T. (ed.) *Patronage and Institutions: Science, Technology, and Medicine at the European Court, 1500–1750*, Suffolk, The Boydell Press, 1991

Emerson, R. 'The Organisation of Science and its Pursuit in Early Modern Europe', in Olby, R.C. Cantor, G.N. Christie, J.R.R. and Hodge, M.J.S. (eds.) *Companion to the History of Modern Science*, London, Routledge, 1990, 960–979.

Fabroni, A. (ed.) *Delle lettere familiari del Conte Lorenzo Magalotti e di altri insigni uomini a lui scritte* (2 vols.), Florence, S.A.R., 1769

Fabroni, A. (ed.) *Lettere inedite d'uomini illustri* (2 vols.), Florence, G. Barbèra Editore, 1775

Favaro, A. *Documenti inediti per la storia dei manoscritti galileiani nella biblioteca Nazionale di Firenze*, Rome, 1886

Favaro, A. (ed.) *Le Opere di Galileo Galilei, Edizione Nazionale* (20 vols.), Florence, Barbèra, 1890

Favaro, A. *Amici e Corrispondenti di Galileo* (3 vols.), (ed. Galluzzi, P.), Florence, Libreria Editrice Salimbeni, 1983

Fermi, S. *Lorenzo Magalotti. Scienziato e Letterato, 1637–1712*, Piacenza, Stab.Tip.-Lit. Bertola & C., 1903

Feyerabend, P. *Against Method: Outline of an Anarchistic Theory of Knowledge*, London, New Left Books, 1975

Findlen, P. 'The Economy of Scientific Exchange in Early Modern Italy', in Moran, B.T. (ed.) *Patronage and Institutions: Science, Technology, and Medicine at the European Court, 1500–1750*, Suffolk, The Boydell Press, 1991

Findlen, P. 'Controlling the Experiment: Rhetoric, Court Patronage and the Experimental Method of Francesco Redi', *History of Science*, Vol. 31, 1993, 35–64

Finley, M.I. Smith, D.M. and Duggan, C. (eds.) *A History of Sicily*, London, Chatto & Windus, 1986

Galilei, G. *On Motion and On Mechanics* (ed. and tr. Drake, S. and Drabkin, I.E.), Madison, University of Wisconsin Press, 1960

Galilei, G. *Discourses and Mathematical Demonstrations Concerning Two New Sciences Pertaining to Mechanics and Local Motion* (ed. and tr. Drake, S.), Madison, The University of Wisconsin Press, 1974

Galluzzi, P. 'Il Mecenatismo Mediceo e le Scienze', in Vasoli, C. (ed.) *Idee, Istituzioni, Scienza ed arti nella Firenze dei Medici*, Florence, Giunti Mantelli, 1980

Galluzzi, P. 'L'Accademia del Cimento: "gusti" del principe, filosofia e ideologia dell'esperimento', *Quaderni Storici*, Vol. 48, 1981, 788–844

Galluzzi, P. 'G.A. Borelli dal Cimento agli Investiganti', in Lomonaco, F. and Torrini, M. (eds.) *Galileo e Napoli*, Naples, Guida Editori, 1987, 339–355

Galluzzi, P. (ed.) *Scienziati a Corte: l'arte della sperimentazione nell'Accademia Galileiana del Cimento, 1657–1667*, Livorno, Sillabe, 2001

Galluzzi, R. *Istoria del Granducato di Toscana sotto il Governo della Casa Medici* (5 vols.), Milan, Anastatica, 1974

Garin, E. 'La cultura filosofica fiorentina nell'età medicea', in Vasoli, C. (ed.) *Idee, istituzioni, scienza ed arti nella Firenze dei Medici*, Florence, Giunti Mantelli, 1980

Gaukroger, S. *Explanatory Structures: A Study of Concepts of Explanation in Early Physics and Philosophy*, Sussex, Harvester Press, 1978

Gaukroger, S. and Schuster, J. 'The Hydrostatic Paradox and the Origins of Cartesian Dynamics', *Studies in History and Philosophy of Science*, Vol. 33, 2002, 535–572

Gaukroger, S. Schuster, J. and Sutton, J. (eds.) *Descartes' Natural Philosophy*, London, Routledge, 2000

Giovannozzi, G. 'La Versione Borelliana dei Conici di Apollonio', in *Memorie della Pontificia Accademia Romana dei Nuovi Lincei*, Rome, 1916, 1–32

Giovannozzi, G. (eds.) *Lettere inedite di Gio. Alfonso Borelli al P. Angelo di S. Domenico sulla versione di Apollonio*, Florence, Scuola Tipografica Calasanzioana, 1916

Grant, E. *Much Ado About Nothing: Theories of Space and Vacuum from the Middle Ages to the Scientific Revolution*, Cambridge, Cambridge University Press, 1991

Grilli, M. and Sebastiani, F. 'Le origini della fisica del calore: le teorie sulla natura del calore da Galileo a Newton', *Physis*, Vol. 24, 1982, 301–356

Guilmartin Jr., J.F. *Gunpowder and Galleys: Changing Technology and Mediterranean Warfare at Sea in the Sixteenth Century*, Cambridge, Cambridge University Press, 1974

Hahn, R. *The Anatomy of a Scientific Institution: The Paris Academy of Sciences, 1666–1803*, Los Angeles, University of California Press, 1971

Hall, A.R. *The Scientific Revolution, 1500–1800: The Formation of the Modern Scientific Attitude*, (2nd ed.) Boston, Beacon Press, 1966

Hall, A.R. *Ballistics in the Seventeenth Century*, Cambridge, Cambridge University Press, 1952

Hall, A.R. and Hall, M.B. (eds. and trs.), *The Correspondence of Henry Oldenburg* (5 vols.), Madison, University of Wisconsin Press, 1965

Heath, T.L. (ed.) *The Thirteen Books of Euclid's Elements*, New York, Dover, Publications, 1956

Heath, T.L. (ed.) *Apollonius of Perga*, Cambridge, W. Heffer & Sons, 1896 (1961 printing)

Heilbron, J.L. *The Sun in the Church: Cathedrals as Solar Observatories*, Cambridge, MA, Harvard University Press, 2001

Henry, J. 'Patrizi's Concept of Space and its Later Influence', *Annals of Science*, Vol. 36, 1979, 549–575

Henry, J. 'Magic and Science in the Sixteenth and Seventeenth Centuries', in Olby, R.C. Cantor, G.N. Christie, J.R.R. and Hodge, M.J.S (eds.) *Companion to the History of Modern Science*, London, Routledge, 1990, 583–596

Henry, J. *The Scientific Revolution and the Origins of Modern Science*, London, Macmillan, 1997

Koenigsberger, H. *The Government of Sicily Under Philip II of Spain*, London, Staples Press, 1951

Koyré, A. *Metaphysics and Measurement: Essays in Scientific Revolution*, London, Chapman & Hall, 1968

Koyré, A. *Galileo Studies* (tr. Mepham, J.), Hassocks, Harvester Press, 1978

Koyré, A. *The Astronomical Revolution* (tr. Maddison, R.E.W.), London, Methuen, 1980

Lefèvre, W. 'Galileo Engineer: Art and Modern Science', in Renn, J. (ed.) *Galileo in Context*, Cambridge, Cambridge University Press, 2001, 11–27

Magalotti, L. *Saggi di naturali esperienze fatte nell'Accademia del Cimento sotto la protezione del serenissimo principe Leopoldo di Toscana*, Florence, 1667

Magalotti, L. *Lettere familiari* (ed. Marchetti, G.), Carnago, SugarCo Edizioni, 1993

Meli, D.B. 'The Neoterics and Political Power in Spanish Italy: Giovanni Alfonso Borelli and his Circle', *History of Science*, Vol. 34, 1996, 57–90

Meli, D.B. 'Shadows and Deception: from Borelli's *Theoricae* to the *Saggi* of the Cimento', *British Journal for the History of Science*, Vol. 31, 1998, 383–402

McClellan III, J.E. *Science Reorganized. Scientific Societies in the Eighteenth Century*, New York, Columbia University Press, 1985

Middleton, W.E.K. *The History of the Barometer*, Baltimore, The Johns Hopkins Press, 1964

Middleton, W.E.K. 'Carlo Rinaldini and the Discovery of the Convection in Air', *Physis*, Vol. 10, 1968, 299–305

Middleton, W.E.K. 'A Cardinalate for Prince Leopoldo de' Medici', *Studi Secenteschi*, Vol. 11, 1970, 167–180

Middleton, W.E.K. *The Experimenters: A Study of the Accademia del Cimento*, Baltimore, The Johns Hopkins Press, 1971

Nastasi, P. 'Una polemica giovanile di Giovanni Alfonso Borelli', *Physis*, Vol. 26, 1984, 215–247

Natucci, A. 'V.Viviani', in Gillispie, C.C. (ed.) *Dictionary of Scientific Biography* (18 vols.), New York, Scribner, Vol. 10, 1990, 48–50

Naylor, R.H. 'Galileo's Experimental Discourse', in Gooding, D. Pinch, T. and Schaffer, S. (eds.) *The Uses of Experiment: Studies in Natural Science*, Cambridge, Cambridge University Press, 1990, 117–134

Nelli, G.B.C. *Saggio di storia letteraria fiorentina del secolo XVII scritta in varie lettere*, Lucca, Patrizio Fiorentino, 1759

Newman, W.R. and Principe, L.M. *Alchemy Tried in the Fire: Starkey, Boyle and the Fate of Helmontian Chymistry*, Chicago, University of Chicago Press, 2002.

Ornstein, M. *The Role of Scientific Societies in the Seventeenth Century* (3rd ed.), Hamden, Archon Books, 1963

Osler, M.J. *Divine Will and the Mechanical Philosophy*, Cambridge, Cambridge University Press, 1994

Pieraccini, G. *La stirpe de' Medici di Cafaggiolo* (3 vols.), Florence, 1925

Pinch, T. 'Towards an Analysis of Scientific Observation: The Externality and Evidential Significance of Observational Reports in Physics', *Social Studies of Science*, Vol. 15, 1985, 3–36

Pinch, T. and Bijker, W.E. 'The Social Construction of Facts and Artefacts: Or How the Sociology of Science and the Technology Might Benefit Each Other', *Social Studies of Science*, Vol. 14, 1984, 399–441

Procissi, A. 'I manoscritti superstiti dell'Accademia del Cimento' in *Celebrazione della Accademia del Cimento nel Tricentenario della Fondazione*, Domus Galileana, Pisa, 1957

Pumfrey, S., Rossi, P.R., Slawinski, M. (eds.) *Science, Culture, and Popular Belief in Renaissance Europe*, New York, Manchester University Press, 1991

Redondi, P. *Galileo Heretic*, New Jersey, Princeton University Press, 1987

Renn, J., Damerow, P., and Rieger, S. 'Hunting the White Elephant: When and How did Galileo Discover the Law of Fall?', in Renn, J. (ed.) *Galileo in Context*, Cambridge, Cambridge University Press, 2001, 29–149

Rose, P.L. 'Galileo's Theory of Ballistics', *British Journal for the History of Science*, Vol. 4, No. 14, 1968, 156–159

Saccenti, M. *Lucrezio in Toscana*, Florence, Leo S. Oslchki, 1966

Schmitt, C.B. 'Experimental Evidence For and Against a Void: The Sixteenth-Century Arguments', *Isis*, Vol. 58, 1967, 352–366

Schofield, C.J. *Tychonic and Semi-Tychonic World Systems*, New York, Arno Press, 1981

Schuster, J.A. 'The Scientific Revolution', in Olby, R.C. Cantor, G.N. Christie, J.R.R. and Hodge, M.J.S. (eds.) *Companion to the History of Modern Science*, London, Routledge, 1990, 217–242

Schuster, J.A. 'Whatever Should We Do with Cartesian Method? – Reclaiming Descartes of the History of Science', in Voss, S. (ed.) *Essays on the Philosophy of Science of René Descartes*, New York, Oxford University Press, 1993, 195–223

Schuster, J.A. 'I domini della conoscenza', in *Storia della Scienza*, Istituto della Enciclopedia Italiana, Vol. 5, 2002, 337–358

Schuster, J. and Taylor, A. 'Seized by the Spirit of Modern Science', *Metascience*, New Series Issue Nine, 1996, 9–28

Schuster, J. 'Blind Trust: The Gentlemanly Origins of Experimental Science', *Social Studies of Science*, Vol. 27, 1997, 503–536

Schuster, J. 'L'Aristotelismo e le sue Alternative', in Garber, D. (ed.) *La Rivoluzione Scientifica*, Istituto della Enciclopedia Italiana, Rome, 2002, 337–357.

Schuster, J.A. and Watchirs, G. 'Natural Philosophy, Experiment, and Discourse: Beyond the Kuhn/Bachelard Problematic', in LeGrand, H.E. (ed.) *Experimental Inquiries: Historical, Philosophical and Social Studies of Experimentation in Science*, Dordrecht, Kluwer Academic Publishers, 1990, 1–48

Schuster, J.A. and Yeo, R.R. (eds.) *The Politics and Rhetoric of Scientific Method*, Dordrecht, D. Reidel, 1986
Segre, M. 'Galileo, Viviani and the Tower of Pisa', *Studies in the History and Philosophy of Science*, Vol. 20, No. 4, 1989, 435–451
Segre, M. 'The Role of Experiment in Galileo's Physics', *Archive for History of Exact Sciences*, Vol. 23, 1980, 227–252
Segre, M. 'Viviani's Life of Galileo', *Isis*, Vol. 80, 1989, 207–231
Segre, M. *In the Wake of Galileo*, New Jersey, Rutgers University Press, 1991
Settle, T.B. 'Galileo's Use of Experiment as a Tool of Investigation', in McMullin, E. (ed.) *Galileo: Man of Science*, New York, Basic Books, 1967, 315–337.
Settle, T.B. 'Giovanni Alfonso Borelli', in Gillispie, C.C. (ed.) *Dictionary of Scientific Biography* (18 vols.), New York, Scribner, Vol. 2, 1990, 306–314
Shapin, S. 'Pump and Circumstance: Robert Boyle's Literary Technology', *Social Studies of Science*, Vol. 14, 1984, 481–494
Shapin, S. 'The House of Experiment in Seventeenth-Century England', *Isis*, Vol. 79, 1988, 373–404
Shapin, S. *A Social History of Truth: Civility and Science in Seventeenth Century England*, Chicago, Chicago University Press, 1994
Shapin, S. *The Scientific Revolution*, Chicago, Chicago University Press, 1996
Shapin, S. and Schaffer, S. *Leviathan and the Air-Pump*, New Jersey, Princeton University Press, 1985
Shea, W.R. *Galileo's Intellectual Revolution*, London, Macmillan, 1972
Shumaker, W. *Natural Magic and Modern Science. Four Treatises: 1590–1657*, New York, Medieval & Renaissance Texts & Studies, 1989
Tamny, M. 'Atomism and the Mechanical Philosophy', in Olby, R.C. Cantor, G.N. Christie, J.R.R. and Hodge, M.J.S. (eds.) *Companion to the History of Modern Science*, London, Routledge, 1990, 597–609
Targioni Tozzetti, G. *Notizie degli aggrandamenti delle scienze fisiche accaduti in Toscana nel corso di anni LX del secolo XVII raccolte dal dottor. Gio. Targioni Tozzetti* (3 vols.), Florence, 1780
Targioni Tozzetti, G. *Atti e memorie inedite dell'Accademia del Cimento e notizie aneddote dei progressi delle scienze in Toscana* (3 vols.), Florence, 1780.
Taton, R. 'Cassini, Gian Domenico' in Gillispie, C.C. (ed.) *Dictionary of Scientific Biography* (18 vols.), New York, Scribner, Vol. 3, 1990, 100–104
Tenca, L. *Le relazioni fra Giovanni Alfonso Borelli e Vincenzio Viviani*, Milan, Istituto Lombardo di Scienze e Lettere, 1956
Tribby, J. 'Dante's Restaurant: The Cultural Work of Experiment in Early Modern Tuscany', in Bermingham, A. and Brewer, J. (eds.) *The Consumption of Culture: 1600–1800*, London, Routledge, 1991, 319–337
Tribby, J. 'Of Conversational Dispositions and the *Saggi*'s Proem', in Cropper, E. Perini, G. and Solinas F. (eds.) *Documentary Culture Florence and Rome from Grand Duke Ferdinand I to Pope Alexander VII: Papers from a Colloquium Held at the Villa Spelman, Florence, 1990*, Bologna, Nuova Alfa Editoriale, 1992, 379–390
Van Duesen, N.C., *Telesio: The First of the Moderns*, New York, 1932
Van Helden, A. 'Eustachio Divini versus Christian Huygens: A Reappraisal', *Physis*, Vol. 12, 1970, 36–50
Van Helden, A. 'The Accademia del Cimento and Saturn's Ring', *Physis*, Vol. 25, 1973, 237–259
Van Helden, A. 'Saturn and His Anses', *Journal for the History of Astronomy*, Vol. 5, 1974, 105–121
Van Helden, A. '"Annulo Cingitur": the Solution of the Problem of Saturn', *Journal for the History of Astronomy*, Vol. 5, 1974, 155–174
Vasoli, C. 'Fondamento e metodo logico della geometria dell'Euclides Restitutus del Borelli', *Physis*, Vol. 11, 1969, 571–598
Viviani, V. 'Racconto Istorico di Vincenzio Viviani', in Favaro, A. (ed.) *Le Opere di Galileo Galilei* (20 vols.), Florence, 1890, Vol. 19, 597–632
Viviani, V. *Vita di Galileo* (ed. Flora, F.), Milano, Rizzoli Editore, 1954
Viviani, V. *Vita di Galileo* (ed. Borsetti, L.), Bergamo, Moretti & Vitali Editori, Bergamo, 1992
Weld, C.R. *A History of the Royal Society* (2 vols.), New York, Arno Press, 1975
Westfall, R.S. *Force in Newton's Physics: the Science of Dynamics in the Seventeenth Century*, London, Macdonald, 1971
Westfall, R.S. *The Construction of Modern Science*, Cambridge, Cambridge University Press, 1977
Yates, F. *Giordano Bruno and the Hermetic Tradition*, London, Routledge & Kegan Paul, 1964

INDEX

Printed in the United States
88279LV00001B/58-60/A